AF783015

Assessment of Inhalation Hazards

ILSI Monographs

Carcinogenicity:
The Design, Analysis, and Interpretation of Long-Term Animal Studies
H.C. Grice and J.L. Ciminera, Editors
1988. 279pp. ISBN 0-387-18301-9

Inhalation Toxicology: The Design and Interpretation of Inhalation Studies and Their Use in Risk Assessment
U. Mohr, Editor-in-Chief
D.L. Dungworth, G. Kimmerle, J. Lewkowski, R.O. McClellan, W. Stöber
1988. 318 pp. ISBN 0-387-17822-8

Radionuclides in the Food Chain
M.D. Carter, Editor-in-Chief
J.H. Harley, G.D. Schmidt, G. Silini, Editors
1988. 518pp. ISBN 0-387-19511-4

Assessment of Inhalation Hazards:
Integration and Extrapolation Using Diverse Data
U. Mohr, Editor-in-Chief
D.V. Bates, D.L. Dungworth, P.N. Lee, R.O. McClellan, F.J.C. Roe, Editors
1989. 382pp. ISBN 3-540-50952-6

ILSI Sponsored by the International Life Sciences Institute
MONOGRAPHS

U. Mohr
Editor-in-Chief

Assessment of Inhalation Hazards

Integration and Extrapolation Using Diverse Data

D.V. Bates D.L. Dungworth P.N. Lee
R.O. McClellan F.J.C. Roe
Editors

Springer-Verlag
Berlin Heidelberg New York
London Paris Tokyo Hong Kong

Ulrich Mohr, Dr. med.
Professor of Experimental Pathology, Medizinische Hochschule Hannover, Institut für Experimentelle Pathologie, Konstanty-Gutschow-Straße 8, 3000 Hannover 61, FRG
David V. Bates, M.D., F.R.C.P., F.R.C.P.C, F.A.C.P., F.R.S.C.
University of British Columbia, 4891 College Highroad, Vancouver, BC V6T 1G6, Canada
Donald L. Dungworth, Ph.D., M.R.C.V.S
Department of Veterinary Pathology, University of California, Davis, CA 95616, USA
Peter N. Lee, M.A.
P.N. Lee Statistics and Computing Ltd., Hamilton House, 17 Cedar Road, Sutton, Surrey SM2 5DA, UK
Roger O. McClellan, D.V.M.
Chemical Industry Institute of Toxicology, P.O. Box 12137, Research Triangle Park, NC 27709, USA
Francis J.C. Roe, D.M., D.Sc., F.R.C.Path.
Consultant in Toxicology, 19 Marryat Road, Wimbledon Common, London SW19 5BB, UK

The contributions to the paper and poster sessions of the meeting have been published in the journal "Experimental Pathology," volume 37, by VEB Gustav Fischer, Jena, GDR.

Library of Congress Cataloging-in-Publication Data
Assessment of inhalation hazards : integration and extrapolation using diverse data / D.V. Bates . . . [et al.].
p. cm. – (ILSI monographs)
ISBN 3-540-50952-6
1. Gases, Asphyxiating and poisonous–Toxicology. 2. Health risk assessment. I. Bates, David V. II. Series.
RA1245.A87 1989
615.9′1–dc20 89-21931

This work is subject to copyright. All rights are reserved, whether the whole or part of materials is concerned, specifically the rights of translation, reprinting, reuse of illustrations, recitation, broadcasting, reproduction on microfilms or in other ways, and storage in data banks. Duplication of this publication or parts thereof is only permitted under the provisions of the German Copyright Law of September 9, 1965, in its version of June 24, 1985, and a copyright fee must always be paid. Violations fall under the prosecution act of the German Copyright Law.

© Springer-Verlag Berlin Heidelberg 1989
Printed in the United States of America

The use of general descriptive names, registered names, trademarks, etc. in this publication does not imply, even in the absence of a specific statement, that such names are exempt from the relevant protective laws and regulations and therefore free for general use.

Product Liability: The publisher can give no guarantee for information about drug dosage and application thereof contained in this book. In every individual case the respective user must check its accuracy by consulting other pharmaceutical literature.

Typeset by Publishers Service, Bozeman, Montana.
Printed and bound by R.R. Donnelley & Sons Company, Harrisonburg, Virginia.
Printed on acid-free paper.
9 8 7 6 5 4 3 2 1

ISBN 3-540-50952-6 Springer-Verlag Berlin Heidelberg New York
ISBN 0-387-50952-6 Springer-Verlag New York Berlin Heidelberg

Series Foreword

The International Life Sciences Institute (ILSI), a nonprofit, public foundation, was established in 1978 to advance the sciences of nutrition, toxicology, and food safety. ILSI promotes the resolution of health and safety issues in these areas by sponsoring research, conferences, publications, and educational programs. Through ILSI's programs, scientists from government, academia, and industry unite their efforts to resolve issues of critical importance to the public.

As part of its commitment to understanding and resolving health and safety issues, ILSI is pleased to sponsor this series of monographs that consolidates new scientific knowledge, defines research needs, and provides a background for the effective application of scientific advances in toxicology and food safety.

Alex Malaspina
President
International Life Sciences Institute

Contents

Section 2. Arsenic

Section 3. Butadiene

Section 4. Benzene

Section 5. Mineral Fibers

Section 6. Environmental Tobacco Smoke

Section 7. Automotive Exhaust Emissions and Polycyclic Aromatic Hydrocarbons

Part IV. Interfaces Between Scientific Judgement and Prudent Environmental Health Policy

Contributors

The complete affiliations for all authors are given as footnotes to the opening pages of their chapters. These page numbers are given in the list below.

J.C. Bailar, M.D., Ph.D., US Dept. of Health and Human Services and McGill University 39

D.V. Bates, M.D., F.R.C.P.C., F.R.C.P.F., F.A.C.P., F.R.S.C., University of British Columbia 365

S. Belinsky, Ph.D., National Institute of Environmental Health Sciences, NIH 85

H.M. Bolt, Dr.med., Dr.rer.nat., Institut für Arbeitsphysiologie, Dortmund 77

J.A. Bond, Ph.D., Lovelace Biomedical and Environmental Research Institute 315

G.A. Boorman, D.V.M., Ph.D., National Institute of Environmental Health Sciences 61

M. Casanova, M.S., Chemical Industry Institute of Toxicology 151

C. Chen, Dr., US Environmental Protection Agency 169

V.J. Cogliano, Ph.D., US Environmental Protection Agency 337

M. Corn, Ph.D., The Johns Hopkins University 233

R.G. Cuddihy, Ph.D., Lovelace Biomedical and Environmental Research Institute 3

H.d'A. Heck, Ph.D., Chemical Industry Institute of Toxicology 151

S. Deutschmann, Dipl-Biol., Institut für Arbeitsphysiologie, Dortmund 189

M.R. ELWELL, D.V.H., PH.D., National Institute of Environmental Health Sciences 61

S.L. EUSTIS, D.V.M., PH.D., National Institute of Environmental Health Sciences 61

V.J. FERON, DR., TNO-CIVO Toxicology and Nutrition Institute 131

M. FISCHER, DR., Institut für Wasser-, Boden- und Lufthygiene des Bundesgesundheitsamtes 327

H. GIBB, PH.D., M.P.H., US Environmental Protection Agency 169

R.C. GRAFSTROM, DR., Institute of Toxicology, Stockholm 103

W.C. GRIFFITH, B.S., Lovelace Biomedical and Environmental Research Institute 3

R.A. GRIESEMER, D.V.M., PH.D., National Institute of Environmental Health Sciences 61

J.R. HARKEMA, D.V.M., PH.D., Lovelace Biomedical and Environmental Research Institute 315

C.C. HARRIS, M.D., National Cancer Institute, NIH 103

U. HEINRICH, DR.RER.NAT., Fraunhofer-Institut für Toxikologie und Aerosolforschung 301

R.F. HENDERSON, PH.D., Lovelace Biomedical and Environmental Research Institute 315

D. HENSCHLER, DR.MED., University of Würzburg 357

J.E. HUFF, PH.D., National Institute of Environmental Health Sciences 177

A. JACOBS, M.D., F.R.C.P., F.R.C.PATH., University of Wales College of Medicine 225

B. JELITTO, DIPL-CHEM., Institut für Arbeitsphysiologie, Dortmund 189

R. KREILING, DR.RER.NAT., Institut für Arbeitsphysiologie, Dortmund 189

R.J. LAIB, DR., Institut für Arbeitsphysiologie, Dortmund 189

J.F. LECHNER, DR., National Cancer Institute, NIH 103

P.N. LEE, MA (OXON), P.N. Lee Statistics and Computing Ltd. 49

G.W. LUCIER, PH.D., National Institute of Environmental Health Sciences, NIH 85

G.M. MATANOSKI, M.D., DR.P.H., The Johns Hopkins University, School of Hygiene and Public Health 201

N. MATSUKURA, DR., National Cancer Institute, NIH 103

J.L. Mauderly, D.V.M., Lovelace Biomedical and Environmental Research Institute 3, 315

R.O. McClellan, D.V.M., Chemical Industry Institute of Toxicology 3, 315, 365

M.A. Medinsky, Ph.D., Lovelace Biomedical and Environmental Research Institute 215

R.L. Melnick, Ph.D., National Institute of Environmental Health Sciences 177

R.A. Miller, D.V.M., Ph.D., Battelle Pacific Northwest Laboratories 177

M. Miyashita, Dr., The First Department of Surgery, Nippon Medical School 103

J. Pauluhn, Dr.rer.nat., Department of Toxicology, Bayer AG 119

G. Pershagen, M.D., Ph.D., Institute of Environmental Medicine, Karolinska Institutet 275

F. Pott, Dr.med., Medical Institute for Environmental Hygiene, Düsseldorf 243

P.W. Preuss, Ph.D., US Environmental Protection Agency 337

F.J.C. Roe, D.M. (Oxon), D.Sc. (Lond.), F.R.C.Path., Consultant in Toxicology, London 69

G. Rose, DM, DSc, FRCP, FFCM, FRCGP, London School of Hygiene and Tropical Medicine 349

J.M. Samet, M.D., University of New Mexico 265

C. Santos-Burgoa, M.D., Ph.D., Istituto Nacional de Salud Publica, Mexico 201

R. Saracci, M.D., International Agency for Research on Cancer (IARC) 255

M.B. Schenker, M.D., M.P.H., Division of Occupational and Environmental Medicine, University of California 293

B. Schneider, Dr.rer.nat., Medizinische Hochschule Hannover 285

L. Schwartz, B.A., The Johns Hopkins University, School of Hygiene and Public Health 201

F. Steinhäusler, Dr.phil., Inst. für Allgemeine Biologie, Biochemie und Biophysik, Salzburg 23

C. Thompson, Ph.D., National Institute of Environmental Health Sciences, NIH 85

B.F. Trump, Dr., Department of Pathology, University of Maryland School of Medicine 103

K. Ulm, Dr.rer.nat., Institute for Medical Statistics and Epidemiology, Technical University Munich 139

M.E. Vahter, Dr.Med.Sc., Institute of Environmental Medicine, Karolinska Institutet 161

R.R. Vangala, Dr., Institut für Arbeitsphysiologie, Dortmund 189

P.D. White, Ph.D., US Environmental Protection Agency 337

J.C. Willey, Dr., National Cancer Institute, NIH 103

J.W.G.M. Wilmer, Dr., TNO-CIVO Toxicology and Nutrition Institute 131

R.K. Wolff, Ph.D., Lilly Research Laboratory, Greenfield 315

R.A. Woutersen, Dr., TNO-CIVO Toxicology and Nutrition Institute 131

S.L. Zeger, Ph.D., The Johns Hopkins University, School of Hygiene and Public Health 201

A. Zwart, Dr., TNO-CIVO Toxicology and Nutrition Institute 131

PART I

Integrative Approach to Assessing Human Health Risk: Two Contemporary Problems

1. Integrating Diverse Data Sets to Assess the Risks of Airborne Pollutants

R.O. McClellan,[1,2] R.G. Cuddihy,[2] W.C. Griffith,[2] and J.L. Mauderly[2]

Introduction

Airborne materials in the workplace and general environment are of concern because of their potential for being inhaled and adversely affecting the health of individuals. The adverse health effects of concern range from acute respiratory distress to chronic diseases like fibrosis, emphysema and cancer. To minimize exposure and, thus, the potential for disease, a variety of guidelines and standards have been developed.

In this paper we briefly describe the approaches used to establish exposure standards to control respiratory diseases and cancer. The introduction emphasizes the need for integrating data from multiple sources such as studies of cells, laboratory animals and people, and understanding mechanisms of carcinogenesis for various exposure situations. This approach is then illustrated using an agent, diesel exhaust, for which occupational and environmental exposures are of concern and abundant data from multiple systems are available. The paper builds on a review of the literature on the health effects of diesel exhaust by McClellan (1987).

The earliest standards were based on human experience and, typically, the presence of overt effects in highly exposed individuals. The standards were designed to minimize the potential for development of acute respiratory effects and typically assumed a threshold existed for exposures causing the signs and symptoms in question. In this situation, a clear demarcation was assumed between high exposure and accompanying respiratory disease and lower exposure that did not produce respiratory disease in individuals and, thus, in the population. If human data were available, it was used directly to establish a no observed effects level (NOEL). The standards were then set at some lower level (typically 1/10th of the NOEL) to provide a margin of safety to protect any potentially sensitive subpopulation.

[1]Chemical Industry Institute of Toxicology, Research Triangle Park, NC 27709, USA.
[2]Lovelace Inhalation Toxicology Research Institute, Albuquerque, NM 87185, USA.

In the absence of human data, information from laboratory animals was used to estimate a NOEL for people. Most frequently the estimated NOEL for people was set at 1/10th of the animal NOEL to account for possible differences in exposure-response relationships among laboratory animals and people. Inclusion of a factor of 10 margin of safety for potentially sensitive subpopulations yielded a standard for people 1/100th of the NOEL for animals.

Using human data, the only extrapolation issue involves the inclusion of a margin of safety to account for potentially sensitive subpopulations. Obviously, an additional extrapolation from laboratory animals to people is involved when human data are not available. However, in both cases, the standard for protection from noncarcinogenic respiratory disease is being set in the general range of actual observations. With either approach, it is assumed that exposure of individuals at levels below the standard does not result in any risk of inducing detectable adverse health effects. Later, we will return to this point.

In addition to continuing concern for controlling exposures that may cause noncarcinogenic effects, interest has increased in recent years regarding airborne materials causing cancer. No doubt, this concern has been heightened by the steady increase in the lung cancer incidence in both males and females in the United States, the vast majority of which is attributed to cigarette smoking.

The process of developing standards for controlling exposure to cancer-causing agents has some characteristics that are different from the process used with agents that cause diseases other than cancer. A key characteristic is the use of models to describe the relationship between exposure and cancer incidence. One use of such models is to extend estimates of risk to exposure levels lower than those for which experimental observations exist. In the absence of a documented exposure-related threshold in incidence or a plausible mechanism for a threshold, it has generally been considered prudent to assume that, even at low levels of exposure, there is a finite probability of cancer being induced. In addition, a linear relationship between exposure and cancer incidence has usually been assumed when extrapolating from actual observations to lower exposure concentrations unless strong evidence exists to the contrary.

Having developed an estimated relationship between exposure and cancer incidence, the next step in the standard setting process is to estimate the level of exposure that will produce a given effect, for example, a 10^{-6} probability of an excess lung cancer occurring in a given exposed population. Obviously, in this situation some individuals will develop cancer while others will not. Thus, attention shifts from individuals to a population for estimating the potential cancer impact of exposure. In addition, in contrast to the situation for acute respiratory effects, exposure-response relations are being extrapolated to estimate the occurrence of cancer at exposure levels that are typically well below the exposure levels at which observations have been made in either people or laboratory animals. It should also be kept in mind that the excess cancer is being observed or estimated against a background incidence of cancer occurring spontaneously or from unidentified cause.

Several difficulties exist in extrapolating high-level human exposure-cancer response data to lower exposure levels. One difficulty is of a statistical nature and relates to the impracticality of testing the validity of the extrapolation. This is not an unusual situation, recognizing that epidemiological studies may require tens of thousands of exposed subjects to observe a statistically significant increase in cancer incidence, for example, a relative risk of 1.5 for exposed versus nonexposed populations. To provide perspective, a population of individuals smoking a pack of cigarettes a day would likely have a relative risk of greater than 5.0 compared to non-smokers. In this hypothetical example, if the specific cancer risk were 1×10^{-2} in the nonexposed population, the excess cancer risk would be 5×10^{-2} in the exposed population. If the exposure-cancer incidence were extended in a linear fashion to an excess risk of 1×10^{-6}, it would represent an excess risk and, thus, an exposure concentration 5×10^{4} lower than the risk and exposure level at which the exposure-cancer relationship was observed. Clearly, the accuracy of the extrapolation cannot be directly validated in an epidemiological study. A second difficulty is that the mechanism(s) of cancer induction at the high exposure level may not be operative with the same effectiveness in producing cancer at lower levels of exposure. The effectiveness could be less or more at the lower level. In either case, it is unlikely that the difference in effectiveness could be established from epidemiological observations.

In the absence of human data or cancer induction by specific materials, as with acute effects, it is necessary to make laboratory animal to man extrapolations. However, it is important to recognize that in the cancer situation, two interrelated extrapolations are occurring—to lower exposure levels and between species. As was the case above for human data, the observations on laboratory animals must involve a relatively large number of animals observed at a high level of exposure to determine a statistically significant increase in cancer attributed to exposure. It is not unusual to compare a hundred or more exposed animals to a like number of nonexposed controls to observe a statistically significant increase, for example, a cancer risk of 1×10^{-1} in the exposed group versus a baseline of $1 - 2 \times 10^{-2}$ in the control group. As in the case of the human data, it is a substantial extrapolation to the risk, and corresponding exposure, levels of interest. Accordingly, it is appropriate to be concerned with whether the mechanisms that gave rise to the excess cancers at the high level of exposure are operative to the same extent at the lower level of exposure. In addition, it is also appropriate to raise the issue of whether the mechanisms operative in the laboratory animals are also likely to be operative in humans at the lower exposure levels.

From the foregoing discussion, it should be apparent that an understanding of the mechanisms of carcinogenesis is absolutely critical to our having scientific confidence in the two extrapolations—from high to low levels of exposure and laboratory animals to people. Our level of knowledge of mechanisms of carcinogenesis is rapidly improving. Unfortunately, it is still far from sufficient for use in directly estimating human cancer risks for exposure to various agents in the environment or workplace. In the absence of such "perfect" knowledge, it is necessary to use whatever information we can obtain from multiple approaches to estimate human health risks. Inevitably, this involves an integration of diverse data sets. Such an integrated approach will be illustrated in this paper using exposure to diesel exhaust as a case study. In the presentation, repeated reference will be made to the relationship between the levels of exposure being studied and the levels ultimately of interest from an occupational or environmental exposure perspective. From this presentation, it will be apparent that the need exists for understanding the mechanisms of diesel exhaust-induced carcinogenesis at various levels of exposure in order to develop scientifically defensible estimates of risk for low levels of exposure.

Multiple Sources of Data

In estimating the carcinogenic risk of exposure to various materials, information from epidemiological, whole animal, and molecular and cellular studies is of use. As illustrated in Fig. 1.1a, too frequently in the past the tendency has been to consider each of these approaches separately and, indeed, sometimes competitively, as they compete for the same scarce financial resources. This has probably occurred in part because each type of study has typically been conducted by individuals trained in different disciplines. Fortunately, as illustrated in Fig. 1.1b, it is increasingly recognized that each type of study can provide complementary information.

Obviously, since our ultimate interest is in estimating the cancer risk for people, it stands to reason that highest priority should be accorded the development and use of epidemiological data. The central element of an epidemiological study is a comparison of two populations, one more heavily exposed than the other population, with a view to determining if an excess of disease exists in the more heavily exposed population. It is important to recognize that in the case of diesel exhaust, both variables, exposure and cancer, occur in both populations. This is so because essentially everyone is exposed to some level of diesel exhaust. Lung cancer, which is the disease of interest, occurs in a relatively high incidence in most populations. As is well recognized, most of the cases of lung cancer are attributable to cigarette smoking. Therefore, the difficulty in any epidemiological study of lung cancer is to determine if there is any excess of lung cancer attributable to the exposure factor in question, in this case, diesel exhaust, over and above that due to cigarette smoking.

Epidemiological investigations can be conducted to show an association, or lack of a demonstrated association, between diesel exhaust and lung cancer. As such, the results are of a descriptive nature linking exposure to response, i.e., diesel exhaust exposure at high levels probably causes lung cancer in people. Rarely are the exposures sufficiently well-characterized to allow estimation of the quantitative relationship between exposure and excess cancer incidence. In addition, because the linkages are inevitably of an empirical nature, they do not

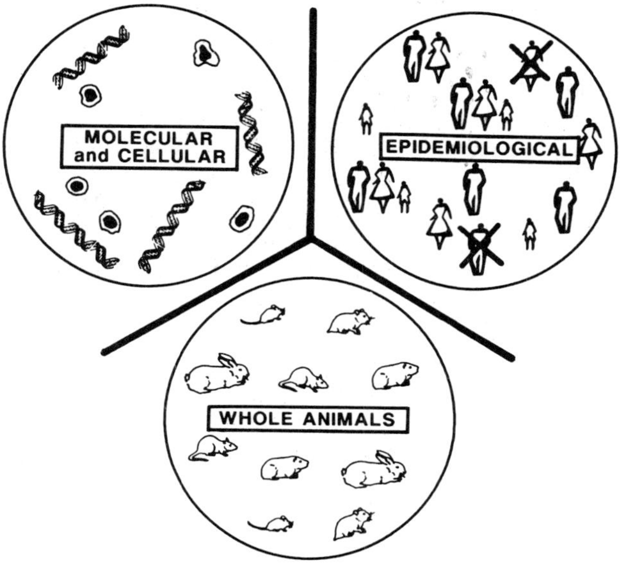

a

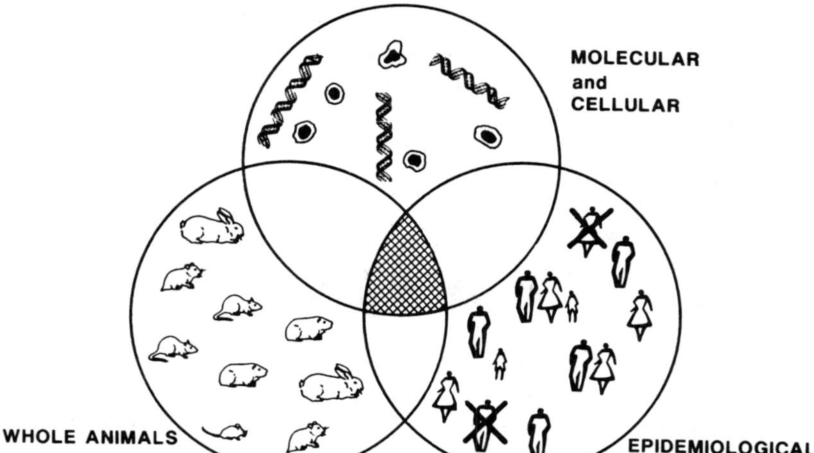

b

Figure 1.1a,b. Role of data from multiple sources of information for estimating the cancer risk for people from occupational or environmental exposure; **a** the competitive approach frequently encountered in the past and **b** the integrated approach advocated for the future

provide a high degree of insight into the mechanisms that may be involved in producing cancer.

In the absence of adequate human data on diesel exhaust-exposed populations, it is necessary to turn to other sources of information. One logical alternative is the use of laboratory animals as surrogates for people with animals chronically exposed to diesel exhaust under controlled conditions. The information from such studies, like that from epidemiological studies, is typically descriptive. However, because such studies are done under highly controlled conditions and animals can be experimentally manipulated, the investigator has a wider range of experimental options than in epidemiological studies. For example, multiple exposure levels can be used. And as will be noted later, it is much easier to introduce "dose" into an exposure-dose-response relationship. This and other experimental options available in the animal studies provide the opportunity to develop more insight into the mechanisms producing cancer.

Before leaving this brief consideration of whole animal studies, it is appropriate to note that for some materials of interest, exposed human populations do not exist because release of the material has been rigorously controlled or it is a new product. In other cases, exposures may have occurred but have been so recent that it is not feasible to conduct an epidemiology study for a disease like cancer that has a long latent period. In these situations, laboratory animal investigations may provide the only opportunity to obtain information on the response of a population of intact mammals.

A third alternative approach to obtaining data to estimate human health risks is to conduct studies using tissues, cells, or macromolecules. This approach has gained a high degree of popularity in recent years with the development of a variety of cell systems for studying mutagenicity and other aspects of toxicity that may serve as indicators of carcinogenicity. These systems have their greatest utility in "qualitative" screening for effects and at the present time have limited utility in quantitatively predicting cancer incidence in laboratory animals or man. This situation exists because at present we have only a rudimentary knowledge of carcinogenicity and how the multiple elements of the carcinogenic process are integrated in the intact mammal. It should be emphasized that studies of macromolecules and cells in vitro, beyond providing screening information, are perhaps of greatest value in providing insight into specific subparts of the carcinogenic process. This underscores their value in complementing whole animal studies and epidemiological investigations. The ultimate goal is to be able to integrate the various subparts into a unifying theory and model that will quantitatively describe the relationship between exposure and excess incidence of cancer in a population.

Epidemiological Studies

A number of epidemiological studies of diesel exhaust-exposed populations have been conducted. The results have been reviewed elsewhere (NIOSH 1986, 1988; McClellan 1987; IARC 1989) and are summarized in Table 1.1. Of these, only

Table 1.1. Major epidemiological studies on the cancer risk of exposure to diesel exhaust.

Population	Source of exhaust	Outcome
• U.S. railroad workers (Kaplan 1959)	Locomotives	Negative
• London transit workers (Raffle 1957; Waller 1980)	Buses	Negative
• Potash miners (Waxweiler et al. 1973)	Mining equipment	Negative
• Canadian railroad workers (Howe et al. 1983)	Locomotives	Suggestive
• Construction equipment operators (Wong et al. 1985)	Construction equipment	Suggestive
• U.S. railroad workers (Garshick et al. 1987 & 1988)	Locomotives	Positive

the two studies of Garshick et al. (1987, 1988) have yielded statistically significant associations between diesel exhaust exposure and excess lung cancer.

The London Transit Worker study (Raffle 1957), although it did not yield a statistically significant increase in lung cancer associated with diesel exposure, was reanalyzed by Harris (1981, 1983) to obtain alternative interpretations. In one analysis, Harris (1981) calculated an upper 95% confidence limit estimate of cancer risk that was still consistent with the original data. In a second analysis, Harris (1983) used an intergroup comparison to yield a maximum likelihood estimate of cancer.

Both Garshick et al. (1987, 1988) studies were of railroad workers enrolled with the U.S. Railroad Retirement Board. A key factor facilitating conduct of the studies was the rapid dieselization of railroads in the United States after World War II, resulting in 95% use by 1959. In the studies, it was assumed that diesel exhaust exposure began in 1959.

In a case-control study, Garshick et al. (1987) tested the hypothesis that lung cancer is associated with exposure to diesel exhaust by studying male workers born on or after January 1, 1900 who had at least 10 years of railroad service and died between March 1, 1981 and February 28, 1982. They obtained death certificates on 87% of 15059 deaths and identified 1256 workers who died from lung cancer. Deceased workers with jobs involving diesel exhaust exposure were compared to deceased workers with jobs that had more limited exposure to diesel exhaust. The relative hazard of lung cancer attributable to diesel exhaust exposure was calculated using a multiple conditional logistic regression to adjust for smoking and asbestos exposure. The results are summarized in Table 1.2. A statistically significant increase in relative odds (1.41 and $p = 0.02$) was observed for lung cancer in the workers aged 64 years or younger at death who had 20 years or more of diesel exhaust exposure. The odds ratio for the workers greater than 65 years of age with 20 years of exposure was not elevated, nor was a statistically significant increase seen in either age group with asbestos exposure. An overwhelming effect of cigarette smoking was seen in both age groups.

Table 1.2. Odds ratio for lung cancer observed in a case-control study of railroad workers (modified from Garshick et al. 1987).

Exposure	Odds ratio	95% Confidence interval	p value
Age ≤ 64 yr			
20-yr diesel exposure	1.41	1.06, 1.88	0.02
Asbestos	1.20	0.87, 1.65	0.27
≤ 50 pack-years	3.29	1.57, 6.93	< 0.01
> 50 pack-years	5.68	2.73, 11.80	< 0.01
Age ≥ 65 yr			
20-yr diesel exposure	0.91	0.71, 1.17	0.47
Asbestos	0.98	0.81, 1.20	0.86
≤ 50 pack-years	4.38	2.90, 6.60	< 0.01
> 50 pack-years	9.14	6.11, 13.70	< 0.01

Reprinted with permission of the *American Review of Respiratory Disease*, 1987; 135:1242–1248.

In a cohort study, Garshick et al. (1988) followed 55407 railroad workers who were aged 40–64 years in 1959 and had started railroad work 10–20 years earlier. They were traced to 1980, yielding 19396 deaths. Death certificates were obtained in 88% of the decedents and resulted in identification of 1694 lung cancer cases. Workers were classified as exposed or unexposed to diesel exhaust based on their job classifications. A proportional hazards model and directly standardized rates were used to calculate the relative risk of lung cancer (Table 1.3). The elevation in relative risk of lung cancer (1.45) was statistically significant for those individuals aged 40–44 years in 1959, i.e., those workers with the longest potential period of diesel exhaust exposure. The relative risk for the older age groups with presumably less diesel exhaust exposure was generally progressively lower.

In evaluating the Garshick et al. (1987, 1988) studies, several considerations are worthy of note. As is frequently the question in lung cancer studies, the extent to which the cigarette smoking history of the subjects has been taken into account is of paramount importance. The extent to which misclassification by smoking history can influence the results is clear from a comparison in Table 1.2 of the odds ratios for cigarette smoking with those for 20 years of diesel exhaust exposure. The added risk of less than 50 pack-years of smoking in the under 64 year age group was more than 5 times that of 20 years of diesel exhaust exposure (2.29 versus 0.41), and for the over 50 pack-year group, more than 11 times that of 20 years of diesel exhaust exposure (4.68 versus 0.41). This, coupled with the potential for survivors tending to underestimate smoking in the face of a potential occupational factor, diesel exhaust exposure, leaves open the issue of whether the ascertainment of smoking history was adequate.

Several other considerations are important in evaluating the Garshick et al. (1987, 1988) studies. In the case-control study, the authors only reported data on the separate age groups. It would have been of interest to have an analysis of the combined groups with varied years of diesel exposure, i.e., less than 10 years, 11–19 years, and more than 20 years. Another concern is the issue of other

Table 1.3. Relative risk of lung cancer observed in a retrospective cohort study of railroad workers (modified from Garshick et al. 1988).

Age in 1959 (yrs)	Relative risk	95% CI
40–44	1.45	1.11, 1.89
45–49	1.33	1.03, 1.73
50–54	1.12	0.88, 1.42
55–59	1.18	0.94, 1.50
60–64	0.99	0.74, 1.33

Reprinted with permission of the *American Review of Respiratory Disease*, 1988; 137:820–825.

occupation-related confounding factors such as the products of combustion of other fuels used before introduction of diesels. And finally, the difficulties in quantifying diesel exhaust exposure must be noted. The related industrial hygiene surveys of Woskie et al. (1988a,b) will still leave open the question of the extent to which exposure measurements made in the 1980s serve as adequate surrogates of the exposures of the 1960s and 1970s. This is a problem potentially in classifying jobs as diesel exhaust-exposed versus nonexposed. It is of even greater importance, as will be noted later, in establishing quantitative cancer risks for diesel exhaust exposures.

Despite any of the potential limitations noted above, it would appear that the Garshick et al. (1987, 1988) studies are the most comprehensive of the diesel exhaust epidemiological studies yet performed. Indeed, considering their size and scope and the apparent unavailability of similar-sized populations with substantial diesel exhaust exposure, it appears unlikely that better diesel exhaust epidemiology studies are likely to be performed soon. This speaks in favor of additional analyses of the extensive data base collected by Garshick et al., including possible analysis by other investigators and an extension of the studies to other time periods.

Considering the results of the Garshick et al. (1987, 1988) studies, it would seem appropriate to view chronic high-level diesel exhaust exposures as carcinogenic to people.

Animal Studies

A number of studies of the effects of exposure of laboratory animals to diesel exhaust or diesel exhaust particle extracts have been performed. These have been reviewed elsewhere (Ishinishi et al. 1986; McClellan 1987; NIOSH 1988; IARC 1989) and are summarized in Table 1.4

The studies of skin tumor induction in mice by repeated painting of diesel exhaust particle extracts indicate the potential for exposure to diesel exhaust causing cancer. However, the quantities used were relatively large and the results of skin tumor induction by particle extracts cannot be quantitatively extrapolated to lung tumor induction by inhaled particles.

Table 1.4. Laboratory animal studies of tumor induction from exposure to diesel exhaust or diesel exhaust particle extracts.

Endpoint	Result
• Skin tumor induction (exposure to diesel exhaust particle extracts)	Positive
• Lung tumor induction	
Exposure to gas phase exhaust only	
Rats, Syrian hamster	Negative-Equivocal
Exposure to total exhaust (gas and particles)	
Rats	Positive
Syrian hamsters	Negative
Mice	Equivocal

The results of the studies that involved inhalation exposure are the most relevant to evaluating the human cancer risk of exposure to diesel exhaust. Within this set of studies, those of greatest interest involved exposure of two strains of rats in five different laboratories to graded levels of diesel exhaust (Stober 1986; Mauderly et al. 1987; Brightwell et al. 1986; Ishinishi et al. 1986; Iwai et al. 1986). In general, an exposure concentration-related increase in lung tumors was observed with exposure to whole diesel exhaust that included particles and gases. The lack of a lung tumor response in rats exposed to diesel exhaust from which the particles were removed implicates the particles in the induction of the lung tumors (Stober 1986).

In evaluating the results of the rat studies, several considerations are worthy of note. One important issue is the dose of toxicant actually experienced by the animals. The Mauderly et al. (1987) study used two measures of dose, lung burden (Wolff et al. 1987) and DNA adducts (Wong et al. 1986) to complement the measurements of exposure. Wolff et al. (1987) demonstrated that lung clearance was impaired at the two high-exposure concentrations (3.5 and 7.0 mg/m^3 for 7 h/day, 5 days/week in the Mauderly et al. study. This impairment in long-term pulmonary clearance led to lung burdens at these levels in excess of that predicted from particle retention kinetics at the lowest exposure concentration (0.35 mg/m^3). At the two highest levels, lung pathology was evident in addition to the presence of lung tumors. At this time, the extent to which pulmonary pathology, consisting of inflammation, fibrosis and metaplasia, is linked to the process(es) producing the lung tumors is not known. Vostal (1986) and Morrow (1988) have previously reviewed information on the lung "over-load" phenomena and noted that exposure of rats to high levels of several particulate materials induced lung tumors in rats. This raises the possibility that the rat lung tumor response in the diesel exhaust-exposed rats was a nonspecific response to the particles.

In the Mauderly et al. (1987) study, an increased level of DNA adducts was measured in the rats exposed to the highest level of diesel exhausts (7.0 mg/m^3) compared to controls (Wong et al. 1986). The adducts represent a nonspecific measure of the interaction of organic constituents of the particles with the DNA and provide support for their potential role in the tumorigenic process.

These interactions themselves might be sufficient to produce lung tumors. Alternatively, the tumors might arise from the interactions of the organic constituents with the DNA, i.e., genetic injury, *and* the nonspecific promotional effects of the lung pathology. If this combined mechanism were involved, it might be possible that the exposure concentration-tumor response relationship might be substantially sublinear at low-exposure concentrations which do not produce pulmonary pathology.

Considering the results of the several rat studies, it would be appropriate to consider chronic high-level exposures to diesel exhaust as carcinogenic to rats. This provides supporting evidence for a carcinogenic response in people chronically exposed to high levels of diesel exhaust.

Molecular and Cellular Studies

Since the initial report in 1977 that extracts of diesel exhaust particles were mutagenic to bacteria, numerous observations have been made using molecular and cellular systems (USEPA 1977). These have been reviewed elsewhere (Ishinishi et al. 1986; McClellan 1987) and are summarized in Table 1.5. The findings in multiple systems give clear evidence of the capacity for organic constituents of diesel exhaust particles to interact with DNA and give rise to mutations, chromosomal alterations and cell transformations, all well-established early steps in the process of carcinogenesis. These observations provide a partial qualitative explanation for the observations of lung cancer in people and rats exposed to high levels of diesel exhaust. However, the present state of our knowledge of carcinogenesis precludes utilizing this fragmentary data to quantitatively estimate in a direct fashion the lung cancer risk for people exposed to diesel exhaust.

It should be noted that Albert et al. (1983), Harris (1983), and Cuddihy et al. (1984) have used a comparative potency approach to indirectly estimate lung cancer risks as an approach first advocated by the U.S. Environmental Protection Agency (Albert et al. 1983). In this approach, the results from several short-term

Table 1.5. Information acquired with molecular and cellular systems relevant to evaluating the cancer risks of exposure to diesel exhaust. Exposures were to extracts of diesel exhaust particles unless otherwise noted.

Endpoint	Results
Mutagenicity	Positive
Bacteria	Positive
Mammalian cells	Positive
Sister chromatid exchanges	Positive
Cell transformation	Positive
DNA adducts*	Positive

*Evaluated in lung tissue of rats exposed to whole diesel exhaust.

cell assays and skin tumor induction bioassays of diesel exhaust particle extracts, cigarette smoke condensate, roofing tar, and coke oven tar are combined with epidemiological data from cigarette smokers and workers exposed to roofing tar or coke oven emissions to indirectly estimate the lung cancer risk to people exposed to diesel exhaust.

Extrapolation Issues

As noted earlier, the estimation of cancer risks for occupational and environmental exposures may involve two extrapolations, between different levels of exposure and between species or levels of biological organization. Both extrapolations are readily apparent from consideration of Fig. 1.2.

Consider first the exposure level extrapolation issue. The range of soot concentrations shown for environmental exposure are taken from Cuddihy et al. (1984). The occupational exposure levels are taken from Woskie et al. (1988a,b), who studied railroad operations and consideration of the likely maximum concentrations to be found in coal mines with diesel equipment if the total particulate standard (2 mg/m^3) is met. The whole animal exposure concentrations are taken from the studies reviewed earlier.

The in vitro and skin painting studies have mostly been conducted with extracts of diesel exhaust particles or with specific constituents of particles. To accommodate the display of such data, a scale for specific constituents in addition to soot concentration has been added to Fig. 1.2. The two scales are offset by a factor of 10^5, recognizing that most individual constituents represent only a small fraction of the soot. For example, particle-phase 1-nitropyrene emissions of 4 μg/min and total particle emissions of 394000 and 607000 μg/min have been reported by Schuetzle and Frazier (1986).

The range of exposure extrapolations is substantial; the concentrations used in the in vitro assays differ from the whole animal, occupational, and environmental exposures by more than a factor of 10^3, 10^5, and 10^6, respectively. The extrapolation range is still substantial in going from animal exposures to environmental exposures, perhaps a factor of 10^2–10^3. Obviously, the epidemiological data on occupationally exposed populations are in the exposure concentration range of immediate concern. However, they are a factor of 10–100 higher concentrations than those of environmental concern. Considering this broad range, it is appropriate to have concern for potential nonlinearity in exposure-response relations.

The second extrapolation issue, between species and levels of biological organization, is also apparent from Fig. 1.2. Further, the extent to which the two extrapolation issues are intertwined is apparent. To ask the obvious questions, what degree of confidence should we have in predicting human responses from the findings of lung cancer induced in rats with accompanying lung pathology at high levels of exposure that have resulted in impaired lung clearance? Is the "true" excess incidence of lung cancer proportionally related to exposure concentration

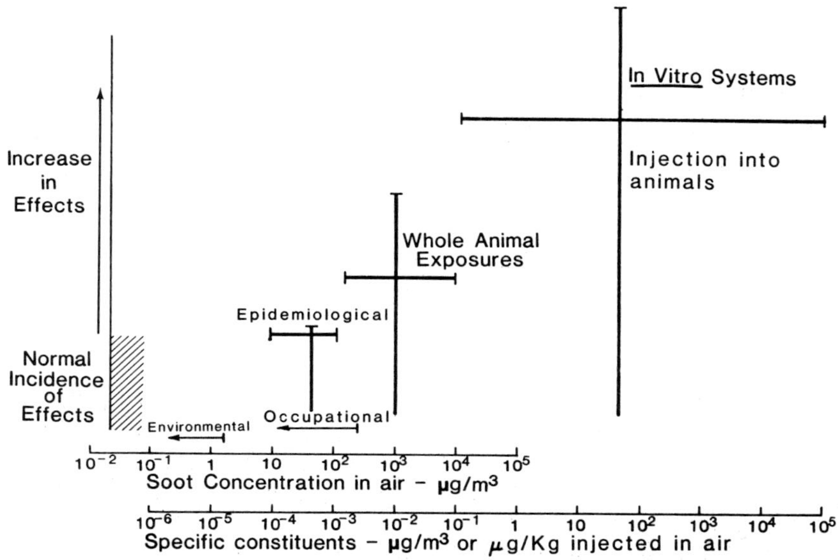

Figure 1.2. Exposure-response relationships relevant to evaluating the cancer risks of diesel exhaust

in occupationally or environmentally exposed people with no exposure-related impairment of lung clearance or lung pathology? The existing data on the mechanisms of diesel exhaust-induced lung cancer are insufficient to answer these questions. The crucial nature of these questions speaks to the need for additional research to address them.

Qualitative and Quantitative Assessments of Lung Cancer Risk

Based on consideration of the epidemiological and animal bioassay evidence, the International Agency for Research on Cancer has classified diesel exhaust as probably carcinogenic to people (IARC 1989). This classification was strongly influenced by the complementary nature of the evidence from epidemiological and animal studies. This qualitative finding is most relevant when considering exposures on the order of those from which the epidemiological and animal data were derived, i.e., hundreds of μg/m³.

It is of interest to consider the several data sets with a view to their value in providing quantitative estimates of lung cancer risk. For comparative purposes, such estimates have been developed using three approaches: the epidemiological data, the rat bioassay data, and comparative potency calculations. To permit comparisons between risk estimates developed using the three approaches, the

Table 1.6. Comparison of lung cancer risk estimates for diesel exhaust exposure (see the text for underlying assumptions).

Approach	Data source	Risk model	Lung cancer risk estimates (cases/yr/μg/m³)
Epidemiological–railroad workers	Garshick et al. 1987	Proportional	3800[a] (400–7400)[b] 950[c] (100–1900)
Epidemiological–London transit workers	Harris 1981	Proportional (upper 95% confidence limit)	3500
	Harris 1983	Proportional (maximum likelihood)	860
Comparative Potency	Harris 1983	Proportional	250
	Cuddihy et al. 1984	Absolute	230
	Albert et al. 1983; Albert and Chen 1986	Absolute	90-light duty 70-heavy duty
Rats	this paper	Relative	270-light duty 70-heavy duty

[a] assuming railroad workers exposed at 125 μg/m³
[b] 95% confidence limit
[c] assuming railroad workers exposed at 500 μg/m³

excess lung cancer incidence will be reviewed in this paper for single exposure scenario. The scenario assumes exposure (at 1 μg/m³) of the United States population (230×10^6 people living an average of 70 years with 10^5 lung cancer deaths per year). The various estimates are shown in Table 1.6.

The calculations for each of the estimates derived using epidemiological data are shown below:

- *Proportional risk model* (Garshick et al. 1987). The first lung cancer risk estimate using this data is developed assuming the railroad workers were exposed to 125 μg/m³ for 40 h per week.

$$0.016 \text{ cancers/yr of diesel} \times \frac{1\ \mu\text{g/m}^3}{125\ \mu\text{g/m}^3} \times \frac{168\ \text{h/wk}}{40\ \text{h/wk}} \times 70\ \text{yr} \times 1$$
$$\times\ 10^5 \text{ cancers/yr} = 3800 \text{ lung cancers/yr}$$
$$(95\% \text{ confidence limit: } 400\text{–}7400)$$

- *Proportional risk model with a calculated 95% upper confidence limit* (Harris 1981).

$$5 \times 10^{-4} \frac{\text{increased lifetime risk}}{\mu\text{g/m}^3 \text{ for 1 yr}} \times 70\ \text{yr} \times 1 \times 10^5 \text{ cancers/yr}$$
$$= 3500 \text{ lung cancers/yr}$$

- *Proportional risk model* (Garshick et al. 1987). The second lung cancer risk estimate using this data is developed assuming the railroad workers were exposed to 500 μg/m³ for 40 h per week.

$$0.016 \text{ cancers/yr of diesel} \times \frac{1\ \mu g/m^3}{500\ \mu g/m^3} \times \frac{168\ \text{h/wk}}{40\ \text{h/wk}} \times 70\ \text{yr} \times 1$$

$$\times\ 10^5 \text{ cancers/yr} = 950 \text{ lung cancers/yr}$$

(95% confidence limit: 100–1900)

- *Proportional risk model based on intergroup comparisons with a maximum likelihood estimate* (Harris 1983).

$$1.23 \times 10^{-4} \frac{\text{increased lifetime risk}}{\mu g/m^3 \text{ for 1 yr}} \times 70\ \text{yr} \times 1 \times 10^5/\text{yr}$$

$$= 860 \text{ lung cancers/yr}$$

The calculations for each of the estimates derived using the comparative potency approach are shown below:

- *Proportional risk model* (Harris 1983)

$$0.35 \times 10^{-4} \frac{\text{increased lifetime risk}}{\mu g/m^3 \text{ for 1 yr}} \times 70\ \text{yr} \times 1 \times 10^5 \text{ cancers/yr}$$

$$= 250 \text{ lung cancers/yr}$$

- *Absolute risk model* (Cuddihy et al. 1984)

$$\frac{0.1 \text{ lung cancer/yr/}10^5 \text{ persons}}{1\ \mu g/m^3} \times 230 \times 10^6 \text{ people}$$

$$= 230 \text{ lung cancers/yr}$$

- *Absolute risk model* (Albert et al. 1983; Albert and Chen 1986)

light-duty diesel

$$0.26 \times 10^{-4} \frac{\text{lifetime risk}}{\mu g/m^3} \times 230 \times 10^6 \text{ people/70 yr}$$

$$= 90 \text{ lung cancers/yr}$$

heavy-duty diesel

$$2 \times 10^{-6} \frac{\text{lifetime risk}}{\mu g/m^3} \times 230 \times 10^6 \text{ people/70 yr}$$

$$= 7 \text{ lung cancers/yr}$$

• *Relative risk model based on logistic regression fit to rat data* (this paper)

light-duty diesel

$$\frac{2.7 \times 10^{-3} \text{ excess lifetime risk}}{\mu g/m^3} \times 1 \times 10^5 \text{ cancers/yr}$$
$$= 270 \text{ lung cancers/yr}$$

heavy-duty diesel

$$\frac{6.7 \times 10^{-4} \text{ excess lifetime risk}}{\mu g/m^3} \times 1 \times 10^5 \text{ cancers/yr}$$
$$= 70 \text{ lung cancers/yr}$$

The estimates using railroad worker case-control data of Garshick et al. (1987) were developed using their coefficient of 0.01648 for cancers per year of diesel exhaust exposure. Two estimates of risk are given: one assuming the railroad workers were exposed to 125 μg/m³ and the other assuming exposure at 500 μg/m³. Unfortunately, the actual levels of exposure during the 1960s and 1970s were not measured and, thus, will never be known.

The risk coefficients used in the calculations for the London Transit Workers were developed by Harris (1981, 1983). It should be borne in mind that the original study by Raffle (1957) did not demonstrate an increase in lung cancer associated with diesel exhaust exposure. Harris developed his estimates by calculating in one case an upper 95% confidence limit consistent with the negative data, and in the second case by calculating a maximum likelihood estimate from intergroup comparisons.

The comparative potency estimates are to a large extent all based on the same extensive data base developed by the U.S. Environmental Protection Agency and summarized by Albert et al. (1983). Thus, the similarity of the risk estimates based on the several interpretations is not surprising.

The lung cancer risk estimates from the rat studies were developed using the data from near lifetime exposure studies conducted by Mauderly et al. (1987), Brightwell et al. (1986), and Iwai et al. (1986), using light-duty diesel engines to produce the exhaust, and by Ishinishi et al. (1986) using a heavy-duty diesel engine to produce exhaust. The excess relative risk was calculated for both situations, light- and heavy-duty diesel engines, using a logistic regression. It should be noted that over the range from 1 to 125 μg/m³, the model departed only slightly from linearity. However, it should be emphasized that the true relationships in this range may be substantially nonlinear because the shape of the curve is largely controlled by the groups exposed at the higher concentrations. It is not practical to precisely define the shape of the curve in this region because to do so would require the exposure of groups of rats that would be much larger than the combined size of all previous studies of diesel exhaust.

In reviewing the lung cancer risk estimates developed using the different methods, one is struck by the range of the estimates: from 7 to 4000 lung cancer cases. This divergence of estimates can be interpreted in several ways. One view is that this range is indicative of the current uncertainties in our knowledge of lung carcinogenesis. It might be tempting to speculate that the estimates derived from epidemiological studies are the most accurate because they are based on human data. Indeed, this may be the case. However, if so, it would be much more satisfying if the biological basis for the difference between the epidemiological-based and comparative potency or laboratory rat-based estimates were understood. This is a point that will be returned to later.

In considering the lung cancer estimates, it is useful to consider the nature of the models. The estimates developed from the railroad worker and rat data were both proportional or relative risk models. For a population containing both smokers and nonsmokers, these models would predict that the vast majority of excess lung cancer cases produced by exposure to diesel exhaust would occur in smokers. For example, it is reasonable to estimate that about 90% of lung cancers occur in smokers even though this group represents well under one-half of the population. Thus, using a proportional risk model and assuming an exposed population with a mix of smokers and nonsmokers similar to the U.S. population, 90% of the excess lung cancer estimated to result from diesel exhaust exposure would occur in smokers. Using an absolute risk model, the excess risk would be distributed evenly across the total population.

Without question, additional research is needed to clarify the relative merits of alternative extrapolation models. Some insights may come from detailed analysis and interpretation of epidemiological data. Other information may be developing in laboratory animal or cell studies. To date, our efforts in this area have been hampered by our lack of good animal or cell system models for studying interactive effects such as the influence of combined effects of cigarette smoking and exposure or other agents. If suitable models were available, it would be appropriate to evaluate the effect of combined exposure to cigarette smoking and diesel exhaust. This might help resolve the question of whether a relative risk model for diesel exhaust based on studies of rats not exposed to cigarette smoke is equally applicable to rats exposed to cigarette smoke and, in turn, applicable to estimating the risks in people smoking cigarettes.

The foregoing discussion emphasizes the importance of obtaining additional information on the mechanisms of cancer induction by diesel exhaust, including co-exposure to other pollutants such as cigarette smoke. However, it is critical that such information be obtained at relevant exposure levels. As noted earlier in the Introduction, two interrelated extrapolation issues are involved in developing mechanistic information at relevant exposure levels in laboratory animals, i.e., from high- to low-exposure concentrations and across species.

As noted earlier, statistical considerations dictate that animal studies of exposure-cancer incidence be conducted at exposure levels well above those of primary interest for risk assessment purposes. Mechanistic studies can be conducted at these exposure levels. However, it follows that mechanisms of carcinogenesis (in company with a demonstrated cancer outcome) cannot be studied at the exposure levels of concern for risk estimation purposes.

Faced with this dilemma, one alternative is to attempt to study surrogate measures or markers of the dose of diesel exhaust to critical biological units, be it macromolecules such as deoxyribonucleic acid, or specific cell types, and then determine the relationship between exposure (concentration and duration) and the measure of dose. Although this approach is conceptually attractive, it is dependent upon knowing that the dose measured is causally associated with the development of the health response of concern, in this case, cancer. Moreover, for this approach to have value, the relationship between the measure of dose and cancer must be known quantitatively if the information is to have significant value for quantitative estimates of health risks. Recent advances in our understanding of the molecular and cellular mechanisms of carcinogenesis offer encouragement that qualitative linkages will be established between the dose to critical biological structures and cancer induction. The situation with regard to quantitation of these linkages is much less optimistic because of the statistical dilemma discussed above. This is an area clearly deserving additional research emphasis.

Summary

Information obtained from diesel exhaust-exposed people, laboratory animals, and cells have been reviewed to illustrate the potential for integrating diverse data sets for purposes of risk estimation. Taken in concert, the integrated data set suggests that with high levels of exposure, diesel exhaust is a pulmonary carcinogen to people. Strengths and weaknesses of the various types of data have been discussed. Despite shortcomings in each type of study, the strength of the evidence for viewing high-level diesel exhaust exposures as carcinogenic is increased because of the complementary nature of the results from different types of studies. A major uncertainty is the utility of the information obtained at high levels of exposure for quantitatively estimating cancer risks at levels of exposure substantially lower than those at which cancer was observed. A critical need exists for acquiring additional information on mechanisms of diesel exhaust-induced carcinogenesis. However, the mechanistic information must be convincingly shown to have relevance at the exposure levels likely to be encountered by people. This is a formidable challenge especially for estimating risks at levels likely to be encountered in the general environment. Because of the substantial data base that already exists on diesel exhaust, it is an attractive model for further studies.

References

Albert RE, Chen C (1986) US EPA diesel studies on inhalation hazards. In: Ishinishi N, Koizumi A, McClellan RO, Stober W (eds) Carcinogenic and mutagenic effects of diesel engine exhaust. Elsevier Science, Amsterdam, pp 411–419

Albert RE, Lewtas J, Nesnow S, Thorslund TW, Anderson E (1983) Comparative potency method for cancer risk assessment: application to diesel particulate emissions. Risk Anal 3:101–117

Brightwell J, Fouillet X, Cassano-Zoppi A-L, Gatz R, Duchosal F (1986) Neoplastic and functional changes in rodents after chronic inhalation of engine exhaust emissions. In: Ishinishi N, Koizumi A, McClellan RO, Stober W (eds) Carcinogenic and mutagenic effects of diesel engine exhaust. Elsevier Science, Amsterdam, pp 471–485

Cuddihy RG, Griffith WC, McClellan RO (1984) Health risks from light-duty diesel vehicles. Environ Sci Technol 18:14A–21A

Garshick E, Schenker MB, Munoz A, Segal M, Smith TJ, Woskie SR, Hammond K, Speizer FE (1987) A case-control study of lung cancer and diesel exhaust exposure in railroad workers. Am Rev Respir Dis 135:1242–1248

Garshick E, Schenker MB, Munoz A, Segal M, Smith TJ, Woskie SR, Hammond SK, Speizer FE (1988) A retrospective cohort study of lung cancer and diesel exhaust exposure in railroad workers. Am Rev Respir Dis 137:820–825

Harris JE (1981) Potential risk of lung cancer from diesel engine emissions. National Academy, Washington

Harris JE (1983) Diesel emissions and lung cancer. Risk Anal 3:83–100

Howe GR, Fraser D, Lindsay J, Presnal B, Yu SZ (1983) Cancer mortality (1965–1977) in relation to diesel fume and coal exposure in a cohort of retired railway workers. J Natl Cancer Inst 70:1015–1019

IARC (1989) Engine exhaust and nitro-aromatic compounds. International Agency for Research on Cancer, Lyon (in press)

Ishinishi N, Kuwabara N, Nagase S, Suzuki T, Ishiwata S, Kohno T (1986) Long-term inhalation studies on effects of exhaust from heavy and light duty diesel engines on F344 Rats. In: Ishinishi N, Koizumi A, McClellan RO, Stober W (eds) Carcinogenic and mutagenic effects of diesel engine exhaust. Elsevier Science, Amsterdam, pp 329–348

Iwai K, Udagawa T, Yamagishi M, Yamaga H (1986) Long-term inhalation studies of diesel exhaust on F344 SPF rats. Incidence of Lung Cancer and Lymphoma. In: Ishinishi N, Koizumi A, McClellan RO, Stober W (eds) Carcinogenic and mutagenic effects of diesel engine exhaust. Elsevier Science, Amsterdam, pp 349–360

Kaplan I (1959) Relationship of noxious gases to carcinoma of the lung in railroad workers. J Amer Med Assoc 171:2039–2043

Mauderly JL, Jones RK, Griffith WC, McClellan RO (1987) Diesel exhaust is a pulmonary carcinogen in rats exposed chronically by inhalation. Fund Appl Toxicol 9:208–221

McClellan RO (1987) Health effects of exposure to diesel exhaust particles. Annu Rev Pharmacol Toxicol 27:279–300

Morrow P (1988) Possible mechanisms to explain dust overloading of the lungs. Fund Appl Toxicol 10:369–384

NIOSH (1986) Evaluation of the potential health effects of occupational exposure to diesel exhaust in underground coal mines. US Department of Health and Human Services, Centers for Disease Control, NIOSH, Cincinnati

NIOSH (1988) Current Intelligence Bulletin No. 50. Carcinogenic effects of exposure to diesel exhaust. US Department of Health and Human Services, Centers for Disease Control, NIOSH, Cincinnati

Raffle P (1957) The health of the worker. Br J Ind Med 14:73–80

Schuetzle D, Frazier J (1986) Factors influencing the emission of vapor and particulate phase components from diesel engines. In: Ishinishi N, Koizumi A, McClellan RO,

Stober W (eds) Carcinogenic and mutagenic effects of diesel engine exhaust. Elsevier Science, Amsterdam, pp 41–63

Stober W (1986) Experimental induction of tumors in hamsters, mice and rats after long-term inhalation of filtered and unfiltered diesel exhaust. In: Ishinishi N, Koizumi A, McClellan RO, Stober W (eds) Carcinogenic and mutagenic effects of diesel engine exhaust. Elsevier Science, Amsterdam, pp 421–439

US Environmental Protection Agency (1977) Precautionary notice in laboratory handling of exhaust products from diesel engines. US Environmental Protection Agency, Washington

Vostal JJ (1986) Factors limiting the evidence for chemical carcinogenicity of diesel emissions in long-term inhalation experiments. In: Ishinishi N, Koizumi A, McClellan RO, Stober W (eds) Carcinogenic and mutagenic effects of diesel engine exhaust. Elsevier Science, Amsterdam, pp 381–396

Waller RE (1980) Trends in lung cancer in London in relation to exposure to diesel fumes. In: Pepelko WE, Danner RM, Clarke NA (eds) Health Effects of Diesel Engine Emissions, US EPA, Cincinnati, pp 1085–1097

Waxweiler R, Wagoner J, Archer V (1973) Mortality of potash workers. J Occupat Med 15:486–489

Wolff RL, Henderson RF, Snipes MB, Griffith WC, Mauderly JL, Cuddihy RG, McClellan RO (1987) Alternations in particle accumulation and clearance in lungs of rats chronically exposed to diesel exhaust. Fund Appl Toxicol 9:154–166

Wong D, Mitchell CE, Wolff RK, Mauderly JL, Jeffrey AM (1986) Identification of DNA damage as a result of exposure of rats to diesel engine exhaust. Carcinogenesis 7:1595–1597

Wong O, Morgan RW, Kheifets L, Larson SR, ,Wharton MD (1985) Mortality among members of a heavy construction equipment operators union with potential exposure to diesel exhaust emissions. Br J Ind Med 42:435–448

Woskie SR, Smith TJ, Hammond SK, Schenker MB, Garshick E, Speizer FE (1988a) Estimation of the diesel exhaust exposures of railroad workers: I. Current exposures. Am J Ind Med 13:381–394

Woskie SR, Smith TJ, Hammond SK, Schenker MB, Garshick E, Speizer FE (1988b) Estimation of the diesel exhaust exposures of railroad workers: II. National and historical exposures. Am J Ind Med 13:395–404

2. Risk Assessment for Radon Inhalation Based on Animal Exposure Data and Human Epidemiology

F. Steinhäusler[1]

The Slow Emergence of a Problem

Large-scale occupational radon exposure in the industrial age started in the 1930s owing to the increasing demand from industry and medicine for radium (^{226}Ra). In the 1940s, owing to the development of nuclear weapons, the scale of the uranium mining industry increased further (Steinhäusler 1989). In the meantime, in Europe there had been indications of a causal relationship between elevated occupational radon exposure and increased risk for lung cancer induction among German and Czechoslovakian miners as early as 1879 (Steinhäusler 1988a). However, possibly because of communication problems, in 1951 the New York Operations Office (NYOO) in the United States still considered an atmospheric radon concentration of 3.7 Bq/l as safe. A survey in 36 uranium mines showed that in reality few operators kept to the NYOO limit: the median radon concentration exceeded the limit by a factor of 19 (Bale 1951).

An additional problem was the general misconception of the significance of inhaled radon decay products. It was only at the beginning of the 1950s that the first dosimetric calculations by BALE (1951) identified radon daughters as the main contributors to the inhalation dose.

In the 1960s and 1970s, national and international organisations frequently lowered the recommended limit. However, they also occasionally increased the values again on grounds of "practicality for industrial applications" (Steinhäusler 1988a). In 1981 the International Commission on Radiological Protection (ICRP) issued its long-awaited recommendations for radon daughters at 4.8 working level months (WLM) per year (ICRP 1981). Figure 2.1 compares the

[1]University of Salzburg, Institute of General Biology, Biochemistry and Biophysics, Division of Biophysics, Hellbrunnerstrasse 34, A-5020 Salzburg, Austria.

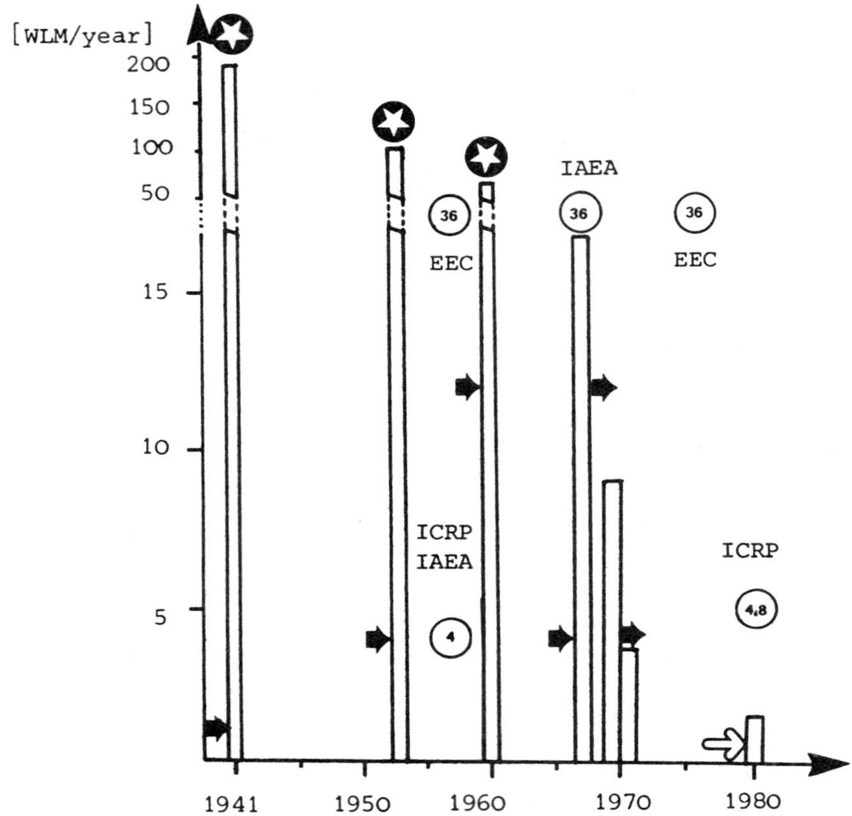

Figure 2.1. Comparison of average exposure of United States underground uranium miners to radon daughters (with recommended limits) between 1941 and 1981. *Stars within circles*, estimated values; *black arrows*, United States limits; *figures within circles*, ICRP, EEC, and IAEA limits; *white arrow*, United States union petition

actual average exposure of underground uranium miners in the United States with the limit recommended at the time. In Europe limits were in excess of ICRP recommendations until the late 1970s. In the United States a significant number of miners exceeded the recommended limits until 1967. Since then average annual exposures have continued to decrease down to the present levels which are equivalent to less than 1% of the values 40 years ago.

In comparison to occupational radon exposure-related hazards, the awareness that radon also causes the highest radiation doses to members of the public is comparatively recent. Although first detailed results regarding indoor exposure in Sweden were already published in 1956 (Hultquist 1956), the international scientific community took little notice. Natural radiation (NR) exposure continued to be associated predominantly with cosmic rays, terrestrial gamma radiation and dietary nuclide uptake. The total dose from all these NR sources remained almost as an "NR constant" equal to 1 mSv per year until 1982. In 1983 the first frequency distribution of individual NR-related risks was published for a typical European urban environment (Steinhäusler et al. 1983). Also the first results of a detailed nationwide indoor radon survey correlated with lung cancer incidence became available (Letourneau et al. 1983). In 1982 the radon contribution alone was estimated by the United Nations Scientific Committee on the Effects of Atomic Radiation (UNSCEAR) to result typically in 2 mSv per year (UNSCEAR 1982). In 1987 the total average effective dose equivalent from NR sources to members of the population of the United States was assessed as 3.0 mSv per year (NCRP 1987).

Scientific Basis for Radon-Induced Lung Cancer Risk Assessment

Assumptions and Facts in Epidemiology

In most industrialized countries death rates due to respiratory diseases have been increasing over the past decades as a result of smoking habits and atmospheric pollution. However, there are large regional differences, reflecting the superposition of many influencing socioeconomic, physiological and environmental variables in the induction of lung cancer. In Europe in the mid-1980s, values of national lung cancer rates differed up to a factor 15 (Österreichisches Statistisches Zentralamt 1987). Since radiation-induced cancer cannot be differentiated from other cancers, any additional lung cancer cases due to elevated radon decay product (Rn-d) exposure have to be detected against a largely variable background of "spontaneous" lung cancer.

Human Rn-d exposure has been studied extensively for uranium miners in the United States and Czechoslovakia (Table 2.1). Relatively large uncertainties are associated with dosimetry and lung cancer diagnosis. As an example of the impact of an influential parameter on the overall outcome of risk assessments, lung cancer mortality is compared for the absolute vs. the relative risk projection model. The absolute model assumes a constant (dose-dependent) risk rate for the complete cancer expression period. The relative model is based on the assumption of proportionality between the radiation-induced and the "spontaneous" cancer rate, thereby necessarily also including a Rn-d contribution from "normal" background levels. Particularly for exposure at younger ages, the results obtained by the two models differ by almost 100% although exposure conditions

Table 2.1. Characteristics of exposure and lung cancer data for uranium miners and observed vs. expected number of lung cancer cases.

Population	Exposure conditions	Characteristics of data on lung cancer	Range of effective dose equivalent (mSv)	O/E
Uranium miners in USA				
High degree of job mobility; some were previously hard-rock miners; 70% smokers	Estimated values in 1950s (based on one measurement per mine per year); by 1967: still 50% of all exposure data estimates; subsequently large number of Rn-d measurements; no quantitative assessment of exposure to non-radiological hazards	About 185 lung cancer cases in 1977; truncated follow-up period; only 82% of original pathological findings confirmed as lung cancer; many miners had "chronic cough"; 4% "possible" lung cancers included but not identified as such; in 33 cases death certificate still outstanding	0–1010 1010–2030	0.8 3.1
Uranium miners in Czechoslovakia				
Unknown whether data include *all* lung cancer cases (e.g. also political prisoners)	120 000 radon measurements, but no data published on radon equilibrium value prior to 1960; since then additional Rn-d measurements; well-documented exposure to toxic metal ore dusts, fungi, arsenic	About 280 lung cancer cases in 1979; truncated follow-up period; histopathological information poor; deficit of epidermoid tumours compared to controls; some tissue samples reported as "unclassifiable"	< 425 425–840 840–1270	2 1.6 2.5

O/E, observed vs. expected number of lung cases.

are the same (Schindel et al. 1988). Using very conservative assumptions for the variability of the input data in order to avoid extremes, it was calculated that the range of uncertainties of the number of lung cancer cases due to Rn-d exposure is at least of the same order of magnitude as the total number expected (Steinhäusler and Hofmann 1985).

Any epidemiological research on radon-induced lung cancer has to quantify the influence of smoking based on: tobacco consumption rate and total amount, the degree of inhalation and data on the age when smoking began (and stopped). Each of these variables can cause the lung cancer mortality rate to change up to a factor of 4 (Fraumeni 1975). Furthermore, the changing ratio of tar to nicotine in tobacco blends during follow-up periods over the decades needs to be taken into account. None of the available radon epidemiologies of miners has been able to provide all of these data. Also the problem of age- and sex-dependent individual "dosimetry for smoke inhalation" has been neglected so far. The topic of synergistic interaction between Rn-d and smoking is unresolved; data range from an additive to a multiplicative type of interaction (Steinhäusler 1988b). At present it can be assumed that—based on a stochastic multistep theory of carcinogenesis—alpha radiation from Rn-d acts as initiator of the cellular damage. Other carcinogenic substances, e.g. those contained in cigarette smoke or ore dust, act as promoters. This results in a reduced latent period for smokers as compared to non-smokers.

In the past, attempts have repeatedly been made to investigate a causal relationship between Rn-d exposure and lung cancer induction, also among non-mining populations. From 1950 to 1980 such studies concentrated on high background areas and radon spas (Dirnagl 1979; Vohra et al. 1982). These studies were able to correlate the exposure to elevated Rn-d levels with biological effects (e.g. atypical cells in sputum analysis and chromosome aberrations in some spa workers). However, epidemiological studies among 1500 workers in radon laboratories and radon spas revealed no increased lung cancer incidence (Bogolyubov and Andrejew 1985). In the 1980s the first results from large-scale indoor studies in Canada, Europe and the United States became available (Clemente et al. 1980; Steinhäusler 1985; Hopke 1987).

In Canada, lung cancer mortality and indoor Rn-d concentration were correlated in 18 cities (Letourneau et al. 1983). Measurements were carried out during summer seasons only. Analysing data on lung cancer, smoking habits, socioeconomic indicators and Rn-d concentration, a carcinogenic hazard could neither be supported nor eliminated. Since the study was also unable to confirm the well-established association between smoking and lung cancer for women, this illustrates the inadequacy of comparing large population groups instead of case-control studies.

In Sweden, lung cancer in women was found to be associated with the number of years lived in dwellings with ground contact and therefore elevated radon concentration indoors (Svensson et al. 1989). The dose-determining Rn-d exposure was not measured, instead annual radon levels were derived from a single 2-week

Table 2.2. Summary of animal inhalation studies at the University of Rochester (USA), COGEMA (France) and Battelle Pacific Northwest Laboratory (USA).

	Animals used	Atmosphere inhaled	Cumulative exposure (WLM)	Biological effect
University of Rochester	80 dogs	Radon conc. (1.8 − 3.7 × 10^7 Bq/m^3)	200–10 000	No carcinogenic dose effect; lifespan shortening; destructive lesions in all lungs
	100 rats	AMAD: 0.2 μm Unattached ^{218}Po: 20%	~ 26 000	
	2000 mice		14 000–72 000	
COGEMA	10 000 rats	Radon conc. (2.8 − 4.6 × 10^7 Bq/m^3 Equilibrium factor: 1 − 100% Cigarette smoke and dust added	20–8500	> 800 lung cancers induced; no apparent threshold; uranium ore dust showed little influence
Pacific Northwest Laboratory	612 hamsters	Radon conc. (3.7 − 11 × 10^6 Bq/m^3;	10 000 (mean)	Lifespan shortening (dogs, rats); 364 lung tumours induced (rats);
	912 rats	Equilibrium factor: 10%–60%	320–10 250	
	69 dogs	Diesel exhaust, cigarette smoke and dust added	12 700–13 500	respiratory tract cancers also in dogs

integrated radon sample per dwelling. Exposure during the previous 5 years was not included. Where information on the type of dwelling for some of the time was lacking, the average exposure during the years for which there were data available was used for the missing years. The results showed that a radon exposure between 4500 and 600 Bq/m³ per year resulted in a relative risk of 1.4%. The impact of the low quality of the individual exposure assessment is indicated by the fact that for radon exposures greater than 6000 Bq/m³ year the relative risk is below 1.

In the United States, average indoor radon levels were compared with the average lung cancer rates for the county in which each measurement was taken (Cohen 1988). Under the assumption of a no-threshold dose-effect relationship, theoretically female lung cancer rates based on the average radon value should have been 25% higher than the national average. Instead, the data indicate a 30% decrease. Since participating persons were home owners concerned about a potential radon problem, the study cannot be considered random. An additional randomized survey comprising the highest and lowest United States lung cancer rates revealed that the radon level for counties with low lung cancer rates was much lower than predicted, and vice versa in counties with high lung cancer rates. The same limitation regarding study design as mentioned for the Canadian example also applies in this case. Neither the Canadian nor the United States study takes into account quantitatively any long-term changes of indoor Rn-d levels (e.g. home improvements) and individual mobility over the past 20–40 years.

In summary, it can be said that epidemiological results from non-mining populations are inconclusive and range from lack of correlation, to positive and even negative correlation. However, they show that any such Rn-d-induced lung cancer increase is so small that it requires more refined epidemiological methods than those applied hitherto.

Despite all the deficiencies mentioned, high cumulative Rn-d exposure of miners resulting in bronchial epithelial doses in excess of 2000 mGy is linked with a statistically significant increase of the lifetime lung cancer risk of about 2×10^{-5}/mGy (NCRP 1984). However, in the dose range important for present conditions, i.e. at bronchial epithelial dose levels below 2000 mGy, the epidemiological results for miners are inconclusive: the ratio of the observed to the expected number of lung cancer cases ranges from 0.8 to 3.1.

On the Validity of Extrapolating from Radon-Induced Cancer in Animals to Man

Animal Rn-d inhalation studies have the inherent advantage of defined exposure conditions. The most extensive of such studies have been carried out in France at the Compagnie Générale des Matières Nucléaires (COGEMA) and in the

Table 2.3. Lifetime risk coefficient (r) for lung tumour induction in rats after Rn-d exposure.

Reference	Cumulative exposure (WLM)	Risk (r) Model A (10^{-5}/mGy)	Risk (r) Model B (10^{-5}/mGy)
Chameaud et al. 1981,1982,1984	23	5.8	1.9
	50	4.5	1.5
	290	2.5	0.8
Cross 1988	320	3.6	1.2
	322	3.5	1.1

Model A, 1 WLM $\hat{=}$ 13 mGy; model B, 1 WLM $\hat{=}$ 40 mGy.

United States at the Pacific Northwest Laboratory and at the University of Rochester. In Table 2.2 the data from these studies are summarized (Cross 1988). In the following, no experiments with hamsters have been used because of the known low lung cancer sensitivity of these species.

In early animal studies, cumulative exposure exceeded occupational exposures by far. This resulted predominantly in cell killing rather than transformation, which explains the reduced efficiency of lung cancer induction and the observed life span shortening at cumulative exposures $\geq 10^4$ WLM.

Table 2.3 shows the lifetime risk coefficient (r) for rats exposed to Rn-d levels comparable to human exposure conditions. Severe dosimetric problems exist with regard to standardizing originally exposure-related rat data to dose values for comparison with data derived from epidemiology. Significant morphometric differences exist between the two respiratory systems: the rat has a monopodial airway branching structure, the human lung represents a symmetrically dichotomously dividing system (Yeh and Schuh 1979). Analysis of experimental morphometric data for both species confirm differences concerning the distribution of branching angles and the termination probability of airways (Koblinger and Hofmann 1988). This causes a species-dependent dose distribution, i.e. the ratio lung dose to mean whole body dose is 40% higher for the rat than for man (Bogolyubov and Andrejew 1986).

Assuming that—despite the above differences—the same exposure-dose conversion factor can be used for rat and man (model A, Table 2.3), r values range dose-dependently from about 3×10^{-5} per milligray to 6×10^{-5} per milligray. However, in an attempt to take into account morphological and physiological differences, a scaling factor (f) for the smaller rat lung can be derived from age-dependent lung dosimetry, i.e. $f = 3$ for newborn humans vs. adults (James 1988; model B, Table 2.3). In this case r values are considerably lower and of the same magnitude as those derived from epidemiology. Further dosimetric studies are needed in the future to resolve this issue.

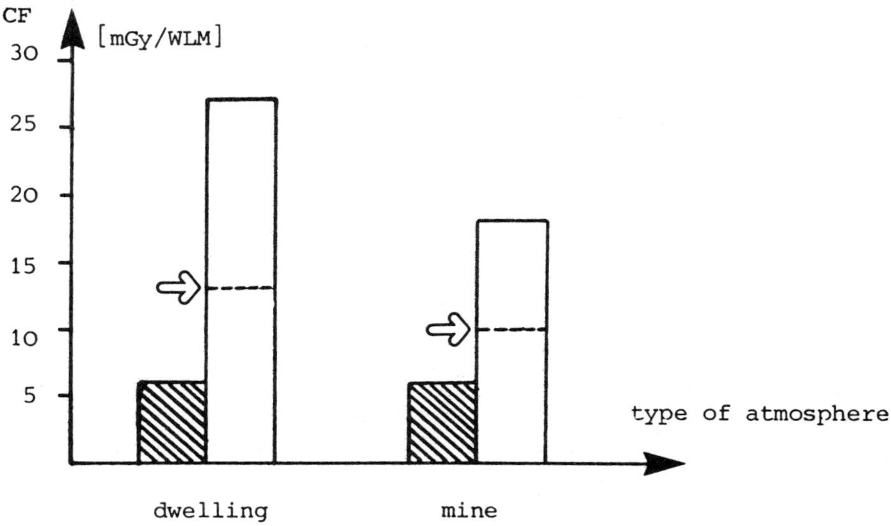

Figure 2.2. Ranges of Rn-d dose-exposure conversion factors (*CF*) for sequental bronchi after inhalation of air indoors and underground. *Hatched columns*, minimal value; *open columns*, maximal value; *arrows*, reference value. From James, Lung Dosimetry © copyright 1988. Reprinted with permission of John Wiley Sons, Inc.

Dosimetric Modelling: Its Usefulness and Limitations

Lung dosimetry is carried out in five stages: choice of physiological lung model; calculation of probability for Rn-d to be deposited in the tracheobronchial (T-B) region of the lung as the site of occurrence of most radon-induced lung tumours; calculation of Rn-d activities resulting from inhalation over a certain period; choice of Rn-d exposure-absorbed dose conversion factor; calculation of effective dose equivalent based on average cell dose, radiation quality and weighting factor. The above steps are dependent on the superimposed nfluence of various physical and biological parameters: particle size distribution of Rn-d-attached aerosols, unattached fraction of Rn-d, alpha radiation quality factor, dose weighting factors, breathing rate, geometry of airways, depth distribution of target cells in the T-B region, and variability of the thickness of the mucus layer. Each of these parameters is a function of either environmental conditions, or differences between individuals (age, sex). Using conservative estimates, this results in a variation in the effective dose equivalent by a factor of at least 2 (Nuclear Energy Agency 1983; Senes 1984). Environmental parameters alone can cause exposure-dose conversion factors to differ by a factor of about 4 depending on the type of atmosphere inhaled (Fig. 2.2; James 1988).

Table 2.4. Estimated number of Rn-d-induced lung cancer cases (*n*) per unit dose to the bronchial epithelium based on risk assessment by different sources.

Reference	$n \times 10^{-5}$/mGy	Remarks
NCRP 1984	1.3	For homes; assumed risk 9.1×10^{-3}/WLM.year
ICRP 1981 (modified)	1.6	For homes; assumed risk 10^{-5}/WLM, year and revised dosimetry (James 1988)
BEIR 1989	2.7	For homes; multiplicative interaction between smoking and Rn-d exposure
UNSCEAR 1982	2–6	For mines
EPA 1986	1.9–7.2	For homes; assumed risk 1.3%–5%

In summary, it can be said that Rn-d dosimetry represents a valuable tool in standardizing doses resulting from Rn-d inhalation in different environments. However, at present the uncertainty associated with the assessment of the effective dose equivalent of an individual can only be estimated.

Risk Assessment in Perspective

Using the available epidemiological data from miners, together with data from dosimetric modelling and animal inhalation experiments, the risk due to low-level Rn-d exposure can only be assessed with large uncertainties (Table 2.4). The estimated number of lung cancer cases ranges from $1.3–7.2 \times 10^{-5}$ per milligray to the bronchial epithelium of the lung, with a "centre value" of about 3×10^{-5} per milligray.

In Western industrialized countries it can be assumed that the lifetime dose to the bronchial epithelium resulting from "normal" Rn-d exposure ranges typically from about 200 to 250 mGy, representing an average risk of about 0.7% under the above assumptions. Applied to Europe and the United States, this would mean that about 6 million lung cancer cases may be induced by Rn-d exposure indoors. This coarse estimate does not take into consideration sex and age distribution and differences in characteristic tobacco consumption. In the following, this risk is compared to other risks society is exposed to. An upper limit of acceptability is provided by the fact that, in the United States, regulatory action has been taken into consideration in all cases if lifetime risk was above 0.4% (Bengtsson 1988). A lower limit can be seen in the recommendation by the Swedish Environmental Protection Authority suggesting an "acceptable lifetime risk" from chemical carcinogens of 0.001%. With regard to the radon problem, a lung cancer risk in excess of 1% (i.e. Rn-d levels > 150 Bq/m^3) requires a sociopolitical decision regarding its acceptability as in the case of limiting the risk of premature death due to, for example, car accidents (~ 2%).

References

Bale WF (1951) Hazards associated with radon and thoron. Memorandum. US Department of Energy, EML Archive, New York

Bengtsson G (1988) Comparison of radiation and chemical risks. SSI-report no. 88-18, ISSN 0282-4434, Statens stralskyddsinstitut, Stockholm

Bogolyubov VM, Andrejew SV (1985) Zaozenkata na riska pri radontherapia. Kurortologia i Fiziotherapia 22/1:27–33

Bogolyubov VM, Andrejew SV (1986) A technique for computation of radiation intensity during radon treatment in experiment (in Russian). Woprosi Kurortologie 4:47

Chameaud J, Perraud R, Lafuma J, Masse R (1981) Contribution of animal experimentation to the interpretation of human epidemiological data. In: Gomez M (ed) Proceedings of the international conference on radiation hazards in mining: control, measurement and medical aspects. Kingsport Press, Kingsport

Chameaud J, Perraud R, Lafuma J, Masse R (1982) Cancer induced by Rn-222 in rat. In: Clemente GF, Nero RV, Steinhäusler F, Wrenn ME (eds) Proceedings of the specialist meeting on assessment of radon and daughter exposure and related biological effects. RD Press, Salt Lake City, p 198

Chameaud J, Masse R, Morin M, Lafuma J (1984) Lung cancer induction by radon daughters in rats (present stage of the data on low-dose exposures). In Stocker H (ed) Proceedings of the international conference on occupational radiation safety in mining. Canadian Nuclear Association, Ontario, p 350

Clemente GF, Nero AV, Steinhäusler F, Wrenn ME (eds) (1980) Proceedings of the specialist meeting on assessment of radon and daughter exposure and related biological effects. RD Press, Salt Lake City

Cohen B (1988) An experimental test of the linear-no threshold theory of radiation carcinogenesis. Proceedings of the American chemical society, 196th national meeting. American Chemical Society, Los Angeles

Cross FT (1988) Evidence of lung cancer from animal studies: In: Nazaroff W, Nero AV Jr (eds) Radon and its decay products in indoor air. Wiley, New York, pp 373–406

Dirnagl IK (1979) Grundlagen der Radontherapie. Z Angew Bäder Klimaheilkd 26/4:1–112

Fraumeni JF (1975) Persons at high risk of cancer. Academic, New York, p 131

Hopke PK (ed) (1987) Radon and its decay products: occurrence, properties and health effects. Am Chemical, Washington DC (ACS Symposium Ser, no. 331)

Hultquist B (1956) Studies on naturally occurring ionizing radiations with special reference to radiation doses in Swedish houses of various types. Kungl Svenska vetenskapsakademius handligar, vol 6/3

International Commission on Radiological Protection (ICRP) (1981) Limits on inhalation of radon daughters by workers. Ann ICRP, vol 6/1

James AC (1988) Lung dosimetry. In: Nazaroff WW, Nero AV Jr (eds) Radon and its decay products in indoor air. Wiley, New York, pp 259–309

Koblinger L, Hofmann W (1988) Comments on the rat lung as a human surrogate in inhalation studies. In: Proceedings of the 7th IRPA Congress, Sydney, vol III. Pergamon, Sydney, pp 1407–1410

Letourneau EG, Mao Y, McGregor RG, Semenciw R, Smith MH, Wigle DT (1983) Lung cancer mortality and indoor radon concentrations in 18 Canadian cities. In: Maxey MN (ed) Proceedings of the Health Physics Society. Albuquerque, pp 470–483

National Council on Radiation Protection and Measurements (NCRP) (1984) Evaluation of occupational and environmental exposures to radon and radon daughters in the United States. NCRP report no 78, Bethesda

National Council on Radiation Protection and Measurements (NCRP) (1987) Ionizing radiation exposure of the population of the United States. NCRP report no 93, Bethesda

Nuclear Energy Agency Group of Experts (NEA) (1983) Dosimetry aspects of exposure to radon and thoron daughter products. OECD, Paris

Österreichisches Statistisches Zentralamt (1987) Bericht über das Gesundheitswesen in Österreich im Jahre 1986. Österreichisches Statistisches Zentralamt, Vienna, p 233

Schindel F, Henrichs K, Paretzke H (1988) Einfluß von Zeit-Projektions-Modellen auf Schätzungen von Krebsrisiken durch Radionuklide. In: Hefner A, Steger F (eds) Proceedings of the 4th European International Radiation Protection Association. Austrian Association for Radiation Protection, Vienna, pp 296–299

Senes Consultants Ltd (1984) Assessment of the scientific basis for existing federal limitations on radiation exposure to underground uranium miners. Report for the American Mining Congress, Toronto

Steinhäusler F (1985) European radon surveys and risk assessment. In: Gammage RB, Kaye SV (eds) Indoor air and human health. Lewis, Chelsea, p 109

Steinhäusler F (1988a) Occupational and non-occupational radon exposure – a story of trial and error. In: Kvasnicka J (ed) Proceedings of the international workshop on radiation protection in mining. Department of Mines and Energy, Darwin, Australia, pp 301–317

Steinhäusler F (1988b) Epidemiological evidence of radon-induced health risks. In: Nazaroff WS, Nero AV Jr (eds) Radon and its decay products in indoor air. Wiley, New York, pp 311–371

Steinhäusler F (1988) Radon spas: source term, doses and risk assessment. Radiat Prot Dosim (24), pp 257–259

Steinhäusler F, Hofmann W (1985) Inherent dosimetric and epidemiological uncertainties associated with lung cancer risk assessment for mining populations. In: Stocker H (ed) Proceedings of the international conference on occupational radiation safety in mining. Canadian Nuclear Association, Toronto, p 327

Steinhäusler F, Hofmann W, Pohl E, Pohl-Rüling J (1983) Radiation exposure of the respiratory tract and associated carcinogenic risk due to inhaled radon daughters. Health Phys 45:331

Svensson C, Pershagens G, Klominek J (1989) Lung cancer in women and type of dwelling in relation to radon exposure. Cancer Res (in press)

United Nations Scientific Committee on the Effects of Atomic Radiation (UNSCEAR) (1982) Ionizing radiation: sources and biological effects. Report to the General Assembly with Annexes, United Nations, New York, USA

US Environmental Protection Agency (EPA) and Department of Health and Human Services (1986) A citizen's guide to radon. Report OPA-86-004

US National Academy of Sciences, Committee of the Biological Effects of Ionizing Radiation (BEIR report) (1989) Health risks of radon and other internally deposited alpha emitters (in press)

Vohra KG, Pillai KC, Mishra UC, Sadavisan S (eds) (1982) Natural Radiation Environment. Wiley, Delhi, India
Yeh HC, Schuh GM (1979) Anatomic models of the tracheobronchial and pulmonary regions of the rat. Anat Rec 195:483–492.

PART II

Types of Evidence: General Strengths and Weaknesses

Section 1. Epidemiological

3. Inhalation Hazards: The Interpretation of Epidemiologic Evidence

J.C. Bailar[1]

Introduction

This paper deals with the interpretation of epidemiologic evidence, with special attention to inferences about causes (inhaled toxicants) and effects (impact on human health). I am quite convinced that epidemiologic evidence cannot be interpreted in isolation from other kinds of evidence. Thus, animal studies and clinical investigations, in particular, will be discussed in detail. I will show one way to integrate all of these forms of information by integrating the great strengths of epidemiology with strengths of other approaches. Finally, I will show how relevant, but possibly flawed, epidemiologic data can be used to resolve some of the important issues of public health and public policy.

Characteristics of Data

Epidemiologic data often become controversial precisely because they are important—that is, because they are seen to be relevant to some major issue of public health. Examples are the epidemiologic studies of the relation between cigarette smoking and health, the effects of chemical food additives, and the benefits of reducing urban air pollution. But epidemiology does not stand alone. For these examples and countless others, data of many types and from a wide variety of sources, including epidemiology, are brought to bear on the problem. One or a few research studies are rarely enough. The collection of data that must be addressed commonly has four features (Bailar and Thomas 1985):

1. It is *vast*, including hundreds or thousands of published scientific papers and countless government reports and other documents.

[1]Office of Disease Prevention and Health Promotion, US Department of Health and Human Services, 468 N Street, S.W., Washington, D.C. 20024, USA, and Department of Epidemiology and Biostatistics, McGill University School of Medicine,* 1020 Pine Ave. West, Montreal, P.Q. H3A 1A2, Canada.

2. It is *complex*, involving a dozen or more scientific disciplines as well as many different technologies; few individuals are equipped to really understand all of the important contributions or (especially) the complex relationships among these contributions.
3. It is largely *of poor quality*, when judged against the standards of present needs—some studies are old, from years when research capabilities were more limited; some are poorly designed, or poorly executed; and some show deficiencies in the protocol, or the published record is silent on critical matters, such as whether subjects were properly randomized, what steps were taken to assure quality control of laboratory measurements, and how the statistical analysis was carried out.
4. It is *off the main target* of interest; common examples include the use of animal studies to estimate effects on humans, extrapolations from high doses to lower doses, and the use of biologic markers to infer the incidence of more serious health problems at a later time.

It is, to me, remarkable that each of these four problems has been prominent in every major health issue I have worked on, including optimal therapy for certain cancers, mammography to detect early breast cancer, the health effects of environmental chemicals, microbial contamination of food products, the late effects of immunization against several diseases, and the effects of chronic loud noise on hearing. We should not be surprised to see that these four features appear in studies of toxic inhalants, too.

If we accept that these features are common in the study of inhalation hazards, or of human health hazards more generally, it may be helpful to prepare for them early, because the most serious problems can be prevented or mitigated by early identification and appropriate planning (Table 3.1). For example, early planning to deal with a large information base can focus on a single, comprehensive review of data sources (rather than several incomplete and overlapping reviews), with concurrent steps to collect the data and make them useful for decisions—that is, to organize, index, abstract, and otherwise prepare the mass of information for use by the scientists and others who must develop a comprehensive analysis and a set of scientific recommendations.

Then, given that the information will be of many quite different types, one should identify those types early, determine what kinds of expertise are needed to work with them, and generally organize these professional efforts to maximal effect. Epidemiologic competence is almost always necessary here. Unfortunately, there is often a shortage of both time and money, and such limitations add to the need for early planning. There are difficult questions at this stage about leadership. In my experience, it is more important that the leader of this integrative effort be just that—a leader—rather than a renowned expert in one of the relevant scientific disciplines.

Difficult questions about data quality cannot be avoided entirely, but many such questions may be dealt with by setting up some organized means to assess

Table 3.1. The organization and process of risk assessment.

Organization	Process
1. Size of information base	Identify sources
	Collect data
	Organize and index
	Abstract
	Convert to machine form, etc.
2. Complexity of information base	Identify needs for expertise
	Find and recruit experts
	Organize joint efforts, within constraints
	Report, etc.
3. Quality of data	Organize quality assessment
	Choice of sources to use, etc.
4. Data not on target	High vs. low doses
	Animal vs. human subjects
	Route of administration
	Schedule and duration
	Demographic subgroups
	Structure-activity, etc.

quality and screen out work that does not meet minimal standards. An example, which included inhalation toxicology, is in a recent report from the United States National Academy of Sciences (1984).

Efforts to reduce problems in the fourth category, data not on the point of direct interest, should begin with a clear understanding of what would be desirable (a "thought experiment"); this is often some epidemiologic study that is simply impossible because of cost, unavailability of enough subjects, ethical concerns about human experimentation, etc. New research can be considered at this stage, but will not be feasible when answers are needed very soon. Available studies are then reviewed to see how close some combination can come to providing what is needed. I will refer later to the need to rely on the *weight* of evidence (all of it, with attention to research quality and relevance) rather than on the *strength* of evidence (one, or just a few, of the "best" studies, regardless of whatever weaknesses they have). It is at this point also that we first come to the general point that plausible epidemiologic findings should not be dismissed as artifact unless there is strong contrary evidence of other kinds. It is true that epidemiologic evidence can rarely prove risk, or give precise estimates of the size of risk, but such evidence can be sufficiently compelling to justify substantial public health action, including law and regulation. Further, epidemiologic data are not symmetric with respect to the interpretation of positive vs. negative findings. Positive findings, even from weak studies, must generally be accorded more weight because they avoid problems about statistical power, have high relevance to human risk, and engage the corresponding asymmetry between the consequences of false-negative and false-positive conclusions.

Table 3.2. Some common problems with all types of human exposure data.

High variability of human exposure, past and present
Time to time
Person to person
Lead times of decades
Synergy
Questions of what to measure:
Peak exposure vs. time-weighted average (short or long)
Short-term vs. lifetime
High correlations
Incomplete and inaccurate monitor systems
Ambient vs. indoor vs. personal monitors
Sample to sample variation
High costs, and small samples

Self-selection, and confounders
Nonresponse, and incomplete follow-up
Reporting errors
Investigator or interviewer bias

Strengths and Weaknesses of Human Data

The defects of epidemiologic studies of inhaled toxicants can be substantial, and the most critical of them span the whole range of epidemiologic approaches: cross-sectional prevalence studies, the usual kinds of case-control and cohort studies, clinical research including intensive investigation in special respiratory or metabolic wards, and the population-based experiments we call randomized controlled trials, or randomized clinical trials. These defects appear on both the input (exposure) and output (health) sides of the cause-effect linkage. Tables 3.2 and 3.3 give examples. Table 3.2 is divided, with topics that I consider largely epidemiologic above the dotted line and biostatistical topics below the line. Thus the table helps to show how each of the listed problems can be attacked by epidemiologic or statistical means, and shows why both kinds of competence are

Table 3.3. Some common problems on the output (health) side.

Long lead times (again)
Multiple causes of most chronic diseases
Inaccurate and incomplete ascertainment—especially for non-lethal and nonquantal effects
Heterogeneity in individual susceptibility
Synergy
Rare outcomes
Creeping center of the information base

Table 3.4. Recommended limits for occupational exposure to benzene in air.

	ppm	
1941	100	Drinker[a]
1947	50 (8-h TWA)	ACGIH[a]
1948	35 (8-h TWA)	ACGIH[a]
1957	25 (8-h TWA)	ACGIH[a]
1963	25 (Ceiling)	ACGIH[a]
1969	10 (8-h TWA)	ANSI[a]
1971	10 (8-h TWA), 25 (acceptable ceiling), 50 (maximum, 10 min)	OSHA[a]
1987	1 (8-h TWA)	OSHA

Modified from Graham et al. 1988. Harvard University Press, Cambridge. Reprinted with permission.
TWA, time-weighted average; ACGIH, American Council of Governmental Industrial Hygienists; ANSI, American National Standards Institute; OSHA, Occupational Safety and Health Act.

essential. (The table refers to disciplines, not individual expertise. There is, in fact, no clear dividing line between the professions of epidemiology and biostatistics; one cannot be good at either of these without being good at the other, too. Efforts to separate the professions, as opposed to the issues, are foolish, destructive, and sure to fail.)

The last point in Table 3.3, which I call the creeping center of the information base, refers to the fact that increasing information about some hazard is much more likely to show that it is bigger than we thought rather than to show that it is smaller. Table 3.4 shows how this has happened for benzene (Graham et al. 1988; J.D. Graham, personal communication). This is illustrated by some unpublished work I did in 1982 with my colleagues Judy Jackson and Emmet Keeler. We tracked the individual exposure recommendations published each year by the American Council of Governmental Industrial Hygienists (ACGIH), a private non-profit group of experts from government, academic, and research institutions, and private industry. These standards are sometimes changed, and reasons for the changes are usually given.

We found that the ACGIH had published standards for 554 chemicals. Most of these were recently published, with little time for possible revisions, but 175 standards had been changed a total of 222 times. Of these, 185 changes, or 83%, were in the direction of reducing allowable exposures. Of these 222 changes, 114 were on the basis of new evidence (33 animal studies, 81 human studies), 46 were on the basis of new interpretations of old evidence, and 35 seemed to be from concern that margins of safety were too small; we could not classify 27 changes. The conclusion seems inescapable that, at least in the experience in the United States, exposure limits tend to become tighter with time, and these changes are driven largely by increases in scientific information. It would seem good public policy to recognize this tendency and to set initial exposure limits somewhat lower than present evidence justifies.

Research Strategy

These various problems and issues lead to two critically important features of broad research strategy on external hazards to human health. The first is to have a variety of approaches in both the methods and the types of data. No category is perfect, and indeed no hierarchy of categories is universally valid. We cannot say, for example, that randomized clinical trials are always better than rodent studies, or vice versa. No single method dominates the others, and one may often find that weaknesses in one area can be overcome, partially or entirely, in another. For example, the strengths of an epidemiologic observation on the species of interest—man—may be affected by lack of precise exposure data, poor definition of dose-response relationships, and incomplete understanding of biologic mechanisms. But occupational and environmental exposures can sometimes be measured more precisely in current workers, and, if processes have not changed, one can assume that past exposures were similar. Thus, present information can strengthen older studies on other workers. Similarly, dose-response relationships can be investigated in several species of laboratory animals, and for related chemical compounds; if these all lead to similar results, one can have at least modest confidence in assuming that the relationship holds for humans. Mechanisms can be investigated in other subjects, human and animal, and extended to the full epidemiologic study group unless there are specific reasons to the contrary. Thus, in general, a variety of approaches in methods and data, with available epidemiologic observations at the core, can provide a stronger base for inference than any one type of study, however well it may be done.

Table 3.5 lists some of the choices and differences in approach. One critical difference is between observational studies (which include most epidemiologic

Table 3.5. The variety of approaches: categories of research subjects and methods.

Human vs. animal
Healthy vs. diseased subjects
Intact subjects vs. biopsy specimens, cell cultures, organ perfusions, etc.
Health endpoints vs. markers
Physiologic variations in exposure vs. extremes
Free-living (or feral) subjects vs. institutional setting (ward or animal laboratory)
Duration and schedule of exposure
Duration of follow-up observations
Experimental vs. observational; experimental often having:
Assignment of treatment by investigator
Structured observations
Internal controls, perhaps of several types
Deliberate selection of homogeneous subject groups
Formalization of prior hypotheses

Table 3.6. Weight of evidence: empirical criteria for inferences of causality.

Strength of association (generally as relative risks or odds ratios)
Dose-response relationships
Temporally correct association

Consistency of association
Specificity of association
Biologic plausibility

From National Academy of Science 1989, National Academy Press, Washington, D.C. Reprinted with permission.

research) and experimental studies (which include most clinical and laboratory research). I have not found any satisfactory definition of the precise difference between these approaches, and perhaps none is needed, but the table does list five items that may help to define "purely" experimental vs. observational research. An experimental study, which generally has a detailed, written protocol, seems to be defined by having most or all of the following features:

- The investigator assigns specific treatments to subjects.
- Observations are structured, in the sense that the protocol calls for certain observations at specified times or at the time of specified events.
- The study itself contains at least one set of observations on the untreated state; this may be by randomization, and it includes parallel, crossover, and self-control designs (Bailar and Mosteller, 1986).
- Another common feature of experimental studies is the deliberate selection of homogeneous study groups. While this is often appropriate in research on why and how things happen, in some qualitative sense, it is often grossly inappropriate for assessing the *size* of risks, because it is not clear how to generalize to types of subjects not tested.
- Finally, experimental studies are more likely than observational studies to have research hypotheses fully developed before any data are collected or examined; this tends to limit the exploitation of serendipitous findings in experimental studies by keeping sample sizes and the range of observations to the minimum needed for the hypotheses at hand.

These features may be roughly, and just a little unfairly, summarized by saying that experimental studies tend to give more precise and more reliable information about the test systems they use, while observational studies, individually less precise and reliable, tend to focus on matters much closer to topics of real interest.

The second critically important feature of broad research strategy is the need to consider the whole weight of the evidence in the determination of whether the assembled information shows a cause-effect relation. Bradford Hill was one of the first investigators to formalize the integration of disparate kinds of information about human health; Table 3.6 shows a recent version of Hill's criteria adapted for studies of diet (National Academy of Sciences 1989), but I believe that these criteria are equally valid for studies of inhaled toxicants and human health. The stronger an association, as expressed by a larger relative risk or odds ratio, the more likely association is to reflect a cause-effect linkage. Confounders can readily produce odds ratios of as much as, say, 1.5 or even 2.0, and it make take very careful epidemiologic study to determine whether an observed relation smaller than 1.5 is one of cause and effect. On the other hand, it is hard to think of real issues with odds ratios of 10 or higher that do *not* reflect cause and effect. (By "real issues" I mean those where there has been genuine debate over the meaning of a large odds ratio.) With advancements in epidemiologic methods, the upper boundary on the range where causality may be doubtful has been declining over the years, and I expect it to decline still more.

Dose-response relationships can be critical to inferences of causality. When it is possible to examine ordered categories of exposure (even when measured exposures are unavailable), one expects higher doses to have more effect, so that the dose-response relationship is monotonic. Certainly, there are well-established phenomena in toxicology where this is not true, but when there is no evidence to the contrary, departures from monotonicity raise questions about causality.

Temporally correct associations also strengthen inferences of causality, though they are not essential. One might find, for example, that current exposure to indoor air pollutants is clearly linked to some pre-existing respiratory condition. Then, although the disease precedes the measured exposure, we might infer that present exposure is a surrogate for exposures prior to the disease, so that the observed (incorrect) temporal association still supports an inference that exposure causes disease. In general, however, one can draw stronger inferences from a "correct" temporal association.

The first three criteria in Table 3.6 can be applied to the evidence as a whole, but are more likely to be useful in evaluating single research studies. The next three apply only to the evidence as a whole, and not to single studies.

Consistency of association refers to the range of study methods and research subjects. If we find some effect in prevalence, case-control, and cohort studies, with supporting evidence from animal experiments, we may feel more confident about inferring cause and effect than if only one of these is positive (even strongly positive) and the rest show no effect.

Specificity, in contrast, refers to the range of endpoints: an exposure linked to just one or two specific health endpoints is more likely to indicate cause and effect than if the same exposure is linked to a large number of unrelated diseases. This is because many kinds of bias are insensitive to specific outcome, and hence raise or lower estimated risks rather broadly.

The last point, biologic plausibility, has two faces. One should look assiduously for evidence both for and against plausibility, then assess the relationship in question objectively.

These six empirical criteria provide the best guide I know for the study of possible causal relationships when the data have the features described above—that

Table 3.7. When is some action appropriate?

Continuum in strength of evidence	Continuum in possible actions
Almost nil	Tell your friends
Possible	Do more studies
	Publish technical papers
Nagging worry	Scientific reviews and editorials
Suggestive	Public statements
	Public information on campaign
Stronger	Labeling requirements
Persuasive	Enforceable tolerance limits
Conclusive	Outright ban

they are vast, complex, of poor quality, and off-target. Indeed, I cannot think of any situation where all six were satisfied but where later evidence argued persuasively against causation.

When Is Some Action Appropriate?

It is clear that the evidence supporting an inference of cause and effect about inhaled toxicants and health can range from nil to conclusive, with continuous gradations along the way. There is another continuum to consider, that of possible actions in response to the evidence. This second continuum is less familiar, and often reaches in several dimensions, but the many broad categories can be put in rough order. Some of these categories are shown in Table 3.7. Each of these, of course, contains its own ranges of more specific actions. A public statement can be strong or weak; labeling of a hazardous product can be minimal to fully informative, or more; tolerance limits can be high or low. The point is that evidence of various kinds and strengths can and should lead to different actions. We need stronger evidence to ban some useful and important product or process than we need to tell the public that there may be a risk.

Table 3.8. Some considerations in developing exposure standards and recommendations.

Size of risk, and size of estimated benefit
Possible unintended consequences, good and bad
Time patterns of exposure (e.g., peak; TWA; occupational; lifetime exposure)
Quantitative vs. qualitative recommendations
Benefits (to whom?) associated with the product or activity
Availability of substitutes
Recommendations for populations vs. those for subgroups or individuals (age, high risk persons, etc.)
Feasibility and likelihood of implementation
Degree of certainty about all of this

TWA, time-weighted average.

This leads directly to a host of questions about how to link the continuum in strength of evidence to the continuum in possible actions. It is important here to recognize that this is not a scientific question but a political one, and the body politic (generally through either consensus or its elected representatives) can quite appropriately consider many things besides the scientific evidence. Some of these are listed in Table 3.8. The response to evidence that some agent harms human health will, and should, reflect the role of economic factors, prevailing philosophy about risk, the societal impact of various kinds of controls, existing law and regulation, and many other things. Still, the science base should be the most critical input to any decision. And that means that we, as scientists, cannot simply do our research, publish our papers, and move on. Some of us, at least, must be willing to get involved in the important public issues that surround our work, and assure that our scientific findings are used correctly and to the public good.

I will close with a few more words about epidemiology. It has become a scientific discipline in its own right, and epidemiologists need and deserve the support of their peers in other disciplines if they are to contribute effectively to the solution of important problems of public health and public policy. Toward this end I offer four observations:

- Epidemiology looks much easier than it is.
- Epidemiology has its own well-developed body of concepts and techniques.
- Epidemiology has a variety of practitioners, some of whom (generally those trained as epidemiologists) are a lot better than others.
- Epidemiology often takes more time, effort, and money than research in other scientific fields.

To put these in a nutshell, if you want a professional job of epidemiology, get a professional and give him or her the needed resources. And do not blame the scientific discipline of epidemiology when untrained amateurs make mistakes.

Public and private decisions about inhalation toxicants can and will be made for the purpose of improving human health; a proper understanding of the full science base can improve those decisions; and the benefits of improved decisions can be very great. Sound epidemiologic research is the most critical component in this process.

References

Bailar JC III, Mosteller F (1986) Medical uses of statistics. New England Journal of Medicine Press, Waltham, MA

Bailar JC III, Thomas SR (1985) What are we doing when we think we are doing risk analysis? Basic Life Sci 33:65–76

Graham JD, Green LC, Roberts MJ (1988) In search of safety. Harvard University Press, Cambridge

National Academy of Sciences (1984) Toxicity testing. National Academy Press, Washington DC

National Academy of Sciences (1989) Diet and health: implications for reducing chronic disease risk. National Academy Press, Washington DC

4. Problems in Interpreting Epidemiological Data

P.N. Lee[1]

Introduction

In most epidemiological studies, subjects are not randomly assigned to the exposure of interest. Problems of causal inference from non-randomized studies have often been discussed but are worth reviewing in view of the prevailing tendency, particularly in the media, to give a causal interpretation to ever weaker associations.

Problems of Inference from a Single Study

In a simple study of the relationship of a factor of interest (F) to a disease (D), results are often presented as a 2×2 table with cell frequencies as shown in Table 4.1. The odds ratio R=ad/bc indicates the strength of the F/D relationship. When positive (for simplicity only positive associations are considered), an association exists. Before causation can be inferred, a number of alternative explanations have first to be excluded.

Chance. Where an experiment tests a single clearly defined prior hypothesis, standard statistical techniques can attach a p value to an odds ratio to aid judgment as to the plausibility of a chance explanation. Most experimenters will then consider values in the range 0.1–0.01 as suggestive, those in the range 0.01–0.001 as quite convincing and those less than 0.001 as very convincing evidence, though their prior beliefs will also be relevant. Where many factors and/or disease endpoints are involved, multiple analyses can be carried out, and care in interpreting p values is required, especially when they relate to hypotheses suggested from the data. Unless the association is very highly significant

[1]25 Cedar Road, Sutton, Surrey, SM2 5DA, England.

Table 4.1. Relationship of a factor of interest to a disease.

Factor	D−	D+
F−	a	b
F+	c	d

indeed, results from data-derived hypotheses should be regarded only as new hypotheses to be confirmed by other studies.

Which Came First? The possibility that the association arose because the disease (or a precursor of it) affects exposure must always be borne in mind. This is most relevant in case-control studies, but cannot be ignored in prospective studies, especially with a short follow-up period.

Misclassification of Exposure and Disease. Misclassification of exposure and disease is a problem in all studies, though its extent will vary with the nature of the disease and exposure. In a 2×2 table, *random* misclassification tends to have a diluting effect on the observed association. As shown in Table 4.2, the downward bias increases with the misclassification rate. For high odds ratios, a relatively small misclassification rate has quite a large effect, but for smaller odds ratios the effect is less. *Differential* misclassification can, however, produce false-positive relationships. There are two possibilities. One is that knowledge of disease (or the disease itself) affects recall of exposure, making diseased subjects more likely to be classified as exposed. This may be a particular problem in case-control studies. The other is that knowledge of exposure may increase the likelihood the disease is diagnosed. McFarlane et al. (1986) showed, in a study of necropsies in which primary lung cancer was found, that the probability of diagnosis in life was significantly higher if the patient smoked. Differential mis-

Table 4.2. Relationship of observed odds ratio (R^*) to true odds ratio (R) based on 2×2 table[a] and varying misclassification rates (α).

		α				
w	R	0.005	0.01	0.02	0.05	0.10
				R^*		
0.9	81	68.27	58.26	43.73	21.92	9.60
0.8	16	14.86	13.84	12.07	8.36	5.05
0.7	5.44	5.24	5.05	4.70	3.84	2.85
0.6	2.25	2.21	2.18	2.11	1.92	1.67
0.55	1.49	1.48	1.47	1.45	1.38	1.29
0.5	1.00	1.00	1.00	1.00	1.00	1.00

[a] w is the proportion of those exposed who are diseased and also the proportion of those not exposed who are not diseased. It is assumed that 50% of the total population are exposed and that 50% of the total population are diseased. Misclassification rates are applied to both exposure and disease and in both directions.

classification can be prevented by collecting data "double-blind," though this is not always practical.

Non-response. Only some subjects approached provide usable data. While the level of non-response may depend on disease status or on the presence of the factor, this does not cause bias. Bias only occurs if there is interaction between disease and factor in the level of non-response, and then maybe in either direction. Thus, clearly, if all the non-responders are in the D+F+ and D−F− groups, the odds ratio will be biased down, while if they are all in the other groups it will be biased up. Where non-response is low, limits to the possible bias can be calculated. A relatively small non-response rate can have quite a large biasing effect.

Confounding. Turning now to explanations of the association involving another factor, confounding is considered first. That smokers have higher cirrhosis rates than non-smokers is not because of a direct effect of smoking, but because smokers drink more than non-smokers, and alcohol predisposes to cirrhosis. Alcohol confounds the smoking/cirrhosis relationship. It is precisely to avoid confounding that randomized studies are carried out. For a variable to be a confounder it must be associated *both* with disease and factor of interest. When judging whether an association might result from confounding, it is important to consider the strength of the associations of the confounding variable with both disease and factor, as well as how often the confounder occurs. If R_c is defined as the relative risk of disease from the confounder, and p_1 and p_2 as the frequency of confounder among people exposed and not exposed to the factor, the *apparent* relative risk R_F in relation to the factor (assuming it actually has no effect) is given by $R_F = [R_c p_1 + (1-p_1)]/[R_c p_2 + (1-p_2)]$. Table 4.3 illustrates how R_F varies with R_c, p_1 and p_2. $U = p_1(1-p_2)/p_2(1-p_1)$ measures the association between the confounder and the factor. Even for very strong associations of confounder and factor, R_F is much less than R_c. R_F is also usually much less than U and is never greater than p_1/p_2. Thus, while confounding can easily generate small spurious relative risks, it requires special circumstances to explain a large relative risk—the confounder must be rare in those not exposed, both associations with the confounder must be stronger than that between factor and disease, and at least one of the associations with the confounder must be very much stronger.

Susceptible Subgroups and Representativeness. In the last section attention changed from a 2×2 table (disease × factor) to a 2×2×2 table (disease × factor × alternative factor). Another situation where study of the 2×2×2 table is needed is where the factor has an effect, but only on a "susceptible" subgroup, the alternative factor here defining susceptibility. Here inference from the 2×2 table is not wrong in the sense that the factor does increase risk, but is wrong in assuming it applies to everyone. While Occam's razor leads one to assume an association to be generally applicable unless there are external reasons not to do so, it is clearly not always the case that what is true for one population is so for another (e.g. effects of sunlight in white- and black-skinned men). One is on firmer

Table 4.3. Apparent relative risk due to a factor (R_F) in relation to the relative risk due to a confounding variable (R_C), the proportion with the confounder amongst people exposed to the factor (p_1) and the proportion with the confounder amongst people not exposed to the factor (p_2).

			R_C						
p_1	p_2	U	1.50	2.00	5.00	10.00	20.00	50.00	∞
			R_F						
1	0.9	∞	1.03	1.05	1.09	1.10	1.10	1.11	1.11
	0.7	∞	1.11	1.18	1.32	1.37	1.40	1.42	1.43
	0.5	∞	1.20	1.33	1.67	1.82	1.90	1.96	2.00
	0.3	∞	1.30	1.54	2.27	2.70	2.99	3.18	3.33
	0.1	∞	1.43	1.82	3.57	5.26	6.90	8.47	10.00
	0	∞	1.50	2.00	5.00	10.00	20.00	50.00	∞
0.9	0.7	3.86	1.07	1.12	1.21	1.25	1.27	1.28	1.29
	0.5	9.00	1.16	1.27	1.53	1.65	1.72	1.77	1.80
	0.3	21.00	1.26	1.46	2.09	2.46	2.70	2.87	3.00
	0.1	81.00	1.38	1.73	3.29	4.79	6.24	7.64	9.00
	0	∞	1.45	1.90	4.60	9.10	18.10	45.10	∞
0.7	0.5	2.33	1.08	1.13	1.27	1.33	1.36	1.38	1.40
	0.3	5.44	1.17	1.31	1.73	1.97	2.13	2.25	2.33
	0.1	21.00	1.29	1.55	2.71	3.84	4.93	5.98	7.00
	0	∞	1.35	1.70	3.80	7.30	14.30	35.30	∞
0.5	0.3	2.33	1.09	1.15	1.36	1.49	1.57	1.62	1.67
	0.1	9.00	1.19	1.36	2.14	2.89	3.62	4.32	5.00
	0	∞	1.25	1.50	3.00	5.50	10.50	25.50	∞
0.3	0.1	3.86	1.10	1.18	1.57	1.95	2.31	2.66	3.00
	0	∞	1.15	1.30	2.20	3.70	6.70	15.70	∞
0.1	0	∞	1.05	1.10	1.40	1.90	2.90	5.90	∞
0	0	∞	1.00	1.00	1.00	1.00	1.00	1.00	1.00

Table 4.4. Distribution of children by agent, social class and maternal smoking in hypothetical example.

Group	Exposure to agent	Social class	Maternal smoking	(n)	Relative risk
1	+	Low	Yes	240	5
2	+	Low	No	160	5
3	+	High	Yes	60	5
4	+	High	No	40	5
5	−	Low	Yes	40	1
6	−	Low	No	60	1
7	−	High	Yes	160	1
8	−	High	No	240	1

ground with data from a variety of populations. Where this shows variation in the association, this might be because the populations are affected differently by confounding with a true cause of disease or because they have a different frequency of susceptible subjects.

Misclassification of Confounding Variables. Often, statistical adjustment for a confounder reduces, but does not eliminate, the association of interest. Only if the confounder is measured without error can it be assumed that it cannot explain the association. Table 4.4 illustrates the false inferences that can be drawn when a confounding variable is misclassified. The table concerns a hypothetical example in which a disease of children is affected by an agent inaccurately measured by social class (20% misclassification in both directions). Maternal smoking does not affect risk of disease but is correlated with the agent. Standardising for social class reduces the apparent increase in risk for children whose mothers smoke only from 1.31 to 1.19. Tzonou et al. (1986) consider misclassification of confounding variables in more detail, with extensive tabulations. They comment that "even misclassification rates as low as 10% can prevent adequate control of confounding" and note the difficulties this can cause in many situations, e.g. dietary studies and studies of sexually transmitted agents. They conclude: "Generally, in the presence of strong risk factors which have potentially confounding effects and which may have been imprecisely measured, the etiologic significance of factors observed to be relatively weakly associated with disease must be treated with caution."

Subgroup Misclassification. In some studies, the association is only investigated among a subset of people reported as not being exposed to an alternative factor known to be very strongly related to the disease. As will be shown below, in the example of environmental tobacco smoke (ETS) and lung cancer, misclassification of exposure to this "contaminating" factor can lead to a spurious association between the disease and the factor of interest.

Plausibility. Though difficult to quantify, plausibility must be considered in judging causality. An observed weak association between a chemical and a risk of disease is more likely to be considered causal if the same disease is known to be caused by a structurally similar chemical. On the other hand, an unexpected association in relation to very low exposure or at an unlikely target site tends to require more confirmatory evidence.

Dose-Response. Often, subjects will be classified not simply as exposed or not exposed, but by degree of exposure. Normally, a true relationship will be stronger with increasing exposure, and emphasis should be given to statistical tests that are powerful for detecting dose-related trend. One should be cautious about accepting as evidence of a causal effect studies which show an association at low doses but not at high doses or which show a very irregular dose relationship. Even where the dose-response is rising smoothly, a causal relationship must not be concluded too readily, since many of the mechanisms which produce bias in a 2×2 table will also produce a spurious dose-response in a 2×k table (k dose levels). Since heavy smoking and heavy drinking are strongly correlated, one would expect to see a dose-related relationship between smoking and liver cirrhosis.

Problems in Inference from Multiple Studies

Meta-analysis. Evidence on a relationship often comes from multiple studies. Particularly where evidence in any one study is based on small numbers, there is an obvious desire to combine the results to reach a more reliable overall assessment. Techniques for such "meta-analysis" are available (Wald et al. 1986); they produce both combined relative risk estimates and also test for heterogeneity between studies. A number of points should be made:

1. Although the overall risk estimate may differ significantly from unity, with no heterogeneity evident, a causal relationship may still not exist. The source of bias may apply to all the studies.
2. The existence of some heterogeneity need not exclude a causal interpretation. The level of exposure may vary between exposed groups in the different studies.
3. The existence of marked heterogeneity suggests either some results may be spurious or that some other important factor is involved. Insight into potential sources of bias may be gained by comparing studies showing an association and those that do not.

Non-reporting Bias. Bias in meta-analysis may arise because the studies combined are not representative of all those carried out. In particular, estimates of relative risk may be biased upwards if scientists tend not to submit or journals tend not to publish studies finding no effect or a difference in an unexpected direction. Chalmers et al. (1987) cite results from a comparison of published and of completed unpublished randomized controlled trials. Of the former, 55% showed a significant ($p < 0.05$) advantage to the new therapy, while of the latter only 15% did. Publication bias is likely to be even greater for non-randomized studies. Often, large epidemiological studies have the potential to answer hundreds of questions, and it is easy to imagine an investigator running numerous statistical analyses, finding a few significant associations of interest and then publishing papers only on these. Pocock et al. (1987), who surveyed 45 recently published clinical trials, point to the problems caused by *inter alia* multiple end points, subgroup analyses, multiple treatments and multiplicity of significance tests. They concluded that "overall, the reporting of clinical trials appears to be biased towards an exaggeration of treatment differences." This must apply also to epidemiological studies.

Avoidance of Non-reporting Bias. Can one avoid the problems of non-reporting bias? Clearly, standard meta-analyses based only on published data may give a false picture. Obviously, the best solution is to get hold of all the data, but this is not always feasible. In the absence of this, more weight might be given in meta-analysis to non-significant than to significant trials. In particular, the first published study might be given little or no weight as, normally, it generates the hypothesis so cannot be used to test it. Journals will, anyway, not normally accept a first study which shows no effect. Another, but more long-term, solution is to carry out "quadruple blind studies." It is well known that those defining disease status should be blind to the exposure, and those defining exposure should be blind to disease so as to avoid bias. To avoid non-reporting bias, one seems to require two further stages of blindness, in which both the scientist writing the paper and the journal refereeing it are blind to its results. The scientist would write two papers, each with a common introduction and methods section and with a common results section with identically structured tables but no data entered. Only the discussion section (and abstract) would vary, illustrating the different conclusions drawn depending on whether a significant relationship was found. The journal would be sent both versions and must agree to accept both or neither. On acceptance, the data would be examined and the tables completed. Ideally, the journal would publish both discussion sections to show how the authors' opinion depended on the findings. This procedure not only eliminates publication bias, but also the possibility that authors select analyses to suit their beliefs.

Environmental Tobacco Smoke and Lung Cancer

Epidemiology is quite satisfactory for isolating true effects provided the association is a strong one. As the association studied becomes progressively weaker, problems caused by non-reporting, misclassification and confounding become relatively more important. Particularly for weak associations, with relative risks of less than 2, it is of vital importance to consider all the possibilities of bias. It may be impossible to make reliable inferences. As an example of the difficulties in interpretation, the weak association between ETS and lung cancer is now discussed in some detail.

Association

The first reports (Hirayama 1981; Trichopoulos et al. 1981) found that non-smokers married to smokers have a higher risk of lung cancer than non-smokers married to non-smokers. Over 20 studies on this issue have now been reported. Although some find no effect or a slight reduction in risk in ETS-exposed groups, the majority find an increase and a number of meta-analyses have reached similar answers: 1.30 (Lee 1984), 1.41 (Wells 1986), 1.2–1.5 (Doll 1986) and 1.35 (Wald et al. 1986). In considering whether this statistically significant association indicates a true causal effect of ETS, various possibilities cited above must be considered.

Plausibility

Because tobacco smoke constituents can be found in the blood, urine and saliva of non-smokers, it is plausible that ETS entails some risk of lung cancer. It does not, however, seem very plausible that ETS might entail a risk of lung cancer anything like as large as indicated by epidemiology. Taken at face value, the epidemiological data indicate risk of lung cancer is proportional to the amount smoked and that risk in relation to passive smoking is 10%–15% of that in relation to active smoking (Lee 1988a). One would expect, therefore, to find that average passive smokers have about 10%–15% of the exposure to relevant smoke constituents of average active smokers. In fact, this is far from the case, with relative exposure about 0.5% for inhaled particulate matter and salivary cotinine and 0.05% for relative retained particulate matter (Lee 1988a). This huge discrepancy underlines the need to examine the epidemiology critically.

Possible Alternative Explanations to be Considered

Chance. This cannot be the total explanation, since the latest meta-analyses show quite a highly significant increased relative risk in relation to ETS exposure for non-smokers, e.g. Wald et al. (1986) give 1.35 and 95% confidence limits 1.19–1.54.

Disease Causes Factor. Nor is it plausible that lung cancer (or a precursor of it) makes non-smokers seek out ETS exposure.

Unrepresentativeness. Wald et al. (1986) found no significant heterogeneity. With studies carried out in the United States, United Kingdom, Greece, Sweden, Japan, China and Hong Kong lack of representativeness is not a serious problem.

Non-reporting Bias. Of 20 relative risks given in Table 1 of Wald et al., only three are statistically significant, and these come from the two studies originally reported in 1981. Even eliminating these studies on the (dubious) grounds that the studies were hypothesis-generating, the overall relative risk remained significant (relative risk = 1.20, limits 1.02–1.40). Since the association between ETS and lung cancer has been big news in recent years, with most researchers and journals aware that any material is of interest, it seems that non-reporting bias cannot fully explain the overall positive relative risk, though it might have biased it upwards slightly.

Confounding. Factors known or suspected to cause lung cancer, such as occupation, air pollution, radioactivity, nutrition and genetic susceptibility, seem unlikely to be an explanation. Where it has been possible to standardise for them, the ETS/lung cancer association has been virtually unaffected, which is not surprising, as none of them seem to satisfy all three requirements: (a) frequent source of exposure; (b) very strongly associated with lung cancer risk; and (c) very strongly associated with ETS.

Table 4.5. An example of bias caused by misclassification of the subject's smoking habits. (From Lee 1988b).

Subject's smoking habits	NS	S	S	S	NS	S
Spouse's smoking habits	NS	S	NS	S	Total	Total
Assumed RR	1	1	20	20	1	20
Assumed effects	Passive=1.0					Active=20.0
Assumed N	65	35	35	65	100	100
Change to N from 5% misclassification	+1.75	+3.25	−1.75	−3.25	+5	−5
Observed N	66.75	38.25	33.25	61.75	105	95
Observed D	100	100	665	1235	200	1900
Observed RR	1.5	2.61	20	20	1.90	20
Apparent effects	Passive=1.75					Active=10.5

NS, non-smoker; S, smoker; RR, risk of lung cancer relative to true non-smoker; N, relative numbers of subjects; D, relative numbers of lung cancer deaths (observed N × true RR, e.g., in column 1, 100 = 65 × 1 + 1.75 × 20).

Underlined numbers are true smokers.

Misclassification of Diagnosis or Exposure. Random misclassification cannot cause upward bias in the relative risk. Differential misclassification of diagnosis seems unlikely to be important, since the doctor is likely to have been blind to the ETS exposure of the patient in most cases, and generally unaffected by it even if he was not. It is more plausible that knowledge of disease may affect reported ETS exposure, but this cannot be a full explanation since it relates only to case-control studies and Wald et al. (1986) found a significant association (relative risk = 1.44, limits 1.20–1.72) for prospective studies.

Misclassification of Confounding Variables. Given correction for those confounders considered above is likely to have no major effect, their misclassification is also probably unimportant.

Contamination by Smokers. Because of the strong correlation between husbands' and wives' smoking habits (Lee 1987), failure to take into account active smoking habits can cause enormous bias. For this reason, all the studies have restricted analysis to non-smokers. However, bias is only avoided if the non-smokers are not "contaminated" by misclassified smokers. Table 4.5 illustrates the potential for bias. Here, 5% of smokers denying smoking has the result of converting true relative risks of 20 for active smoking and 1 for passive smoking into observed relative risks of 10.5 for active smoking and 1.75 for passive smoking. As discussed at length by Lee (1988b) the size of the apparent passive smoking effect depends crucially on the assumed levels of misclassification and of concordance of spouse's smoking habits. Based on very limited data on contamination levels, Wald et al. (1986) concluded this source of bias would have little effect, causing upward bias by a factor of 1.04. Review of all available data by Lee (1988b, 1989) indicates that this conclusion is incorrect. The data from the larger studies

Table 4.6. Misclassification of regular smokers in six large studies.

Study	Cotinine cut-off used	Self-reported non-smokers studied	Found to be regular smokers (%)	Found to be regular smokers (*n*)	N as percentage of all regular smokers (%)
Pierce et al. 1987	69 ng/ml saliva	622	2.4	15	5.0
Haddow et al. 1987	70 ng/ml saliva	2871	2.4	68	6.2
Coultas et al. 1987	100 ng/ml saliva	896	2.7	24	9.1
Heller, personal communication	100 ng/ml serum	1296	0.9	12	2.7
Lee 1987	100 ng/ml saliva	808	1.4	11	2.0
Wald et al. 1986	30% smokers concentrations	705	0.4	3	0.5

The data of Wald et al. was based on four small studies, one based on nicotine. Subjects were considered regular smokers if the cotinine level was above cut-off. The calculation of the last column for Coultas et al. and Wald et al. assumes 80% of self-reported smokers are regular smokers – the percentages were not available but were 81%, 76%, 72%, and 86% for the other studies. Data for Coultas et al. exclude those aged < 13 and use additional information provided by the authors.

(Table 4.6) show contamination is likely to be important and sufficient to explain all or virtually all the observed relationship between ETS and lung cancer.

Conclusion on ETS and Lung Cancer

No existing studies on ETS and lung cancer have collected objective data on uptake of smoke constituents to monitor self (or surrogate) reports of active smoking status (or of ETS exposure). Evidence of levels of misreporting of active smoking from other studies shows clearly that it is premature to accept the conclusion of Wald et al. (1986) that "people known to have increased exposure to ETS seem to have an excess risk of lung cancer, which is not explained satisfactorily by bias."

General Conclusions

In any non-randomised epidemiological study with a relative risk less than 2, great care must be taken before inferring causality. The closer the relative risk is to 1, the more severe the problems of interpretation due to one or more of the various sources of bias. Epidemiologists are asked to search for ever smaller risks. Though clearly some can never be detected, a partial step in the right direction can be made if epidemiologists improve their procedures. Objective means should be used to quantify disease, exposure and confounding variables wherever possible. Where subjective answers are used, these should be collected blind. Non-response should be minimised. Statistical procedures should be defined in advance, and journals should accept papers on the quality of the study and not on the results obtained. Until more attention is paid to these points, it will remain likely that many reports of statistically significant but weak associations will be false-positives.

References

Chalmers TC, Levin H, Sachs HS, Reitman D, Berrier J, Nagalingam R (1987) Meta-analysis of clinical trials as a scientific discipline. I. Control of bias and comparison with large co-operative trials. Stat Med 6:315–323

Coultas DB, Howard CA, Peake GT, Skipper BJ, Samet JM (1987) Salivary cotinine levels and involuntary tobacco smoke exposure in children and adults in New Mexico. Am Res Respir Dis 136:305–309

Doll R (1986) Lung cancer: observed and expected changes in incidence from active and passive smoking. Paper presented at the 14th UICC conference, Budapest

Haddow JE, Knight GT, Palomaki GE, Kloza EM, Wald NJ (1987) Cigarette consumption and serum cotinine in relation to birthweight. Br J Obstet Gynecol 94:678–681

Hirayama T (1981) Non-smoking wives of heavy smokers have a higher risk of lung cancer: a study from Japan. Br Med J 282:183–185

Lee PN (1984) Passive smoking. In: Cumming G, Bonsignore G (eds) Smoking and the lung. Plenum, New York, pp 187–216

Lee PN (1987) Passive smoking and lung cancer. Association a result of bias? Hum Toxicol 6:517–524

Lee PN (1988a) An alternative explanation for the increased risk of lung cancer in non-smokers married to smokers. In: Perry R, Kirk PW (eds) Indoor and ambient air quality. Selper, London

Lee PN (1988b) Misclassification of smoking habits and passive smoking. A review of the evidence. Springer, Berlin Heidelberg New York

McFarlane MJ, Feinstein AR, Wells CK (1986) Necropsy evidence of detection bias in the diagnosis of lung cancer. Arch Int Med 146:1695–1698

Pierce JP, Dwyer T, DiGiusto E et al. (1987) Cotinine validation of self-reported smoking in commercially run community surveys. J Chronic Dis 40:689–695

Pocock SJ, Hughes MD, Lee RJ (1987) Statistical problems in the reporting of clinical trials. A survey of three medical journals. N Engl J Med 371:426–432

Trichopoulos D, Kalandidi A, Sparros L, MacMahon B (1981) Lung cancer and passive smoking. Int J Cancer 27:1–4

Tzonou A, Kaldor K, Smith PG, Day NE, Trichopoulos D (1986) Misclassification in case-control studies with two dichotomous risk factors. Rev Epidemiol Sante Publique 34:10–17

Wald NJ, Nanchahal K, Thompson SG, Cuckle HS (1986) Does breathing other people's tobacco smoke cause lung cancer? Br Med J 293:1217–1222

Wells AJ (1986) Misclassification as a factor in passive smoking risk. Lancet 2:638

Section 2. Animal

5. Rodent Carcinogenicity Studies: Their Value and Limitations

G.A. Boorman,[1] S.L. Eustis,[1] M.R. Elwell,[1] and R.A. Griesemer[1]

Introduction

While we have known since the late 1700s that certain chemicals cause cancer in man, large programs for screening the carcinogenic potential of chemicals have been initiated only in recent decades. The emphasis for cancer testing in the United States occurred with the passage of the National Cancer Act of 1971 (Page 1977). The demonstration that vinyl chloride caused cancer in man provided further impetus for testing, since 4 years earlier vinyl chloride had been shown to cause cancer in rats (Viola et al. 1971).

In recent years, the scientific and technical quality of the long-term carcinogenicity studies has improved, but unfortunately too little effort is devoted to understanding the mechanisms and relevance of positive responses found in these studies. Despite many limitations, rodent carcinogenicity studies remain the major and most practical experimental procedure for identifying potential carcinogens without exposing people (Tomatis 1979; Wilkinson 1987).

History of Long-term Rodent Studies

Experimental induction of cancer in animals was achieved by Yamagiwa and Ichikawa in 1915 by painting rabbit ears with coal tar, later Yoshida and Sasaki induced hepatomas in rats (Sugimura 1986), and the work on nitrosamines in the 1930s and 1940s provided impetus for additional animal carcinogenicity studies. In the 1960s, it was estimated that a vast majority of human cancer was environmentally related and thus preventable (Page 1977). The National Cancer Act of 1971 in the United States provided funds for carcinogenicity testing of chemicals in rodents. It was conceived as a multistep process with initial tests in rodents for carcinogenicity, anticipating that industry or appropriate regulatory/research

---

[1]P.O. Box 12233, Research Triangle Park, North Carolina 27709, USA.

organizations would conduct additional research when a positive response was found (Page 1977). Carcinogenicity in two species or the induction of a tumor type rarely seen in the animals might be expected to prompt immediate regulatory action or further evaluation: however, it was recognized that the animal results could not always be directly extrapolated to predict with precision the hazards to humans (Page 1977). By 1974, numerous compounds were being tested in long-term rodent studies by the National Cancer Institute. Limited resources, however, resulted in few additional studies on chemicals found to be positive.

In 1978, the United States National Toxicology Program combined government testing of environmental chemicals under one program to increase the quality and amount of toxicology data obtained from each study. Animal studies were designed with more emphasis on other toxicity endpoints including chemical distribution studies, reproductive studies, genotoxicity evaluations, and assessing toxicity endpoints at more time periods using interim sacrifices and stop-treatment groups. Careful attention to pathology review and quality assessment resulted in more complete evaluations. However, current studies are seldom designed to provide data on mechanisms, and extrapolation of data directly from these studies to human risk assessment is sometimes questioned (Ames et al. 1988).

Value of Long-term Rodent Studies

Predictability for Human Carcinogens

Chemicals that cause cancer in humans cause cancer in other mammals under appropriate experimental conditions. For several human carcinogens [bis(chloromethyl) ether, mustard gas, vinyl chloride], evidence for carcinogenicity was first found in experimental animals, but the majority were identified through epidemiological studies (Griesemer 1988). Of the 44 agents for which there is sufficient or limited evidence of carcinogenicity to humans, 37 that have been adequately tested all produced cancer in one or more species of animals (Wilbourn et al. 1986). This association does not firmly establish that chemicals that cause cancer in animals can cause cancer in humans. However, in the absence of data in humans to the contrary, it is prudent to regard agents which cause cancer in animals as having the potential to cause cancer in man (IARC 1988). The International Agency for Research on Cancer (IARC) requires evidence of malignant, or an appropriate combination of benign and malignant, neoplasms in two species of animals or in two or more independent studies to consider a chemical carcinogenic in animals (IARC 1988).

Occasional lack of concordance between the results in rats and mice would suggest that carcinogenicity studies may not always be predictive for humans. However, when the species correlation was examined for 266 long-term carcinogenicity studies, the overall concordance between rats and mice exposed to the same chemical was 74% (Haseman and Huff 1987). Even more impressive were the results for 250 studies from three sources where 98 chemicals were negative for both rats and mice and 109 were positive for both for a concordance of 83% (Purchase 1980). Von Wittenau and Estes (1983) have suggested that the correlation between rats and mice is so good that mouse results are redundant. It should be noted that 69/266 studies reported by Haseman and Huff (1987) and 43/250 studies reported by Purchase (1980) were positive in only one of the two rodent species under the conditions of the test. Different doses or survival problems may explain some of the discordance, but for other cases the explanation is not yet known.

Practical Considerations for Use of Rodents

Rodents were chosen in part because of their low cost, small size, ease of care, and limited lifespan. In principle, there is no basis for choosing mice rather than elephants (Paget 1987), but for practical considerations rats and mice have become the most commonly used species for carcinogenicity studies. A major advantage for their continued use is the vast amount of accumulated biological data. The spontaneous diseases and naturally occurring neoplasms have been extensively described for the more common strains of rats and mice. There is also extensive knowledge on the metabolism and pharmacokinetics, and more limited data for DNA binding and oncogene activation, for numerous carcinogens in rodents. This information is of value in extrapolating the relative hazard of a chemical to humans.

It is ironic that, on the basis of retrospective animal studies which reproduced the manifestations of drug toxicity in man, prospective toxicological studies were developed: nevertheless, the predictive value of these studies is now being questioned (Griffin 1986).

Limitations of Rodent Carcinogenicity Studies

Insensitivity

A major limitation of rodent carcinogenicity studies as presently conducted is their relative insensitivity. Even with group sizes of 50 and low background incidence of tumors at target sites, it is difficult to draw conclusions from an increase in tumor incidence of less than 10%–15%. Increasing the group size to several hundred animals increases the sensitivity by relatively little (Gart and Tarone 1987).

Cost and Time to Conduct Studies

So-called two-year carcinogenicity studies using both rats and mice cost 1–2 million dollars per test substance and take 4–6 years to complete all phases of testing and evaluation.

Problems Distinguishing Between Initiators and Promoters

Some chemicals have the potential to initiate the carcinogenic response, while other chemicals do not appear to cause tumors alone but can markedly affect later stages of the carcinogenic response induced by the initiator. Many of the initiators are genotoxic, whereas promoters often are not (Pereira 1985). Unfortunately, standard long-term rodent studies do not distinguish between these two types of responses. The National Academy of Sciences Report on Carcinogenicity of Cyclamate (1985) noted that the relevance for humans of using animals studies to detect promoting or enhancing effects is yet to be determined and that research to identify relevant models and methods of validation is required.

Problems with Extrapolation to Humans

One limitation of rodent carcinogenicity studies is that they provide insufficient information to enable precise extrapolation to humans. Anatomical and physiological differences of test species may result in differences of absorption, distribution, metabolism, and pharmacokinetics of the chemicals which affects the nature and magnitude of the carcinogenic response. For example, concentrations of cytochrome P450 present in the various portions of the respiratory tract and the capacity of the respiratory tract tissues to metabolize xenobiotics vary with species (Dahl 1988). Since reactive metabolites may be the ultimate carcinogen, the capacity of a species to metabolize a particular compound and the site of the metabolism determines if and where a toxic effect is seen. Knowledge of metabolic pathways in a species is necessary for extrapolation to expected effects in humans (Dahl 1988).

In addition to cellular and metabolic differences, rodents are usually exposed to higher doses for shorter periods of time than humans. Extrapolation of the results of rodent cancer tests done at high doses to effects on humans exposed at low doses is done routinely despite difficulties with this type of extrapolation (Ames et al. 1988).

Rodent Studies as Part of Risk Assessment

Rodent studies are now part of the public health-based regulatory activities known as risk assessment and risk management. Risk assessment is the scientific process by which we estimate the probability that some harm will come to an individual or a population as a result of exposure to a substance or situation. Risk assessment may comprise four steps: (a) hazard identification which includes epidemiological data, animal bioassay data, in vitro effects, and structure activity comparisons of chemicals; (b) dose-response assessment; (c) exposure assessment; and (d) risk characterization (Ris and Preuss 1988). Positive results in animals may be taken as evidence that the substance may pose a hazard for humans. Other information such as genotoxicity, metabolism, or structural similarity to chemicals with known hazards may also be used with animal data to support or further explain the hazard potential to humans. Risk management includes the political, economic, and social aspects of decision making regarding action on a potential human hazard (Ris and Preuss 1988).

Future Needs in Rodent Studies

Pharmacokinetic and Metabolism Studies

Additional comparative pharmacokinetic and metabolic research following a positive response in a rodent long-term study is not often performed. For a pharmaceutical firm, further studies may not be cost effective since a carcinogenic response frequently eliminates further consideration of a product. Further studies are usually restricted to compounds already in the market place which have commercial value and a degree of public acceptance.

Government testing agencies are under pressure to test additional chemicals. Worker exposure to chemicals in the production of microchips for computers, extensive groundwater contamination around hazardous dump sites, and ambient ozone levels in urban environments are only a few of the many situations for which the need for immediate testing competes with the need for additional follow-up studies on chemicals already shown to be carcinogenic in rodents.

The intense political pressures for regulatory agencies to act quickly with limited resources forces a standardized approach. While carcinogens found in rodent studies are experimental carcinogens only (Weisburger and Williams 1983), there is pressure to regulate rather than to wait for additional pharmacokinetic and metabolism studies.

Need for Molecular Biology

There has been rapid progress in the understanding of the molecular biology of cancer with the demonstration of oncogenes in both human and rodent tumors. Oncogene studies have included potent carcinogens and also evaluation of rodent tumors from standard long-term studies (Reynolds et al. 1987). Evaluation of mouse liver tumors induced by furan and furfural suggests that these chemicals act by induction of novel weakly activating point mutations in the *ras* genes (Reynolds et al. 1987). Using molecular biology techniques, it may be possible to distinguish between chemically induced and spontaneously occurring tumors of the mouse liver. This has important implications for risk assessment (Reynolds et al. 1987). Currently, the National Toxicology Program is collecting tumor tissue from rodents in carcinogenicity and toxicity studies for evaluation of oncogene activation. This additional information may be helpful in human risk assessment.

Transgenic mice are new models that offer an alternate, more rapid screen of chemicals in rodents. While the value of this model is yet to be determined, the transgenic mouse may markedly alter screening of chemicals for carcinogenicity and has the potential to enhance our understanding of the mechanisms of the cancer response at specific target sites in rodents and humans.

Future Needs in Risk Assessment

The desire for a risk-free society may hamper our ability to reduce risk from chemical exposure since a chemical determined to be carcinogenic in rodents may not be studied further but replaced by an untested chemical. Methods available for risk assessment do not permit us to assess human risks with a degree of reliability and accuracy that is acceptable to all segments of the scientific community (Monro 1987). The National Academy of Sciences (1985), in their report on saccharin, drew attention to this problem when they pointed out that the estimates of human risk of bladder cancer based upon quantitative extrapolation of studies in rodents ranged from 0.2–1144000 cancer cases in the United States population over the next 70 years. The extent to which results from a particular rodent study are relevant to humans may be one of the greatest sources of uncertainty regarding animal tests in risk assessment (Hart and Turturro 1987).

Some substances clearly produce tumors in rodents which are of questionable relevance to humans, including melamine-induced bladder cancer in male rats, forestomach tumors induced by butylated hydroxyanisole (BHA), and possibly pheochromocytoma induction by xylitol (Monro 1987). Bladder cancer in male rats was only found in rats that had bladder stones consisting of melamine precipitates. The reaction of agencies to conduct risk assessment evaluations on every substance found to induce tumors in animals needs to be first tempered by a thoughtful evaluation of the relevance of the animal results to humans (Monro 1987).

The saccharin case makes clear that the public is willing to accept certain risks from food additives if it perceives that the benefits are great enough and the risks are low enough (Flamm and Lorentzen 1988). Future policies on carcinogen regulation need to reflect the fact that chemicals of low carcinogenic potency which appear to induce tumors by virtue of their ability to cause physiological, toxicological, or biochemical perturbations at high doses may not be considered as hazardous as carcinogens which produce an abundance of tumors at low doses (Monro 1987). There need to be more trans-species studies and more studies in humans. In the past, the conceptual framework was to apply risk estimate models to the most sensitive target organ in the most sensitive species. A positive carcinogenic response in a rodent study should lead to studies of pharmacokinetics, metabolism in various species, genotoxicity, DNA binding of the parent compound or metabolites, evaluation for oncogene expression in the induced tumors and perhaps additional long-term studies in different strains and species. Once the findings are considered relevant to humans, it is appropriate to develop suitable risk assessment criteria and plan risk management.

Conclusions

Education of our colleagues in the acceptance of uncertainty and to avoid stating possibilities as established facts is necessary (Paget 1987). As these caveats are more clearly understood by the public, there should be increased confidence in the ability of regulatory agencies to distinguish risks for humans. Rapid evolution in our understanding of the molecular biology of tumors provides hope that future cancer risk decisions will be based on scientific logic to a greater extent than in the past. When regulations must be promulgated on inadequate data, they should be interim regulations subject to adjustment as more information becomes available.

References

Ames BN, Magaw R, Gold LS (1988) Ranking possible carcinogenic hazards. Adv Mod Environ Toxicol 15:65–90

Dahl AR (1988) Comparative metabolic basis for the disposition and toxic effects of inhaled materials. In: Mohr U, Dungworth D, Kimmerle G, Lewkowski J, McClellen R, Stober W (eds) Inhalation toxicology. Springer, Berlin Heidelberg New York, pp 41–65

Flamm WG, Lorentzen RJ (1988) Quantitative risk assessment (QRA): a special problem in the approval of new products. Adv Mod Environ Toxicol 15:91–108

Gart JJ, Tarone RE (1987) On the efficiency of age-adjusted tests in animal carcinogenicity experiments. Biometrics 43:235–244

Griesemer RA (1988) Carcinogenicity protocols. In: Mohr U, Dungworth D, Kimmerle G, Lewkowski J, McClellan R, Stober W (eds) Inhalation toxicology. Springer, Berlin Heidelberg New York, pp 119–137

Griffin JP (1986) Predictive value of animal toxicity studies. In: Walker SR, Dayan AD (eds) Long-term animal studies; their predictive value for man. MTP, Boston, pp 107–116

Hart RW, Turturro A (1987) Selection and extrapolation – the task of implementation. In: Roloff MV (ed) Human risk assessment; the role of animal selection and extrapolation. Taylor and Francis, New York, pp 269–273

Haseman JK, Huff JE (1987) Species correlation in long-term carcinogenicity studies. Cancer Lett 37:125–132

IARC (1988) IARC monographs on the evaluation of the carcinogenic risk of chemicals to humans, man-made mineral fibres and radon, vol 43. International Agency for Research on Cancer, Lyon, pp 15–30

Monro IC (1987) International perspectives on animal selection and extrapolation. In: Roloff MV (ed) Human risk assessment; the role of animal selection and extrapolation. Taylor and Francis, New York, pp 23–30

National Academy of Sciences (1985) Evaluation of cyclamate for carcinogenicity. National Academy, Washington, DC
Page NP (1977) Concepts of a bioassay program in environmental carcinogenesis. In: Kraybill HF, Mehlman MA (eds) Advances in modern toxicology environmental cancer, vol 3, Wiley, New York, pp 87–171
Paget GE (1987) Animal selection and extrapolation – the problem defined. In: Roloff MV (ed) Human risk assessment; the role of animal selection and extrapolation. Taylor and Francis, New York, pp 1–7
Pereira MA (1985) Mouse liver tumor data: assessment of carcinogenic activity. Toxicol Indust Health 1:311–333
Purchase IFH (1980) Inter-species comparison of carcinogenicity. Br J Cancer 41:454–468
Reynolds SH, Stowers SJ, Patterson RM, Maronpot RR, Aaronson SA, Anderson MW (1987) Activated oncogenes in B6C3F1 mouse liver tumors: implications for risk assessment. Science 237:1309–1316
Ris CH, Preuss PW (1988) Risk assessment and risk management: a process. Adv Mod Environ Toxicol 15:1–21
Sugimura T (1986) Studies on environmental chemical carcinogenesis in Japan. Science 233:312–318
Von Wittenau MS, Estes PC (1983) The redundancy of mouse carcinogenicity bioassays. Fund Appl Toxicol 3:631–639
Tomatis L (1979) The predictive value of rodent carcinogenicity tests in the evaluation of human risks. Ann Rev Pharmacol Toxicol 19:511–530
Viola PL, Bigotti A, Caputo A (1971) Oncogenic response of rat skin, lungs, and bones to vinyl chloride. Cancer Res 31:516–522
Weisburger JH, Williams GM (1983) The distinct health risk analyses required for genotoxic carcinogens and promoting agents. Environ Health Perspec 50:233–245
Wilbourn J, Haroun L, Heseltine E, Kaldor J, Partensky C, Vainio H (1986) Response of experimental animals to human carcinogens: an analysis based upon the IARC monographs programme. Carcinogenesis 7:1853–1863
Wilkinson CF (1987) Being more realistic about chemical carcinogenesis. Environ Sci Technol 21:843–847

6. The Quality and Relevance of Data from Studies in Laboratory Rodents

F.J.C. Roe[1]

The Rodent as a Model for Predicting Chronic Toxicity and Enhancement of Ageing-Related Diseases in Humans

The rationale for conducting toxicity and carcinogenicity tests of chemicals in laboratory animals assumes that the information obtained will be useful in the prediction of how humans will respond to exposure to the same chemicals. This assumption is reasonably well based in terms of response to short-term high-dose exposure to chemicals, but much less so in terms of later responses to lower doses. As animals grow older, it becomes more and more difficult to distinguish between toxic effects and changes attributable to ageing. Indeed, in many long-term experiments in rodents, most of the differences between exposed and control groups are simply in the incidence and severity of ageing-related diseases. Since the spectra of the most common ageing-related diseases which afflict humans and laboratory rodents are quite different, it is only to be expected that the actual manifestations of chronic toxicity in rodents are quite different from those to be expected in humans.

It is nowadays generally accepted as reasonable to distinguish between *genotoxic* and *non-genotoxic* mechanisms in carcinogenesis (Butterworth and Slaga 1987; Roe 1988a). In the case of the former, there is often a close similarity between the responses of different species. Furthermore, where differences occur, they can sometimes be explained by differences in metabolism and/or distribution of cells with detoxifying or metabolically activating enzymes. By contrast, non-genotoxic mechanisms—of which many are known and doubtless many more are awaiting discovery—are often seemingly species specific. Such mechanisms often involve disturbances of endocrine or other homeostatic control systems, and in many cases the effects of exposure to non-genotoxic carcinogens seen in the late stages of carcinogenicity tests seem to be simply enhancements of spontaneously arising, ageing-related neoplasia. It has long been recognized that, in rodent studies on chemicals, it is easier to enhance the

[1] 19 Marryat Road, Wimbledon Common, London SW19 5BB, UK.

incidence of a commonly occurring tumour than that of a rare tumour. It may well be that this is because the underlying disturbances of physiological and endocrine status involved in non-genotoxic carcinogenesis are ageing related. In other words, conditions which either specifically or non-specifically cause ageing will *pari passu* enhance the risk of non-genotoxic carcinogenicity.

In any event, the important points for epidemiologists to appreciate are: (a) *that many manifestations of chronic toxicity relate more to the ways in which ageing affects a species than to the nature of chemicals which predispose to premature ageing*; and (b) *that there can be no mathematical formula for predicting non-genotoxic carcinogenicity for humans from the results of laboratory tests.*

Influence of Caloric Intake on Ageing-Related Disease and Neoplasia in Rodents

The amount of food a laboratory rat or mouse consumes during a carcinogenicity test on a chemical may have a highly significant influence on how long it lives and on the age-standardised incidence of ageing-related and neoplastic lesions (Conybeare 1988). In the past, far too little attention has been paid by experimentalists to this fact, and it is not surprising, therefore, that many statisticians extrapolating from animal data to humans have totally ignored it. Experimentalists tend to argue that, since test and control animals are offered the same diet, one is able to compare "like" with "like." However, if for one reason or another the test chemical increases or reduces the food intake of animals, then one ends up being unable to compare like with like. Epidemiologists argue that humans also overeat and are therefore like overfed rats. However, this is not strictly true. Humans are really not like rodents, confined to a small cage and given nothing else to do but eat a diet which nutritionally far exceeds their needs. In the case of rats, the manifestations of overnutrition include severe, and sometimes fatal, renal disease and many kinds of endocrine disturbance (see Table 6.1). Thus, one may encounter almost 100% incidences of pituitary, mammary and interstitial cell tumours of the testis in control as well as in test groups.

In a recently briefly reported study (Roe 1988b), my colleagues and I reported that the incidence of life-threatening, or actually fatal, malignant neoplasms in rats was significantly higher in rats given access to food throughout the 24 h of each day than in rats offered the same diet but for only 6.5 h per day. Common sense dictates that one should regard findings in animals rendered grossly abnormal because of overfeeding as very dubious predictors of chronic toxicity in humans.

Doubtless, many humans also consume more calories than they need, and there is good evidence that this predisposes to premature death and earlier onset of life-threatening diseases including some forms of cancer. However, the effects of overnutrition in rats on the incidence of the diseases listed in Table 6.1 seemingly have little or no parallel in humans.

Table 6.1. Overnutrition-related diseases in laboratory rats.

Non-neoplastic
Chronic progressive nephropathy
Cortico-medullary and pelvic nephrocalcinosis
Mesenteric and pancreatic hypertrophic periarteritis
Radiculo-neuropathy affecting the cauda equina
Inflammatory skin lesions
Parathyroid hyperplasia
Widespread metastatic calcification
Neoplastic
Benign and malignant tumors of
Pituitary
Mammary gland
Adrenal medulla
Leydig cells of testis
C-cells of thyroid
Islet cells of pancreas

Route of Exposure and Dose

Before the distinction between genotoxic and non-genotoxic carcinogenicity began to be understood, a positive result in a carcinogenicity test was taken to indicate that the agent was "intrinsically" carcinogenic irrespective of the route or dose of administration. We now know somewhat better. Sarcoma induction in a rat following the subcutaneous injection of a chemical to which humans are only exposed by the inhalation route would not nowadays be accepted as adequate evidence of carcinogenic risk for humans since, although the cancer may have arisen because the test substance is a genotoxic carcinogen, there is also a big chance that a non-specific mechanism is responsible (Grasso and Golberg 1966).

However, there remain areas of uncertainty. The relevance of intratracheal instillation and intrapleural or intraperitoneal injection in rodents for distinguishing carcinogenic from non-carcinogenic inhalable dusts is a subject for debate among investigators. The sheer inelegance of these methods for exposure offends those who are concerned with the importance of dose at the tissue and cellular level. Also, increasing evidence of lack of correlation between (a) the results of tests based on these methods of exposure; (b) the results of tests involving exposure by inhalation; and (c) epidemiological data suggests that, while such tests may have a place in screening for possible carcinogenicity, they are wholly unreliable for quantitative risk assessment.

The science and technology of inhalation toxicity have advanced considerably during the last 2 decades. Nevertheless, there is one obstacle to the development of realistic animal models which is seemingly insuperable. Despite the strong epidemiological evidence of an association between cigarette smoking and lung cancer risk in humans, there is no acceptable animal model for this. Three factors contribute to the problem. First, when rats and mice are exposed to cigarette smoke, it is via the nose and not, as in smokers, via the mouth. Second, a heavy smoker may expose him- or herself off or on throughout 16 h or more per day, whereas it is difficult and impossibly expensive to expose laboratory animals to smoke for more than a few hours per day. Third, the nicotine and carbon monoxide components of tobacco smoke are acutely toxic. The human smoker can avoid overdosing by stopping smoking as soon as he or she feels symptoms indicative of toxicity, but the animal under experiment cannot do this. For these reasons, it is not possible to expose laboratory rodents to doses equal to or above those achieved by smokers. Apart from this, one does not really know whether rodents would develop human-type lung cancers even if one could get adequate doses of smoke into their lungs.

The lack of a realistic model for lung carcinogenesis in relation to smoking hampers not only the development of putatively less hazardous cigarettes but also mechanistic studies with inhalable agents to which both smokers and non-smokers may be exposed.

It is in fact interesting to note that, in a study in which rats were exposed to tobacco smoke, the main effect of exposure was not an increase in lung cancer incidence but a reduction in mammary tumour incidence (Davis et al. 1975).

Extrapolation from High-to-Low Dose

It is not unreasonable to expect to find a positive dose-response relationship for a direct-acting toxin or carcinogen. However, dead cells cannot give rise to clones of cancer cells so that, if the test doses are pushed up to a point where cell death occurs, then the dose-response curve may flatten out or actually fall. In the case of substances which only act as toxins or carcinogens after they have been metabolically activated by tissue enzymes, and in the case of toxins or carcinogens for which the body has only a limited capacity for detoxification, the shape of the dose-response curve may depart from linearity at either end of the dose range. For non-genotoxic agents there can be no general rule whereby one can predict the shape of the dose-response curve. Threshold dose levels below which nothing happens may exist, and only if one has detailed information about how they bring about their observed effects should one presume to be able to extrapolate either upwards or downwards from the dose levels that have actually been studied. Those responsible for making risk assessments should always be mindful of the considerable uncertainties which surround high-to-low dose extrapolation.

Need for New Techniques for Assessing Disturbances of Physiological Status

Since disturbances of physiological and endocrine status are clearly so important and to some extent antedate manifestations of chronic toxicity and of non-genotoxic carcinogenicity, it would obviously be advantageous to have clinical techniques for detecting such disturbances and monitoring their severity while animals are still alive. In the rat, since hyperplasia and neoplasia of endocrine tissues are such prominent manifestations of premature ageing and are so commonly associated with overnutrition, it would be very helpful to have methods for measuring the levels of circulating hormones such as prolactin, growth hormone and 17β-oestradiol, etc. in small samples of blood. At present some of the available methods require so much blood that the animal has to be killed for the measurements to be made. Also, assay methods specifically for rat-type hormones as distinct from human-type hormones have not been developed and/or are not generally available.

Thus the depressing picture is that contract research laboratories and other laboratories tend to measure a large number of parameters for which methods are readily available but which are of no more than marginal interest from a toxicological viewpoint, whilst making no measurements on parameters which could provide early evidence of changes indicative of premature ageing and increasing endocrine imbalance.

The development of new methods for measuring such parameters could revolutionise the predictive value of rodent tests. First, one might be able to devise a system of animal husbandry by which rats could be maintained into old age in more or less normal endocrine status. Second, against a background in which untreated control animals remained physiologically normal, one could far more easily detect effects of test substances on endocrine tissues and on the development of ageing-related disease.

Unrealistic Aspects of Rodent Toxicology

Rodents in the wild differ from laboratory rodents in many ways. Apart from not being confined to a cage, unrestricted sexual activity, a need to forage for food, and a need to avoid predators, wild rodents carry a wide range of parasites and pathogens from which laboratory rats are largely or wholly free. Except in this last particular, the life style of wild rodents is much more like that of humans than is that of laboratory rodents. Although humans are not maintained behind barriers, they are immunized against serious diseases, they are treated for worm and ectoparasitic infestations whenever necessary and they are protected by food hygiene laws. Thus, like laboratory rodents, they are more or less specified pathogen free. The lack of any need to forage for food, the lack of exercise, the lack of opportunity to fulfil sexual urges, the overprovision of unnecessarily nutritious food and general boredom render laboratory rodents exceedingly poor models for humans. Over and above this is another potentially serious and usually unnecessary defect with the model. Rodents commonly eat at night and sleep during the day. Nevertheless, if they are exposed to chemicals other than in the food (e.g. by inhalation or by gavage), such exposure normally takes place during the daytime when they are not eating. Also, blood sampling and other measurements are made during the day. In the case of the safety evaluation of a drug to which humans are exposed during the day when they are eating, it simply does not make sense to dose an animal by gavage during the daytime when it is not eating. Absorption, metabolism and excretion of ingested test chemicals and many haematological and serum chemistry parameters are influenced by the time of dosing in relation to feeding patterns.

Tumour Incidence Data in Rodents as a Predictor of Cancer Mortality Data in Humans

Undoubtedly the most serious point that needs to be made concerns the important difference between the data collected by experimentalists conducting carcinogenicity tests in animals and the data collected by cancer epidemiologists. In the laboratory tests, animals are subjected to careful systematic macroscopic examination at necropsy and to routine sectioning of a long list of tissues. The findings are expressed as incidences of benign and malignant tumours in different tissues. Animals which survive to the end of studies may be killed when seemingly quite healthy and yet be found to have one or more small benign tumours in internal organs. Cancer therapy is not offered to rodents in carcinogenicity tests and, for humane reasons, few animals are allowed actually to die from neoplasms. Instead, sick animals are killed. Apart from this, experimentalists are generally very reluctant to diagnose the cause of death in animals that do die. Thus, there are no reliable *cancer mortality* data for laboratory animals in carcinogenicity tests, only *tumor incidence* data.

By contrast, because of general low necropsy rates, epidemiologists have to rely on rather inaccurate cancer mortality data for humans and have virtually no reliable tumour incidence data for internal organs.

A slowly growing benign tumour has a totally different significance from a more rapidly growing, metastasing malignant tumour, and therefore it is simply nonsensical to presume to be able to extrapolate from tumour incidence data in animals to cancer mortality risk in man. Nevertheless, this is presently often done by statisticians making risk assessments.

Main Conclusions

1. In some ways animals tests for carcinogenicity can provide better data than can be collected by cancer epidemiologists. The effects of high and accurately measured doses can be observed, and pathogenesis and possible mechanisms can be studied because interim sacrifice is considered ethical. However, it is no easier to study the effects of exposure to low doses over long periods in animals than it is in man because of limitations on the numbers of animals that can be studied.

2. The practice of carrying out thorough and systematic necropsies on all animals in carcinogenicity tests provides data that are of great value in the determination of dose-response relationships. However, *cancer mortality* data are not

generated by experimental oncologists – only *tumour incidence* data. Since many of the tumours observed are small, benign and slowly growing, and since comparable tumour incidence data are not available for humans, one cannot make realistic estimates of cancer risk for humans from animal test data.

3. Most carcinogenicity tests in laboratory rats and mice involve *ad libitum* access to overnutritious feedstuffs. Consequently, obesity, premature death, renal disease, polyarteritis, radicular nephropathy and a host of endocrine diseases complicate the interpretation of carcinogenicity studies, particularly in rats. Also, the background incidence of neoplasms in both species is significantly increased by overfeeding such that incidences of lung, liver and lymphoreticular neoplasms may each exceed 30% or even 50% in untreated mice, and pituitary, mammary, adrenal medullary, testicular and other endocrine tumours may approach 100% incidence in untreated rats. Common sense dictates that the results of carcinogenicity tests conducted in such animals cannot possibly be relied upon for the prediction of cancer risks in man.

4. There are many different non-genotoxic mechanisms by which the risk of tumour development may be increased. Many of these mechanisms are seemingly species specific, and many of them entail prolonged disturbance of physiological and/or endocrine functions, and, in the case of some test substances, such disturbances only occur with very high doses. These facts must in future be taken into account by those who presume to be able to calculate risk to humans from animal carcinogenicity test data.

References

Butterworth BE, Slaga TJ (eds) (1987) Nongenotoxic mechanisms in carcinogenesis. In: Banbury report, vol 25. Cold Spring Harbor Laboratory, Cold Spring Harbor

Conybeare G (1988) Modulating factors: challenges to experimental design. In: Grice HC, Ciminera JL (eds) Carcinogenicity. The design, analysis and interpretation of long-term animal studies. Springer, Berlin Heidelberg New York, pp 149–172

Davis BR, Whitehead JK, Gill ME, Lee PN, Butterworth AD, Roe FJC (1975) Response of rat lung to inhaled tobacco smoke with or without prior exposure to 3,4-benzpyrene (BP) given by intratracheal instillation. Br J Cancer 31:469–484

Grasso P, Golberg L (1966) Subcutaneous sarcoma as an index of carcinogenic potency. Food Cosmet Toxicol 4:297–320

Roe FJC (1988a) Let us forget tumour promotion and start thinking about mechanisms of non-genotoxic carcinogenicity. Acta Pathol Microbiol Immunol Scand 96 [Suppl 2]:91–99

Roe FJC (1988b) Toxicity testing: some principles and some pitfalls in histopathological evaluation. Hum Toxicol 7:405–410

Section 3. Dosimetric

7. Identification of the Effective Dose of Inhaled Toxicants: General and Specific Problems

H.M. Bolt[1]

Introduction

The establishment of a dose-effect relationship is based on a parameter of dose. This may be difficult to assess when an inhaled toxicant is being considered. The problem is generally known for particles, and for inhaled gases and vapors. A classical example of the latter is vinyl chloride. Vinyl chloride is a gaseous compound and known as a carcinogen in man and in experimental animals; its biochemical activation mechanism has been clarified. Vinyl chloride requires metabolic transformation to its epoxide, which is the ultimate genotoxic carcinogen at the DNA level (for review, see Bolt 1988).

In 1978, Gehring et al. indicated that the experimental dose-response curve (the relation between the external vinyl chloride concentration and the number of experimentally induced hemangiosarcomas) is atypical because the metabolic capacity of the test animals (rats) is saturated at elevated atmospheric vinyl chloride concentrations (i.e., above 500 ppm). If, however, the metabolized amount of toxicant per unit of time is calculated and used as a surrogate of the dose, a classical "dose"-response relationship is obtained (i.e., a straight line in a log "dose"-probit incidence plot).

This is illustrated in Fig. 7.1 which presents the dose-response (response = hemangiosarcoma incidence) of two different long-term studies with vinyl chloride in rats, under consideration of the rate of metabolism as a parameter of internal "dose."

The results of both studies are generally consistent as the treatment period in the experiment of Feron et al. (1981) was much longer than in that of Maltoni and Lefemine (1975). Maltoni and Lefemine had exposed rats for 1 year (4 h a day, 5 days a week) to the gas, whereas Feron et al. treated their rats orally for 140 weeks (5 days a week). It is also apparent from Fig. 7.1 that excessively high external doses are not accompanied by excessively high effects or target doses

[1]Institut für Arbeitsphysiologie an der Universität Dortmund, Ardeystr. 67, D-4600 Dortmund 1, FRG.

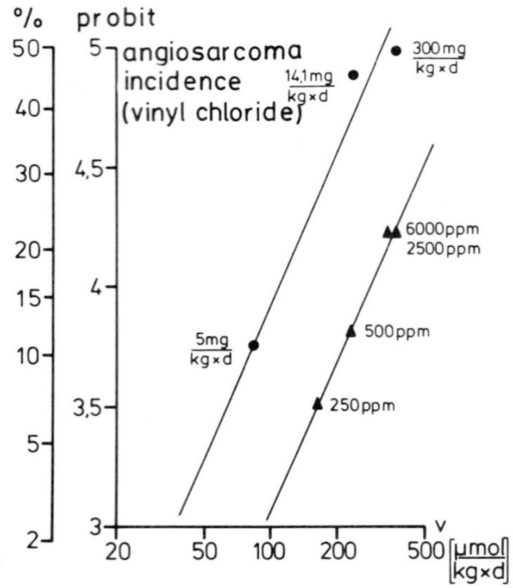

Figure 7.1. Dose-response relationship of vinyl chloride in rats. *Abscissa*, log amount (μmol/kg body weight) which was metabolized during 1 day of the exposure period. *Ordinate*, probit angiosarcoma incidence. The tumor data were taken from two long-term bioassays: Maltoni and Lefemine (1975), inhalation exposure (*triangles*); and Feron et al. (1981), oral exposure (*circles*)

(when the metabolic rate is taken as a means of the target dose per unit of time). This discrepancy is due to the metabolic saturation of the activating enzyme system. It appears, therefore, that the problem of defining target doses must be tackled individually on a compound-by-compound basis.

In this paper, focus will be on those carcinogenic compounds which have been chosen for consideration in the present volume. Particular emphasis will be placed on the target site and possible/probable mechanisms of tumorigenesis.

General Problems of Dosimetry

Table 7.1 shows that some of the inhalable compounds considered in this volume show effects only (or mostly) locally, while others show systemic effects. Approaches to dosimetry must be focused on local deposition patterns of the locally active compounds, and on the general disposition (in the whole organism) of the toxicants acting systemically.

Very recently, specific problems of species-specific deposition of chemicals in the upper airways, especially the nose, have been extensively discussed at a symposium held in October 1988 in The Netherlands (see Feron and Bosland 1989). Particular preference was given there to the discussion of formaldehyde and

Table 7.1. Compounds selected for the present volume: local versus systemic effects.

	Effects	
	Local	Systemic
Chemicals		
Formaldehyde	+ (Nose)	−
Arsenic	+ (Lung)	+
1,3-Butadiene	−	+
Benzene	−	+
Particulates		
Mineral fibers, including asbestos	+ (Lung, pleura)	−
Environmental tobacco smoke	+ (Lung)	?
Automobile exhaust	+ (Lung)	?

related aldehydes. It was emphasized that the inhaled concentration of such compounds does not parallel the tissue dose; regional deposition is obviously important, not only for particles, but also for locally active gases and vapors. For directly active carcinogens (and for those which are locally metabolized to their ultimately active principle), it was pointed out that DNA alkylation and cytotoxicity were major determinants of differential organotrophy, and recently considerable progress has been made in experimentally determining target dose parameters.

Formaldehyde

Formaldehyde acts locally on the respiratory tract, particularly on the nose epithelium. For an assessment of general exposure, the urinary excretion of formate and of thiazolidine metabolites has been proposed (Hemminki 1984), but both parameters are difficult to assess because of intrinsic variations in excretion of these products.

On the other hand, it is generally accepted that a locally delivered dose is the important parameter on which the cancer risk depends in experimental animals exposed to formaldehyde (Starr and Buck 1984). A characteristic biochemical initial effect of formaldehyde at the nose epithelium is formation of DNA-protein crosslinks which can be quantitated experimentally (Casanova-Schmitz et al. 1984). This is regarded as an important lesion in the pathogenesis of formaldehyde-induced tumors (Lam et al. 1985). The nonlinearity with external dose of both the DNA modification (DNA-protein crosslinks) and the experimental nasal tumor formation has been discussed (see Swenberg et al. 1983; Bolt 1987). It appears that the (local) metabolism of formaldehyde by (GSH-dependent) formaldehyde dehydrogenase is an important inactivating mechanism which becomes saturated at higher external dose levels (Heck and Casanova 1987; Casanova and Heck 1987).

In view of these data, the concept of *administered dose versus delivered dose* has been developed (Starr and Buck 1984; Starr and Gibson 1985) as follows (Starr and Gibson 1985):

The *administered dose* is an external measure of exposure directly controlled in laboratory studies of toxicity. For inhalation studies, it refers to the concentration of a test chemical in the inhalation chamber air. In contrast, the *delivered dose* is an internal measure of exposure referring to the quantity or concentration of the biologically active form of a test chemical present in specific target tissues. The latter measure is presumed to be the direct causative variable in mechanistic descriptions of the carcinogenic process at the cellular and molecular levels.

Arsenic

The carcinogenicity of arsenic is based on human epidemiological data and clinical case studies. Lung cancers were observed in workers occupationally exposed to inorganic arsenic, especially in mining and copper smelting industries. Oral exposures have resulted in systemic carcinogenesis, particularly with the skin as target organ (see IARC, 1987). There are indications that arsenic might act via an inhibition of DNA repair enzymes, rather than via a direct genotoxic effect. This aspect is discussed by Laib et al. (this volume).

As far as dosimetry is concerned, arsenic must be regarded as a systemically active compound with multiple potential targets.

The internal dose in man may be assessed on the basis of urinary arsenic excretion. The Deutsche Forschungsgemeinschaft (DFG 1988) has evaluated a correlation between ambient exposure concentration of As_2O_3 and urinary As (based on the analytical hydrogenation method). This is used for biological monitoring of persons occupationally exposed to arsenic.

1,3-Butadiene

1,3-Butadiene is systemically carcinogenic in experimental animals. Of interest are pronounced quantitative species differences in carcinogenicity between mice (high carcinogenicity) and rats (low carcinogenicity). Metabolism of the compound proceeds via its epoxide, 1,2-epoxybutene (vinyl oxirane), as demonstrated by several groups in vitro and in vivo (Bolt et al. 1983; Filser and Bolt 1984; Schmidt and Loeser 1985; Kreiling et al. 1986; Bond et al. 1986). Epoxybutene is a reactive metabolite which is distributed within the entire organism (Bond et al. 1986) and which is also exhaled to some extent (Filser and Bolt 1984). It is carcinogenic on mouse skin (van Duuren et al. 1966) and is directly mutagenic in *Salmonella typhimurium* (Gervasi et al. 1985). Hence, this primary epoxide of 1,3-butadiene is viewed as being the (or at least one important) carcinogenic metabolite of the parent compound. In experimental animals, dosimetry is possible by determination of 1,2-epoxybutene in blood or in exhaled air (see above). It has turned out that mice convert 1,3-butadiene at a higher rate to 1,2-epoxybutene than rats, and that rats inactivate the epoxide much better than mice do (Kreiling et al. 1987). This seems to be a key factor in explaining the pronounced species differences in the carcinogenicity of 1,3-butadiene.

Methods that can be used for biological monitoring of persons occupationally exposed to 1,3-butadiene have not been published so far.

Benzene

Benzene has a unique biological target because of its effects on the human bone marrow. Experimentally (in mice) tumorigenesis at other sites has also been noted (Maltoni et al. 1983; Snyder et al. 1988). It is systemically carcinogenic. The bone marrow is apparently capable of converting benzene to biologically reactive intermediates.

Under inhalation conditions, the intrinsic dose of benzene can be assessed from the levels of benzene in blood (DFG 1988) or in exhaled air (ACGIH 1988). The major metabolite, phenol, is also produced endogenously (mainly due to the metabolism of aromatic amino acids by gut bacteria) and is therefore not a useful indicator of exposure at low exposure concentrations (DFG 1988). Experimentally (and also in industrial practice) the urinary excretion of *S*-phenylmercapturic acid has been used as a parameter of intrinsic exposure. This metabolite has the advantage of not being a product of endogenous metabolism (Stommel et al. 1987).

For biological monitoring of persons occupationally exposed to benzene, correlations have been published by the DFG between external exposure, benzene blood level, and phenol excretion rates (DFG 1988).

Environmental Tobacco Smoke

Parameters which may be used for an assessment of the internal dose of (environmental) tobacco smoke have been recently reviewed by IARC (1986). Mostly, the determination of nicotine is recommended since it is a specific agent of tobacco smoke. The nicotine metabolite cotinine has the advantage of a longer biological half-life (plasma $t_{1/2}$ in man: 1-2 h for nicotine, 16-20 h for cotinine; Benowitz et al. 1982, 1983), and is therefore widely used as a marker for nicotine uptake.

Sensitive methods for the determination of cotinine are available. Some are based on immunochemical techniques, although difficulties may be encountered in discriminating between some closely related nicotine metabolites (Schepers and Walk 1988). Furthermore, a mathematical model has been presented for nicotine and cotinine as biological markers for environmental tobacco smoke exposure (Schwartz et al. 1987). Such methods are mostly used in epidemiological field studies on the effects of tobacco smoke.

Automobile Exhaust

The question of determination of a biologically relevant target dose for exposure to automobile exhaust is very complex. Experimentally, automobile exhaust may produce lung cancer, but further systemic effects cannot be ruled out.

An interesting approach to the dosimetry problem has recently been published by the group of Ehrenberg (Törnqvist et al. 1988). These authors drew blood samples from rats and hamsters experimentally exposed to automotive engine exhausts and determined the hemoglobin adducts caused by the exhaust constituents ethylene and propylene (i.e., N-terminal 2-hydroxyethylvaline and 2-hydroxypropylvaline in hemoglobin). The adduct levels corresponded to a 5%–10% metabolic conversion of the alkenes inhaled during the experimental exposure. Whether this method can be used also in humans under realistic exposure conditions, remains to be determined.

The few examples presented have demonstrated that questions of dosimetry, especially the determination of delivered dose, internal dose, or target dose are of considerable importance for all the inhalable toxicants which are discussed in the present volume.

References

ACGIH (1988) TLVs, threshold limit values and biological exposure indices for 1988–1989. American Conference of Governmental Industrial Hygienists, Cincinnati, OH, USA

Benowitz NL, Jacob P, Jones RT, Rosenberg, J (1982) Inter-individual variability in the metabolism and cardiovascular effects of nicotine in man. J Pharmacol Exp Ther 221:368–372

Benowitz NL, Kuyt F, Jacob P, Jones RT, Osman AL (1983) Cotinine disposition and effects. Clin Pharmacol Ther 34:604–611

Bond JA, Dahl AR, Henderson RF, Dutcher JS, Mauderly JL, Birnbaum LS (1986) Species differences in the disposition of inhaled butadiene. Toxicol Appl Pharmacol 84:617–627

Bolt HM (1987) Experimental toxicology of formaldehyde. J Cancer Res Clin Oncol 113:305–309

Bolt HM (1988) Roles of etheno-DNA adducts in tumorigenicity of olefins. CRC Crit Rev Toxicol 18:299–309

Bolt HM, Schmiedel G, Filser JG, Rolzhäuser HP, Lieser K, Wistuba D, Schurig V (1983) Biological activation of 1,3-butadiene to vinyl oxirane by rat liver microsomes and expiration of the reactive metabolite by exposed rats. J Cancer Res Clin Oncol 106:112–118

Casanova M, Heck H d'A (1987) Further studies of the metabolic incorporation and covalent binding of ^{3}H- and ^{14}C-formaldehyde in Fischer-344 rats: effects of glutathione depletion. Toxicol Appl Pharmacol 89:105–121

Casanova-Schmitz M, Starr TB, Heck H d'A (1984) Differentiation between metabolic incorporation and covalent binding in the labeling of macromolecules in the rat nasal mucosa and bone marrow by inhaled ^{14}C- and ^{3}H-formaldehyde. Toxicol Appl Pharmacol 76:26–44

Deutsche Forschungsgemeinschaft (DFG) (1988) Maximum concentrations at the workplace and biological tolerance values for working materials. Report no XXIII. VCH, Weinheim

Feron VJ, Bosland MC (eds) (1988) Nasal carcinogenesis in rodents. Symposium, October 24–28, 1988, Veldhoven, NL

Feron VJ, Hendriksen CFM, Speek AJ, Til HP, Spit BJ (1981) Lifespan oral toxicity study of vinyl chloride in rats. Food Cosmet Toxicol 19:317–333

Filser JG, Bolt HM (1984) Inhalation pharmacokinetics based on gas uptake studies. VI. Comparative evaluation of ethylene oxide and butadiene monoxide as exhaled reactive metabolites of ethylene and 1,3-butadiene in rats. Arch Toxicol 55:219–223

Gehring PJ, Watanabe PE, Park CN (1978) Resolution of dose-response toxicity data for chemicals requiring metabolic activation: example—vinyl chloride. Toxicol Appl Pharmacol 44:581–591

Gervasi PG, Citti L, del Monte M, Longo V, Benetti D (1985) Mutagenicity and chemical reactivity of epoxide intermediates of the isoprene metabolism and other structurally related compounds. Mutat Res 156:77–82

Heck H d'A, Casanova M (1987) Isotope effects and their implications for the covalent binding of inhaled ^{3}H- and ^{14}C-formaldehyde in the rat nasal mucosa. Toxicol Appl Pharmacol 89:122–134

Hemminki K (1984) Urinary excretion of formaldehyde in the rat. Chem-Biol Interact 48:243–248

IARC (1986) Tobacco Smoking. IARC Monogr Eval Carcinog Risk Hum 38:189–194

IARC (1987) Arsenic and arsenic compounds. IARC Monogr Eval Carcinog Risk Hum [Suppl] 7:100–106

Kreiling R, Laib RJ, Filser JG, Bolt HM (1986) Species differences in butadiene metabolism between mice and rats evaluated by inhalation pharmacokinetics. Arch Toxicol 58:235–238

Kreiling R, Laib RJ, Filser JG, Bolt HM (1987) Inhalation pharmacokinetics of 1,2-epoxybutene reveal species differences between rats and mice sensitive to butadiene-induced carcinogenesis. Arch Toxicol 61:7–11

Lam CW, Casanova M, Heck H d'A (1985) Depletion of nasal mucosal glutathione by acrolein and enhancement of formaldehyde-induced DNA-protein cross-linking by simultaneous exposure to acrolein. Arch Toxicol 58:67–71

Maltoni C, Lefemine G (1975) Carcinogenicity assays of vinyl chloride: current results. Ann NY Acad Sci 246:195–224

Maltoni C, Conti B, Cotti G (1983) Benzene: a multipotential carcinogen. Am J Ind Med 4:589–630

Schepers G, Walk RA (1988) Cotinine determination by immunoassays may be influenced by other nicotine metabolites. Arch Toxicol 62:395–397

Schmidt U, Loeser E (1985) Species differences in the formation of butadiene monoxide from 1,3-butadiene. Arch Toxicol 57:222–225

Schwartz SL, Ball RT, Witorsch P (1987) Mathematical modelling of nicotine and cotinine as biological markers of environmental tobacco smoke exposure. Toxicol Lett 35:53–58

Snyder CA, Sellakumar AR, James DJ, Albert RE (1988) The carcinogenicity of discontinuous inhaled benzene exposures in CD-1 and C57 Bl/6 mice. Arch Toxicol 62:331–335

Starr TB, Buck RD (1984) The importance of delivered dose in estimating low-dose cancer risk from exposure to formaldehyde. Fundam Appl Toxicol 4:740–753

Starr TB, Gibson JE (1985) The mechanistic toxicology of formaldehyde and its implications for quantitative risk assessment. Ann Rev Pharmacol Toxicol 25:745–767

Stommel P, Schöbel S, Müller G, Norpoth K (1987) Bestimmung des Benzolmetaboliten S-Phenylmercaptursäure im Harn von Benzol-exponierten Arbeitnehmern. Verh Dtsch Ges Arbeitsmed 27:289–292

Swenberg JA, Barrow CS, Boreiko CJ, Heck H d'A, Levine RJ, Morgan KT, Starr TB (1983) Non-linear biological responses to formaldehyde and their implications for carcinogenic risk assessment. Carcinogenesis 4:945–952

Törnqvist M, Kautiainen A, Gratz RN, Ehrenberg C (1988) Hemoglobin adducts in animals exposed to gasoline and diesel exhausts. J Appl Toxicol 8:159–170

Van Duuren BL, Langseth L, Orris L, Teebor G, Nelson N, Kushner M (1966) Carcinogenicity of epoxides, lactones and peroxy compounds. JNCI 37:825–835

8. Molecular Dosimetry of Chemical Carcinogens: Implications for Epidemiology and Risk Assessment

G.W. Lucier[1], S. Belinsky[1], and C. Thompson[1]

Introduction

It is becoming increasingly evident that molecular approaches can contribute a great deal toward clarifying some of the central issues of the risk assessment process for chemical carcinogens (Perera 1987). For example, these approaches can help explain some of the underlying mechanisms responsible for epidemiological observations. Secondly, it may be possible to accurately identify populations or individuals at risk to the toxic effects of selected classes of environmental chemicals. Finally, biochemical data should remove some of the uncertainty in the risk assessment process that is inherent when gross biological endpoints in animals, such as tumor incidence, are used to estimate low dose human risks.

The framework for incorporating molecular data or "biomarkers," as they are frequently called, into the risk assessment process is illustrated in Fig. 8.1. The essence of this illustration is that there are numerous biological and biochemical events that ultimately determine an adverse health effect following exposure to a toxic chemical. The chemical must be internalized leading to the presence of the chemical in blood or tissues. There are numerous cases where toxic chemicals can be detected at extraordinarily low concentrations in blood. For example, recent developments in analytic methodology have lowered the limits of detection for 2,3,7,8-tetrachlorodibenzo-p-dioxin (TCDD) to the low ppt range allowing for detection of this compound in the general population (Young and Shepard 1984). Once internalized, many chemicals are metabolized by a wide variety of drug-metabolizing enzymes present in virtually every tissue of the body (Lucier et al. 1979). These enzymes include the multiple forms of cytochrome P-450, epoxide hydrolase, glutathione transferase, glucuronyltransferase, sulfotransferase, and others. Depending on the chemical being metabolized, each of these enzymes may function as an activation or detoxication pathway. For example, glutathione transferase detoxifies electrophilic arene oxides of polycyclic aro-

[1]National Institute of Environmental Health Sciences, PO Box 12233, Research Triangle Park, North Carolina 27709, USA.

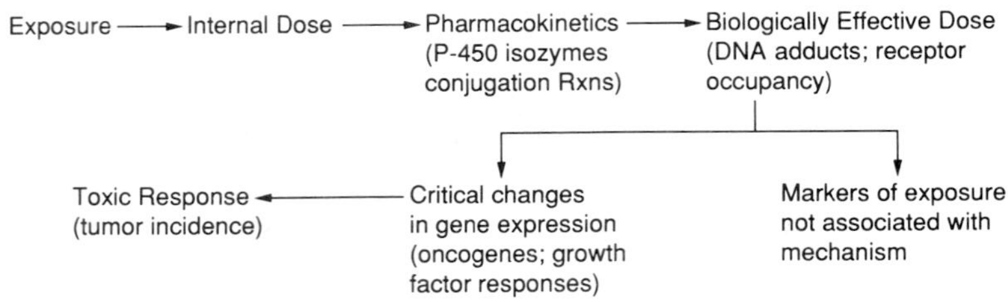

Figure 8.1. Mechanistic basis for the use of biomarkers in epidemiology and risk assessment

matic hydrocarbons (Jerina and Bend 1975), whereas it forms a DNA reactive metabolite of ethylene dibromide (Guengerich et al. 1987).

The balance between activating and detoxifying enzyme systems governs the rate of delivery of bioactive metabolites to the macromolecular target site. The macromolecular interaction of interest could be a DNA adduct for carcinogens which are initiating agents and receptor occupancy for chemicals which are tumor promoters. It should be noted that "initiation" and "promotion" are operational terms not stages which have clearly defined mechanisms. In any event, the concentration of DNA adduct or occupied receptor has been termed the "biologically effective dose." The biologically effective dose leads to changes in gene expression which may or may not be associated with the mechanism of carcinogenesis. Changes not involved in the carcinogenesis process may provide a reliable "marker of exposure" such as occurs for the induction of arylhydrocarbon hydroxylase following exposure to polycyclic aromatic hydrocarbons (for example, benzo(a)pyrene and TCDD). Other changes in gene expression such as activation of protooncogenes and alterations in cell proliferation pathways may have direct links to mechanisms of chemical carcinogenesis (Reynolds et al. 1987; Swenberg et al. 1987) and represent possible candidates of "markers of effect" for chemical carcinogens. A detailed review and evaluation of biomarkers has recently been published by a National Research Council Committee (1987).

One of the most compelling issues in the risk assessment process is the estimation of dose-response relationships. For reasons of economic necessity, high doses are used in animal bioassays for carcinogenicity of environmental chemicals. Extrapolation of tumor incidence arising from high dose exposures in animals to low-dose risks in humans is a highly uncertain exercise as illustrated in Fig. 8.2. Biomarkers, such as activated metabolites of DNA adducts, can usually be measured at much lower doses than those needed to detect increased tumor incidence. Therefore, it seems reasonable to assume that a careful evaluation of biomarkers in experiments covering a wide dose range offers an opportunity to remove some of the uncertainty in the risk assessment process by a more reliable estimation of the shape of the dose-response curve at low doses. These kinds of studies have been termed "molecular dosimetry."

In this presentation, we will address some of the central issues involved in the application of molecular dosimetry information to components of risk assess-

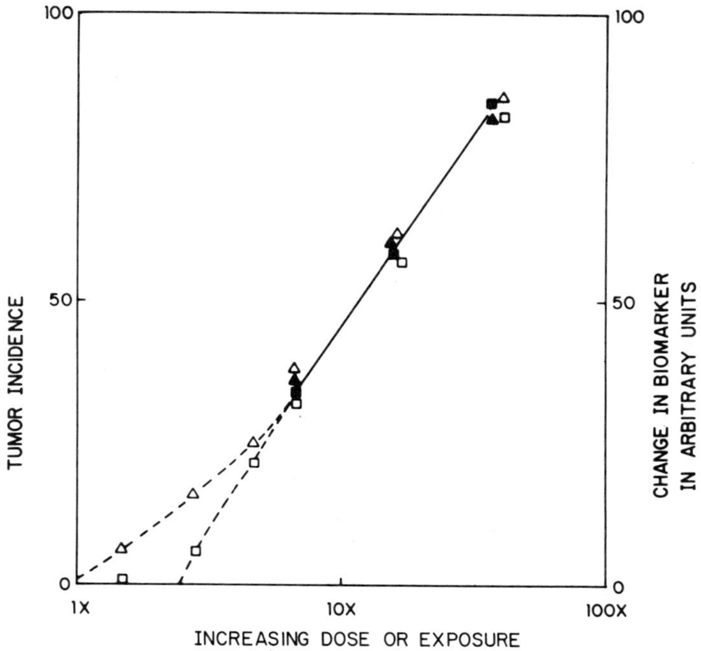

Figure 8.2. Schematic representation of tumor incidence and change in biomarkers following exposure to varying doses of two chemical carcinogens which exhibit the same dose response relationships for tumor incidence and change in biomarkers at high doses (*solid symbols*, tumor incidence for compounds A (*triangles*) and B (*squares*); *open symbols*, change in biomarkers for compounds A (*triangles*) and B (*squares*), respectively). However, dose-response relationships, at low doses, for changes in biomarkers differs between compounds A and B. If the biomarker is involved in the mechanism of action for compounds A and B, then use of the biomarker data in low-dose extrapolation of cancer risks to humans should lessen the uncertainty in the risk assessment process

ment through the use of environmentally relevant compounds as examples. These examples include benzene, cigarette smoke, and TCDD. We will also discuss advantages and difficulties in quantifying DNA adducts in human populations. Finally, we will attempt to evaluate the scientific foundation for conducting molecular dosimetry studies for agents which are carcinogenic because of effects on cell proliferation pathways.

Benzene Metabolism

Benzene has been shown to be carcinogenic in humans (Aksoy and Erdem 1978; Goldstein 1989) as well as in animal bioassays (Huff et al. 1989). High occupational exposures to benzene have occurred in a number of settings although the exposures have been decreased by continued regulatory actions. Exposure to benzene also occurs via gasoline, mainstream tobacco smoke, and environmental tobacco smoke (Wallace 1987).

In the chronic toxicity study on benzene conducted by the National Toxicology Program (NTP), male and female $B6C3F_1$ mice and F344 rats were exposed to benzene by gavage. Species differences in the toxic responses of the hematopoietic system were evident, with mice being considerably more sensitive than rats (Huff et al. 1989). Our studies (in collaboration with NTP and Lovelace Inhalation Toxicology Research Institute) were designed to evaluate the role of metabolism in the observed species differences and to determine if there are quantitative and/or qualitative differences in benzene metabolism between inhalation or oral exposures. Benzene was administered to $B6C3F_1$ mice and F344 rats over a wide dose range and key metabolites were quantified at various time points during and after exposure. There are numerous reports in the literature concerning benzene metabolism and toxicity in rats and mice (reviewed by Medinsky, this volume; Kalf 1987; Snyder et al. 1989). These studies provide strong evidence that metabolism to phenol and subsequent conjugation with glucuronic acid and sulfate represents detoxication pathways. Likewise, conjugation with glutathione to form a mercapturic acid metabolite also appears to be a detoxified product. In contrast, muconaldehyde, a consequence of ring opening of benzene, is capable of reaction with cellular macromolecules and is thought to be responsible for some of the toxic effects of benzene (Latriano et al. 1986). Muconaldehyde is unstable and difficult to quantify, so muconic acid is a more reliable marker of metabolism through the muconaldehyde pathway. Benzoquinone is also considered to be a toxic metabolite of benzene (Kalf 1987) and hydroquinone conjugates provide reliable markers for metabolism through this pathway (Medinsky et al., this volume). Therefore, we quantified hydroquinone conjugates and muconic acid as markers of metabolic activation whereas phenol conjugates and mercapturic acid conjugates were considered as detoxication pathways (Table 8.1). In our dose response studies, we used 5, 50, and 600 ppm (6 h) for inhalation exposure and 1, 10, and 200 mg/kg for oral exposures. Metabolite concentrations were quantified by the area under the curve (AUC).

The data on species differences in benzene metabolism have been published (Sabourin et al. 1987, 1988) and reveal that mice form more of the toxic metabo-

Table 8.1. Activation and deactivation pathways for benzene metabolism.

Activation pathways
muconic acid
hydroquinone conjugates
Deactivation pathways
phenol conjugates
mercapturic acid conjugates

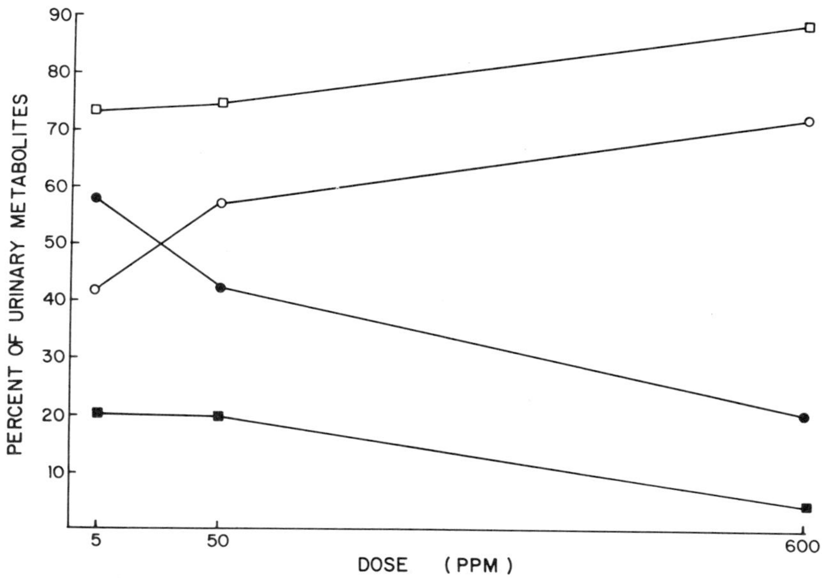

Figure 8.3. Dose-response relationships for activation/deactivation pathways for benzene in B6C3F$_1$ mice and F344N rats following inhalation exposures. *Open squares*, rat deactivation pathways; *open circles*, mouse deactivation pathways; *solid circles*, mouse activation pathways; *solid squares*, rat activation pathways

lites than rats at any dose examined, and this species difference was evident in metabolites present in urine, blood, liver, lung, and bone marrow. Furthermore, mice formed more of the toxic metabolites following either inhalation or oral exposures. One way of evaluating dose-response relationships is to determine the percentage of the total metabolites which are detected as activation pathways. Analysis of our metabolism data in this way revealed a higher percentage of benzene metabolism via toxic pathways than nontoxic pathways at low doses (5 ppm inhalation and 1 mg/kg orally), whereas the nontoxic pathways predominate at the higher doses (Henderson et al. 1989; Sabourin et al. 1989a) (Fig. 8.3). The same relationships were seen in all tissues and fluids examined.

The effects of different exposure rates on benzene metabolism were determined by exposing rodents over different time intervals to the same total amount of benzene. Dose rates were 600 ppm × 0.5 h, 150 ppm × 2 h, and 50 ppm × 6 h. In general, the rate of inhalation exposure to benzene did not affect the AUC of metabolites in tissues of rats. However, a lower concentration of toxic metabolites was detected in mice receiving the high dose rate (600 ppm × 0.5 h) compared to the lower dose rates (Sabourin et al. 1989a).

The shift toward greater relative formation of less toxic metabolites at higher inhalation or oral exposures could result in underestimating the toxicity of low exposures of benzene when extrapolating from rodent studies conducted with high-level dosing regimens. These data were used to develop a physiological model to describe the uptake and metabolism of benzene in mice (Medinsky et al. 1989; Medinsky, this volume). Simulations of human inhalation exposures were also conducted using metabolic rate constants derived from mice. These simulations reveal that for exposures below 25 ppm, hydroquinone conjugates would be the predominate metabolite produced, whereas for higher exposures phenol conjugates predominate. This preferential production of putative toxic metabolites at low exposure concentrations may have important implications in risk assessment for benzene. The muconaldehyde and benzoquinone pathways presumably lead to DNA adduct formation, but DNA adducts are not detected following in vivo exposures unless the dose is extraordinarily high. Therefore, DNA adducts are probably not a good molecular dosimeter of benzene, although hemoglobin adducts offer some promise (Sabourin et al. 1989b).

DNA Adducts

DNA adducts represent one of the most promising molecular dosimeters and are frequently used to estimate genotoxic potential of chemicals. However, several issues need to be addressed in order to evaluate the biological significance of DNA adducts. These include adduct heterogeneity, cell specificity for adduct formation, adduct persistence and repair, adducts in lymphocytes, and adducts arising from dietary constituents or endogenous compounds.

Adduct Heterogeneity

Some DNA adducts are mutagenic whereas others are not. Experiments on mutational specificity (*lac*I gene of *Escherichia coli*), produced by a series of alkylating agents, provide evidence that DNA adduct location governs site-specific mutations (Glickman et al. 1987). Moreover, some DNA sequence changes may lead to selectable phenotypic alterations, whereas others may not.

Cell Specificity of Adducts

Most tissues are comprised of multiple cell types and cell types vary considerably in their capacity to metabolically activate/deactivate environmental chemicals to DNA-reactive species. For example, lung is comprised of multiple cell types some of which have high concentrations of cytochrome P-450 isozymes, whereas others have low or non-detectable mixed function oxidase activity. Moreover, the relative concentrations of various P-450 isozymes and enzymes are cell-type dependent. Therefore, one compound may produce high concentrations of promutagenic adducts in one cell type but not another, whereas the opposite might occur for a compound which is activated by a different P-450 isozyme. For example, exposure to some nitrosamines leads to high concentrations of alkylated DNA adducts in the Clara cell (Belinsky et al. 1987), whereas no cell specificity was seen for benzo(a)pyrene-derived DNA adducts (Stowers and Anderson 1985). In any event, DNA adduct concentrations derived from a whole tissue homogenate may grossly underestimate adduct concentrations in a given cell type.

Adduct Persistence and Repair

Some DNA adducts are repaired quickly, and some adducts are hardly repaired at all, with adduct loss correlating with cell turnover. Therefore, the concentration and gene location of DNA adducts will change with time after exposure to a genotoxic chemical. Furthermore, the existence of nonrandom repair in the genome makes it difficult to utilize total DNA repair capacity as an indicator of cell susceptibility to carcinogens (Hanawalt 1987). It is especially important in human studies to know the duration and timing of exposure for proper evaluation of the biological significance of a given adduct concentration.

Lymphocyte Adducts

In human studies, it is rarely possible to obtain nontumorigenic target tissue for the quantitation of DNA adducts. Lymphocytes are a readily accessible source of human cells that are known to contain DNA adducts. However, there is little information on the reliability of using lymphocyte adduct concentrations for the estimation of target cell or tissue adduct concentrations. In some cases, lymphocytes may be good surrogates and in other cases they may be inappropriate.

Adducts of Dietary Constituents or Endogenous Hormones

Experiments in control animals have revealed that as animals age DNA adducts are detected in increasing amounts (Randerath et al. 1986). Although the relationship of these adducts to tumor development in control animals is unclear, they must be derived from dietary constituents or endogenous chemicals such as hormones. In another experiment, treatment of Syrian hamsters with either steroidal or nonsteroidal estrogens resulted in the formation of DNA adducts in kidney, the major site of estrogen-induced tumors in this species (Liehr et al. 1986). Surprisingly, these adducts were not the consequence of direct adduction of the estrogen molecule. These kinds of studies raise the possibility that a chemical may have genotoxic potential even if its own chemical structure does not bind directly to DNA.

DNA Adducts of Some Nitrosamines

Many mutagenic and carcinogenic compounds have been detected in tobacco and tobacco smoke (Loeb et al. 1984). Several nitrosamines present in both main stream and sidestream smoke and in unburned tobacco (Hoffmann et al. 1979)

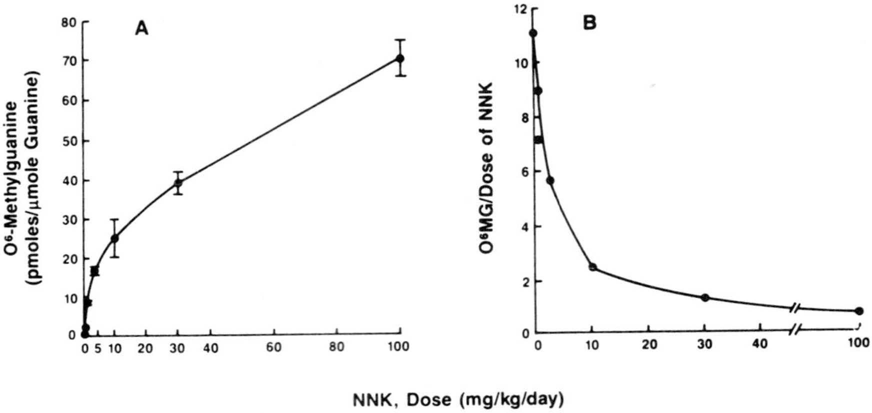

Figure 8.4. Dose response (**A**) and efficiency of O^6-MG formation (**B**) in lung. **A** O^6-MG was determined in lungs from rats treated for 12 days with 0.1, 0.3, 1.3.30, or 100 mg/kg NNK. *Points* are means from four rats; *bars*, SE. **B** Concentration of O^6MG in lungs after 12 days of treatment with NNK was divided by the dose of carcinogen and plotted against dose. (From Belinsky et al. 1987)

have been identified as major constituents of tobacco products. The tobacco-specific nitrosamine 4-(*N*-methyl-*N*-nitrosamino)-1-(3-pyridyl)-1-butanone (NNK) is a potent carcinogen in laboratory animals. Treatment of rats with NNK results in the induction of tumors in the nasal cavity, lung, liver, and pancreas (Hoffmann et al. 1984; Rivenson et al. 1988). At lower doses of NNK, the prevalence of malignant lung tumors was much greater than tumors of other target tissues. Pulmonary tumors are also induced in mice and hamsters following either acute or chronic exposure to this carcinogen (Castonguay et al. 1983; Hecht et al. 1983). The induction of lung tumors from NNK appears to stem from its activation via α-hydroxylation resulting in the formation of the promutagenic adduct O^6-methylguanine (O^6MG) (Belinsky et al. 1986).

O^6MG was found to accumulate in lungs from rats following treatment over 12 days with doses of NNK ranging from 0.1 to 100 mg/kg (Belinsky et al. 1987). The dose response for NNK was nonlinear; the O^6MG to dose ratio, an index of alkylation efficiency, increased dramatically as the dose of carcinogen was decreased (Fig. 8.4). These results have important implications toward assessing risk for exposure to NNK and indicate that extrapolation using high dose alkylation data would greatly underestimate risk. Probably an equally important contributing factor to the potent induction of pulmonary tumors by NNK is the cell specificity observed for O^6MG formation. The Clara cell, although accounting for only 1% of the pulmonary cells in the lung of the rat (Crapo et al. 1983), was found to possess the highest concentration of O^6MG, 28-fold higher than alkylation levels detected in DNA from whole lung, following treatment with NNK (Belinsky et al. 1987). Moreover, as the dose of carcinogen was decreased from 100 to 0.3 mg/kg, the alkylation efficiency in this cell population increased 38-fold.

Further support for cell specificity toward O^{6}MG formation stems from recent studies with *N*-nitrosodimethylamine (NDMA), a weak carcinogen in the rodent lung (Hecht et al. 1986). DNA methylation by NDMA does not exhibit any cell selectivity and adduct concentrations in the Clara cell were 50-fold higher after treatment with NNK than NDMA at equimolar doses (Devereux et al. 1988). Differences in DNA methylation patterns in lung for NNK and NDMA appear related to metabolic activation by the cytochrome P-450 isozymes. Using antibodies to cytochrome P-450 isozymes from lung, the major isozyme involved in activation of NNK was P-450b, while activation by NDMA occurred solely by P-450c. Thus, differences in O^{6}MG, specifically the high affinity for activation of NNK in the Clara cell appears to be an important factor in explaining why at low doses NNK is a potent pulmonary carcinogen while NDMA is not.

Another factor which appears important in the induction of neoplasia by NNK is the repair of O^{6}MG. Constitutive activity of the repair protein O^{6}MG-DNA methyltransferase was two fold greater in macrophages and type II cells than alveolar cells and Clara cells (Belinsky et al. 1988). Treatment with NNK for 4 days had no effect on activity in macrophages, but decreased activity in alveolar small cells and type II cells 57% and 84% respectively, while activity in Clara cells was reduced to below limits of detection. The effect of NNK on O^{6}MG-DNA methyltransferase activity was consistent with rates of removal of O^{6}MG. For example, the loss of O^{6}MG from DNA of macrophages was rapid ($t_{1/2}$ = 48 h) while very little loss of this adduct was observed in Clara cells over an 8-day period following cessation of carcinogen treatment. Taken together, these data indicate that the high concentration of O^{6}MG in Clara cells and the low rate of repair of this promutagenic adduct may be critical factors in the potent pulmonary carcinogenicity induced by the tobacco specific carcinogen NNK.

DNA Adducts in Human Samples

It is obviously more difficult to evaluate dose response relationships for chemical carcinogens in humans compared to animal models. First, radioactive carcinogens cannot be administered to humans, and secondly, the accessibility of tissues and cells is appropriately limited. With regard to the first issue, several approaches to detect DNA adducts in human samples have been evaluated (Wogan and Gorelick 1985; Santella 1988). The two most commonly used methods are immunoassays and ^{32}P-postlabeling. In general, immunoassays are both specific and sensitive for alkylated adducts such as O^{6}MG (see review by Wogan and Gorelick 1985). However, these methods are not easily applied to quantitation of adducts for bulky aromatic hydrocarbons such as benzo(a)pyrene-derived adducts. The main problem relates to a lack of specificity of the antibodies used in the assay which cross react with a number of PAH-related adducts (Santella et al. 1988; Harris et al. 1985a). Therefore, use of these antibodies may be restricted to quantitation of total antigenicity and not specific DNA adducts. The

Table 8.2. Detection of DNA adducts in human samples by ^{32}P-postlabeling.

Population	Tissue	Fraction positive	Reference
1 smokers	white blood cells	4/4	Phillips et al. 1986
nonsmokers	white blood cells	6/6	
2 Controls	oral mucosa	10/19	Dunn and Stith 1986
3 smokers	placenta	16/17	Everson et al. 1986
nonsmokers		3/14	
4 smokers	placenta	28/30	Everson et al. 1988
nonsmokers		4/23	
5 smokers	white blood cells	15/15	Jahnke et al. 1989
nonsmokers		11/11	

second assay frequently used to quantify DNA adducts in humans is the ^{32}P-postlabeling technique. A complete description of this assay is described elsewhere (Reddy and Randerath 1987; Gupta et al. 1988). It offers the advantage of extraordinary sensitivity, being capable of detecting 1 adduct in 10^{10} normal nucleotides when appropriate modifications are made to the procedure. This assay permits detection of DNA adducts in samples from control animals (Randerath et al. 1986) as well as chemically exposed and unexposed humans (Table 8.2). The assay is particularly useful for detecting adducts of nonpolar polycyclic aromatic hydrocarbons such as 7,8-diol-9,10-oxide benzo(a)pyrene deoxyguanosine (BPDE). It is not sensitive for the detection of water soluble or alkylated DNA adducts. This problem arises from the limitations of the chromatographic systems (thin-layer chromatography or high-pressure liquid chromatography) used to separate modified from normal nucleotides.

The ^{32}P method has been used in our laboratories to examine lymphocyte DNA adducts from smokers and nonsmokers (Jahnke et al. 1989). Similar to the findings of others, we could not consistently detect smoking-related adducts in lymphocytes; several distinct adducts were detected in smokers as well as nonsmokers (Table 8.3). The most provocative finding from our studies is that there was a great deal of interindividual variation in the amount or kind of DNA adduct detected on the thin-layer chromatograms. We were uncertain whether this variation represented experimental, intraindividual or true interindividual variation. In order to evaluate these possibilities, several individuals were resampled after a 1-month interval. Results revealed that the postlabeling DNA adduct pattern did not change appreciably in lymphocytes from these subjects. These data are consistent with the idea that a great deal of interindividual variation exists in human populations in relation to the cellular machinery that governs the rates of DNA adduct formation and repair.

The mechanisms responsible for human interindividual variation in responsiveness to chemical carcinogens seem to be unfolding. For example, polymorphisms are present for some metabolic activation/deactivation enzyme systems including specific cytochrome P-450 isoenzymes (Nebert 1988; Harris et al. 1985b) and one form of glutathione transferase (Seidegard and Pero 1985). Individuals deficient in the polymorphic glutathione transferase are more at risk

Table 8.3. Detection of DNA adducts by ^{32}P-postlabeling in human lymphocytes from smokers and nonsmokers.

DNA adducts (unidentified)	Number of individuals in which a specific adduct was detected			
	Nonsmokers $n = 15$		Smokers $n = 11$	
	(n)	(%)	(n)	(%)
a	8	53	8	73
b	9	60	8	73
c	0	0	2	18
d	3	20	2	18
e	7	47	7	64
f	6	40	7	64
g	1	7	2	18
h	5	33	4	36
i	3	20	2	18

Data summarized from Jahnke et al. (1989)

for contracting lung cancer from cigarette smoke (Siedegard et al. 1986). Likewise, individuals having high concentrations of the P-450 isoenzyme, which metabolizes debrisoquine, are more at risk for lung cancer (Ayesh et al. 1984). We are currently investigating the relationship between activities of metabolic activation/deactivation reactions and the DNA adduct pattern detected in human lymphocytes. One component of these studies has involved in vitro incubation of human lymphocytes with chemical carcinogens such as benzo(a)pyrene. These studies quantify rates of formation of individual metabolites, DNA adduct concentration, and frequency of sister chromatid exchanges (SCEs). Results revealed a large amount of interindividual variation in all these parameters. However, none of the parameters correlated with the quantity or kind of DNA adduct detected by ^{32}P-postlabeling of the same lymphocyte samples prior to incubation with ^{3}H-benzo(a)pyrene. The reasons for this finding are not clear but could reflect the fact that chemicals other than benzo(a)pyrene are producing the DNA adducts detected in unincubated lymphocytes. Another possible explanation is that lymphocyte DNA adducts are a consequence of metabolic activation in cells other than lymphocytes, such as liver or lung cells. Likewise, there were no strong correlations between different parameters in our in vitro studies (Thompson et al. 1989). For example, DNA adduct concentrations did not correlate with benzo(a)pyrene-induced increases in SCE frequency. Our results emphasize the contention that one needs to be cautious when extrapolating between commonly used parameters for estimating DNA damage in humans.

Molecular Dosimetry of Tumor Promoters

The development and variation of pharmacokinetic models to explain the metabolism, DNA binding, DNA repair, and associated gene mutation rate have accompanied our increasing knowledge regarding the mechanisms of genotoxic

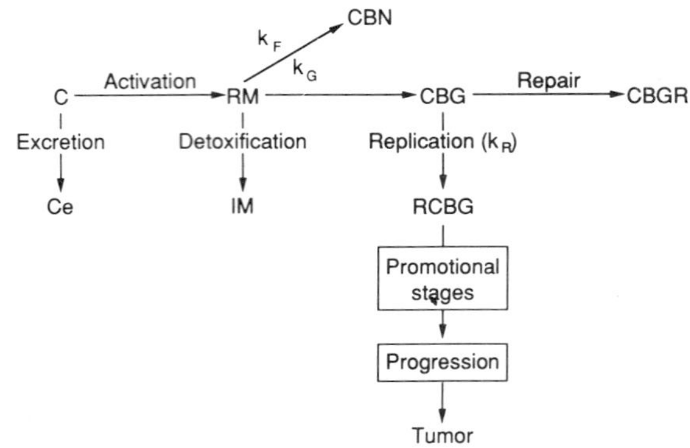

Figure 8.5. Pharmacokinetic model for the metabolic fate and carcinogenic effects of some carcinogens. *C*, nonactivated chemical; *RM*, reactive metabolite; *Ce*, excreted chemical; *IM*, inactive metabolite; *CBN*, covalent binding, nongenetic; *CBG*, covalent binding genetic; *CBGR*, repaired covalently bound genetic material; *RCBG*, retained genetic damage, critical and noncritical. In this model activation, detoxification, and repair are assumed to follow Michaelis-Menten kinetics, while the other reactions follow first-order kinetics. The Michaelis-Menten parameters for activation, detoxification, and repair are denoted as ($V_M{}^A$, $k_M{}^A$), ($V_M{}^D$, $k_M{}^D$), and ($V_M{}^R$, $k_M{}^R$), respectively. The symbols for the first-order rate constants are k_F, k_G, and k_R. (From Hoel et al. 1983)

carcinogens. Such a model is illustrated in Fig. 8.5. However, a lack of mechanistic information hinders modeling of the various components of tumor promotion and progression which are likely to be composed of several events (Barrett and Wiseman 1987). Current theories of chemical carcinogenesis implicate multiple and sequential somatic mutation and cell proliferation events in tumor development.

Compounds which readily form DNA adducts, are positive in short-term tests for genetic toxicity and are positive for carcinogenicity in chronic bioassays are generally considered to be genotoxic carcinogens. In a recent compilation of NTP data designed to test the concordance between short-term tests and in vivo carcinogenicity, it was revealed that nearly 40% of rodent carcinogens are negative for genetic toxicity (Tennant et al. 1987). The prevailing opinion was that promotional mechanisms were responsible for the carcinogenicity of nongenotoxic carcinogens.

In general, the overall process of tumor promotion involves clonal expansion of genetically altered cells. The selective proliferative actions of nongenotoxic carcinogens may be caused by chemically induced cytotoxicity or by receptor-mediated cell proliferation. The former mechanism is probably a high-dose phenomenon, whereas the latter mechanism may occur at high or low doses. Therefore, the mechanism of tumor promotion needs to be considered in the development and application of pharmacokinetic models to predict risk from exposure to a given tumor promoter. Evaluation of dose-response relationships for tumor promoters is further complicated by the existence of multiple receptor-

Table 8.4. Effects of TCDD on receptor systems.

Receptor	Effect	Reference
Estrogen receptor	Decreased binding capacity	Romkes et al. 1987
Glucocorticoid receptor	Decreased binding capacity	Nelson et al. 1989
Epidermal growth factor receptor (EGF)	Decreased binding capacity	Hudson et al. 1985
EGF-stimulated phosphorylation	Decrease	Sunahara et al. 1987
Ah receptor	Increase	Sloop and Lucier 1987

mediated effects produced by the structurally-diverse chemicals which are tumor promoters in animal models. For example, TCDD interacts with a cytosolic/nuclear receptor system that is specific for TCDD and its structural analogs (Poland et al. 1987; Safe 1986). This receptor system possesses all the properties expected of receptor proteins, functions in an analogous manner to steroid receptors and is required for the toxic effects of TCDD (Poland and Knutson 1984) (Table 8.4). TCDD and its structural analogs also can modify production of growth factors and growth factor receptor concentrations, thereby influencing the production of second messengers possibly involved in the regulation of cell division (Hudson et al. 1985; Madhukar et al. 1984; Lucier et al. 1987). TCDD also effects glucocorticoid receptor action, and it exerts profound influences on the activities of metabolic activation/deactivation reactions (Lucier et al. 1986; Nelson et al. 1989) involved in the production of electrophilic moieties which may produce additional DNA damage.

Estrogens, like TCDD, possess their own receptor systems, but this class of tumor promoter produces a different spectrum of receptor-mediated events than TCDD (Nelson et al. 1988). The phorbol acid esters, which are classical promoters in rodent skin, bind to protein kinase C and this interaction is thought to be required for the carcinogenic actions of the phorbol esters (Johnson et al. 1987). Phenobarbital, unlike the previously discussed tumor promoters, does not bind selectively to a receptor system, and exposure to this compound produces a different biochemical sequelae than TCDD, estrogens or phorbol esters. Taken together, these data provide convincing evidence that there is not a common underlying mechanism for all tumor promoters. Therefore, estimation of exposure based risk estimates for tumor promoters will be difficult. Although some models have been developed to predict human risks from exposure to tumor promoters (Moolgavkar 1987), their widespread application awaits a better understanding of the multiple and diverse sequence of biochemical events which comprise the processes of tumor promotion.

References

Aksoy M, Erdem S (1978) Follow-up study on the development of leukemia in 44 panxytopenic patients with chronic exposure to benzene. Blood 52:285–292

Ayesh R, Idle JR, Ritchie JC, Crothers MJ, Hetzel MR (1984) Metabolic oxidation phenotypes as markers for susceptibility to lung cancer. Nature 312:169–170

Barrett CJ, Wiseman RW (1987) Cellular and molecular mechanisms of multistep carcinogenesis: relevance to carcinogen risk assessment. Environ Health Perspect 76: 63–70

Belinsky SA, White CM, Boucheron JA, Richardson FC, Swenberg JA, Anderson MW (1986) Accumulation and persistence of DNA adducts in respiratory tissue of rats following multiple administrations of the tobacco specific carcinogen 4-(N-methyl-N-nitrosamino)-1-(3-pyridyl)-1-butanone. Cancer Res 46:1280–1284

Belinsky SA, White CM, Devereux T, Swenberg JA, Anderson MW (1987) Cell selective alkylation of DNA in rat lung following low dose exposure to the tobacco specific carcinogen 4-(N-methyl-N-nitrosamino)-1-(3-pyridyl)-1-butanone. Cancer Res 47: 1143–1148

Belinsky SA, Dolan ME, White CM, Maronpot RR, Pegg AE, Anderson MW (1988) Cell specific differences in O^6-methylguanine-DNA methyltransferase activity and removal of O^6-methylguanine in rat pulmonary cells. Carcinogenesis 9:2053–2058

Castonguay A, Lin D, Stoner GD, Radok R, Furuya K, Hecht SS, Schut HAJ, Klainig JE (1983) Comparative carcinogenicity in A/J mice and metabolism by cultured mouse peripheral lung of N′-nitrosonornicotine, 4-(methylnitrosamino)-1-(3-pyridyl)-1-butanone (NNK) and the effect of smoke inhalation. Carcinogenesis 4:1287–1290

Crapo JD, Young SL, Fram EK, Pinkerton KE, Barry BE, Crapo RO (1983) Morphometric characteristics of cells in the alveolar region of the mammalian lungs. Am Rev Respir Dis 128:S42–S46

Devereux TR, Anderson MW, Belinsky SA (1988) Factors regulating activation and DNA alkylation by 4-(N-methyl-N-nitrosamino)-1-(3-pyridyl)-1-butanone and nitrosodimethylamine in rat lung and isolated lung cells, and the relationship to carcinogenicity. Cancer Res 48:4215–4221

Dunn BF, Stith HF (1986) ^{32}P Postlabeling of aromatic DNA adducts in human oral mucosal cells. Carcinogenesis 7:1115–1120

Everson RB, Randerath E, Santella RM, Cefalo RC, Avitts TA, Randerath K (1986) Detection of smoking related covalent DNA adducts in human placenta. Science 231:54–57

Everson RB, Randerath E, Santella RM, Avitts TA, Weinstein IB, Randerath K (1988) Quantitative association between DNA damage in human placenta and maternal smoking and birth weight. JNCI 80:567–575

Glickman BW, Horsfall MJ, Gordon AJE, Burns PA (1987) Nearest neighbor affects G:C to A:T transitions induced by alkylating agents. Environ Health Perspect 76:29–32

Goldstein BD (1989) Introduction: Occam's razor is dull. Environ Health Perspect 82:3–6

Guengerich FP, Peterson LA, Cmarik KL, Koga N, Inskeep PB (1987) Activation of D_1 haloalkanes by glutathione conjugation and formation of DNA adducts. Environ Health Perspect 76:15–18

Gupta RC, Earley K, Sharma S (1988) Use of human peripheral blood lymphocytes to measure DNA binding capacity of chemical carcinogens. Proc Natl Acad Sci 85:3513–3517

Hanawalt PC (1987) Preferential DNA repair in expressed genes. Environ Health Perspect 76:9–14

Harris CC, Vahakangas K, Newman MJ, Trivers GE, Shamsuddin A, Sinopoli N, Mann DL, Wright WE (1985a) Detection of benzo(a)pyrene-diol epoxide-DNA adducts in peripheral blood lymphocytes and antibodies to the adducts in serum from coke oven workers. Proc Natl Acad Sci USA 82:6672–6676

Harris CC, Vahakangas K, Autrup H, Trivers GE, Shamsuddin AKM, Trump BF, Boman

BM, Mann DL (1985b) Biochemical and molecular epidemiology of cancer risk. In: The pathologist and the environment. Williams and Wilkins, Baltimore MD, pp 140–167 (International Academy of Pathology Monograph No 26)

Hecht SS, Adams JD, Numoto S, Hoffmann D (1983) Induction of respiratory tract tumors in Syrian golden hamsters by a single dose of 4-(methylnitrosamino)-1-(3-pyridyl)-1-butanone (NNK) and the effect of smoke inhalation. Carcinogenesis 4:1287–1290

Hecht SS, Trushin N, Castonguay A, Rivenson A (1986) Comparative tumorigenicity and DNA methylation in F344 rats by 4-(N-methyl-N-nitrosamino)-1-(3-pyridyl)-1-butanone and N-nitrosodimethylamine. Cancer Res 46:498–502

Henderson RF, Sabourin PJ, Bechtold WE, Griffith WC, Medinsky MA, Birnbaum LS, Lucier GW (1989) The effect of dose, dose rate, route of administration and species on tissue and blood levels of benzene metabolites. Environ Health Perspect 82:9–18

Hoel DG, Kaplan NL, Anderson MW (1983) Implication of nonlinear kinetics on risk estimation in carcinogenesis. Science 219:1032–1037

Hoffmann D, Adams JD, Brunnemann KD, Hecht SS (1979) Assessment of tobacco specific N-nitrosamines in tobacco products. Cancer Res 39:2505–2509

Hoffmann D, Rivenson A, Amin S, Hecht SS (1984) Dose-response study of the carcinogenicity of tobacco-specific N-nitrosamines in F344 rats. J Cancer Res Clin Oncol 108:81–86

Hudson LG, Toscano WA, Greenlee WF (1985) Regulation of epidermal growth factor binding in human keratinocyte cell line by 2,3,7,8-tetrachlorodibenzo-p-dioxin. Toxicol Appl Pharmacol 77:251–259

Huff JE, Haseman JK, DeMarini DM, Eustis S, Maronpot RR, Peters A, Persing R, Chrisp CE, Jacobs AC (1989) Multiple site carcinogenicity of benzene in Fischer 344 rats and B6C3F$_1$ mice. Environ Health Perspect 82:142–184

Jahnke GD, Thompson CL, Walker MP, Gallagher JE, Lucier GW, DiAugustine RP (1989) Multiple DNA adducts in lymphocytes of smokers and nonsmokers determined by ^{32}P-postlabeling analysis. Cancer Res (to be published)

Jerina DM, Bend JR (1977) Glutathione S-transferase. In: Jollo DJ, Koesis JJ, Snyder R, Vainio H (eds) Biological reactive intermediates: formation, toxicity, and inactivation. Plenum, New York, pp 207–236

Johnson MD, Housey GM, O'Brian CA, Kirschmeier PT, Weinstein IB (1987) Role of protein kinase C in regulation of gene expression and relevance to tumor promotion. Environ Health Perspect 76:89–95

Kalf GF (1987) Recent advances in the metabolism and toxicity of benzene. CRC Critic Rev Toxicol 18:141–159

Latriano L, Zaccaria A, Goldstein BD, Witz G (1986) Formation of mucondialdehyde, an open-ring metabolite of benzene, in mouse liver microsomes: an additional pathway for toxic metabolites. Proc Natl Acad Sci USA 83:8356–8360

Liehr JG, Avitts TA, Randerath E, Randerath K (1986) Estrogen-induced endogenous DNA adduction: possible mechanism of hormonal cancer. Proc Natl Acad Sci 83:5301–5305

Loeb LA, Ernster VL, Warner KE, Abbotts J, Laslo J (1984) Smoking and lung cancer, an overview. Cancer Res 44:5940–5958

Lucier GW, Lamartiniere CA, Dieringer CA, Lui E (1979) Metabolic activation/deactivation during development. Environ Health Perspect 29:17–26

Lucier GW, Rumbaugh RC, McCoy Z, Hass R, Harvan D, Albro P (1986) Ingestion of soil contaminated with 2,3,7,8 tetrachlorodibenzo-p-dioxin (TCDD) alters hepatic function in rats. Fundament Appl Toxicol 6:364–371

Lucier GW, Nelson KG, Everson RB, Wong TK, Philpot RM, Tiernan T, Taylor M, Sunahara GI (1987) Placental markers of human exposure to polychlorinated biphenyls and polychlorinated dibenzofurans. Environ Health Perspect 76:79–87

Madhukar BV, Brewster DW, Matsumura F (1984) Effects of in vivo administered 2,3,7,8-tetrachlorodibenzo-p-dioxin on receptor binding of epidermal growth factor in the hepatic plasma membrane of rat, guinea pig, mouse and hamster. Proc Natl Acad Sci USA 81:7407–7411

Medinsky MA, Sabourin PJ, Lucier GW, Birnbaum LS, Henderson RF (1989) A physiological model for simulation of benzene metabolism by rats and mice. Toxicol Appl Pharmacol (in press)

Moolgavkar SH, Knudson AJ (1981) Mutation and cancer: a model for human carcinogenesis. J Natl Cancer Inst 66:1037–1052

National Research Council Committee on Biological Markers (1987) Biological markers in environmental health research. Environ Health Perspect 74:3–9

Nebert DW (1988) Genes encoding drug-metabolizing enzymes: possible role in human disease. In: Woodhead AD, Bender MA, Leonard RC (eds) Phenotypic variation in populations: relevance to risk assessment. Plenum, New York

Nelson K, Vickers A, Sunahara G, Lucier GW (1989) Receptor and DNA ploidy changes during promotion of rat liver carcinogenesis. In: Barrett JC, Langenbach R, Elmore E (eds) Tumor promoters: biological approaches for mechanistic studies and assay systems. Raven, New York (in press)

Perera F (1987) The potential usefulness of biological markers in risk assessment. Environ Health Perspect 76:141–145

Phillips DH, Hewer A, Grover PL (1986) Aromatic DNA adducts in human bone marrow and peripheral blood leukocytes. Carcinogenesis 7:2071–2075

Poland A, Knutson JC (1984) Reflections on the mechanism of action of halogenated aromatic hydrocarbons. In: Biological mechanisms of dioxin action. Banbury report no 18. Cold Spring Harbor Laboratory, Cold Spring Harbor, New York

Poland A, Glover E, Kende AS (1976) Stereospecific high affinity binding of 2,3,7,8-tetrachlorodibenzo-p-dioxin by hepatic cytosol. J Biol Chem 251:4936–4946

Randerath K, Reddy MV, Disher RW (1986) Age- and tissue-related DNA modifications in untreated rats: detection by ^{32}P-postlabeling assay and possible significance for spontaneous tumor induction and aging. Carcinogenesis 7:1615–1617

Reddy MV, Randerath K (1987) ^{32}P-Postlabeling assay for carcinogen-DNA adducts: nuclease P_1-mediated enhancement of its sensitivity and applications. Environ Health Perspect 76:41–47

Reynolds SH, Stowers SJ, Patterson RM, Maronpot RR, Aaronson SA, Anderson MW (1987) Activated oncogenes in B6C3F$_1$ mouse liver tumors: implications for risk assessment. Science 237:1309–1316

Rivenson A, Hoffmann D, Prokopczyk B, Amin S, Hecht SS (1988) Induction of lung and exocrine pancreas tumor in F344 rats by tobacco-specific and areca derived N-nitrosamines. Cancer Res 48:6912–6917

Romkes M, Piskorska-Pliszczyneka J, Safe S (1987) Effects of 2,3,7,8-tetrachlorodibenzo-p-dioxin on hepatic and uterine estrogen receptor levels in rats. Toxicol Appl Pharmacol 87:306–314

Sabourin PJ, Chen BT, Lucier GW, Birnbaum LS, Fisher E, Henderson RF (1987) Effect of dose on the absorption and excretion of 14C-benzene administered orally or by inhalation in rats and mice. Toxicol Appl Pharmacol 87:325–336

Sabourin PJ, Bechtold WE, Birnbaum LS, Lucier GW (1988) Difference in the metabolism of inhaled 3H-benzene by F344/N rats and B6C3F$_1$ mice. Toxicol Appl Pharmacol 94:128–140

Sabourin PJ, Bechtold WE, Griffith WC, Birnbaum LS, Lucier GW, Henderson RF (1989a) Effect of exposure concentration, exposure rate, and route of administration on metabolism of benzene by F344 rats and B6C3F$_1$ mice (in press)

Sabourin PJ, Sun JD, Birnbaum LS, Lucier GW, Henderson RF (1989b) Effect of repeated benzene inhalation exposures on subsequent metabolism of benzene. Exp Pathol (in press)

Safe SH (1986) Comparative toxicology and mechanism of action of polychlorinated dibenzo-p-dioxins and dibenzofurans. Ann Rev Pharmacol Toxicol 26:371–379

Santella RM (1988) Applications of new techniques for the detection of carcinogen adducts to human population monitoring. Mutat Res 205:271–282

Santella RM, Weston A, Perera FR, Trivers GT, Harris CC, Young TL, Nguyen D, Lee BM, Poirier MC (1988) Interlaboratory comparison of antisera and immunoassays for benzo(a)pyrene-diol-epoxide-I-modified DNA. Carcinogenesis 9:1265–1269

Seidegard J, Pero RW (1985) The hereditary transmission of high glutathione transferase activity towards trans-stilbene oxide in human mononuclear leukocytes. Hum Genet 69:66–68

Seidegard J, Pero RW, Miller DG, Beattie EJ (1986) A glutathione transferase in human leukocytes as a marker for the susceptibility to lung cancer. Carcinogenesis 7:751–753

Sloop TC, Lucier GW (1987) Dose-dependent elevation of Ah receptor binding by TCDD in rat liver. Toxicol Appl Pharmacol 88:329–337

Snyder R, Dimitriadis E, Guy R, Hu P, Cooper K, Bauer, Witz G, Goldstein BD (1989) Studies on the mechanism of benzene toxicity. Environ Health Perspect 82:31–36

Stowers SJ, Anderson MW (1985) Formation and persistence of benzo(a)pyrene metabolite-DNA adduct. Environ Health Perspect 62:31–39

Sunahara GI, Nelson KG, Wong TK, Lucier GW (1987) Decreased human birth weights after in utero exposure to PCBs and PCDFs are associated with placental EGF-stimulated receptor autophosphorylation capacity. Mol Pharmacol 32:572–578

Swenberg JA, Richardson FC, Boucheron JA, Deal FH, Belinsky SA, Charbonneau M, Short BG (1987) High- to low-dose extrapolation: critical determinants involved in the dose response of carcinogenic substances. Environ Health Perspect 76:57–63

Tennant RW, Margolin BH, Shelby MD, et al. (1987) Prediction of chemical carcinogenicity in rodents from in vitro genetic toxicity assays. Science 236:933–941

Thompson CL, McCoy Z, Lambert JM, Andries MJ, Lucier GW (submitted) Relationships between benzo(a)pyrene metabolism, benzo(a)pyrene diol epoxide DNA adduct formation and sister chromatid exchanges in human lymphocytes from smokers and non-smokers. Cancer Res

Wallace LA (1987) The total exposure assessment methodology (TEAM) study. EPA Project Summary, EPA/600/S6-87/002, Washington DC, pp 1–14

Wogan GN, Gorelick NJ (1985) Chemical and biochemical dosimetry of exposure to genotoxic chemicals. Environ Health Perspect 62:5–18

Young AL, Shepard BM (1984) Determinations and measurement of human exposure to the dibenzo-p-dioxins. Bull Environ Contam Toxicol 33:702–709

Section 4. Mechanistic

9. Pathobiological Effects of Fibers and Tobacco-Related Chemicals in Human Lung Cells In Vitro

C.C. Harris,[1] J.C. Willey,[1] N. Matsukura,[1] J.F. Lechner,[1] M. Miyashita,[2] R.C. Grafstrom,[3] and B.F. Trump[4]

Introduction

Respiratory carcinogenesis is a multistage process in which tobacco smoke is the most important etiological agent. Several lines of evidence suggest that tobacco smoke has both tumor-initiating and tumor-promoting activities. Tobacco smoke has been shown to contain both initiating and promoting substances in the mouse skin model of carcinogenesis (Hoffmann et al. 1983b; Wynder and Hoffmann 1979). Analysis of epidemiological data is also consistent with the notion that tobacco smoke-induced bronchogenic cancer fits the initiation-promotion model (Doll and Peto 1978). The epidemiological evidence for promoting activity is twofold. First, risk of lung cancer is proportional to the fifth power of the duration of smoking, and second, the effect of smoking is substantially reversible so that cessation of smoking results in declining risk. It therefore follows that a smoker's respiratory tract is repeatedly exposed to both tumor initiators and promoters; this needs to be taken into account in the design of experimental protocols.

A major chemical class of carcinogens in tobacco smoke is the *N*-nitrosamines including those derived from nicotine and, to a lesser extent, from minor tobacco alkaloids (Hoffmann et al. 1982). Nicotine, therefore, may have a role in lung cancer beyond its habituating properties. The *N*-nitrosamines present in tobacco products occur at the highest reported concentrations outside of industrial environments. Interestingly, these carcinogens may also be formed in the oral cavity during tobacco chewing and snuff dipping (Hoffmann and Adams 1981), practices associated with oral cancer (Winn et al. 1981).

[1]Laboratory of Human Carcinogenesis, National Cancer Institute, National Institutes of Health, Building 37, 9000 Rockville Pike, Bethesda, MD 20892, USA.
[2]The First Department of Surgery, Nippon Medical School, 1-1-5 Sendagi, Bunkyo-ku, Tokyo 113, Japan.
[3]Institute of Toxicology, P.O. Box 60 208, S-104 01 Stockholm, Sweden.
[4]Department of Pathology, University of Maryland School of Medicine, 10 South Pine Street, Baltimore, MD 21201, USA.

The most important group of cocarcinogens in tobacco smoke, assayed in mouse skin, are reported to be the catechols (Hecht et al. 1981). It appears that catechols are formed by burning polyphenols and polysaccharides, including cellulose, that are present in tobacco (Hoffmann et al. 1985). A possible mechanism of their cocarcinogenic action is suggested by the observation that they block the detoxification pathway of benzo(a)pyrene metabolism; this presumably increases the concentration of activated carcinogen available for binding to DNA (Hoffmann et al. 1985). Catechols from cigarette smoke are also reported to generate hydrogen peroxide and superoxide free radicals (Nakayama et al. 1984).

Tobacco smoke condensate, aldehydes, peroxides, and other tumor promoters have been found to have a differential effect on normal human bronchial epithelial (NHBE) cells, compared with transformed cells of the same lineage (Miyashita et al. 1989). These substances caused squamous differentiation of normal cells and an inhibition of their clonal growth rate but had significantly less effect on human lung carcinoma cells. This suggests a mechanism whereby tumor promoters could cause selective clonal expansion of initiated cells in bronchial epithelium (Harris et al. 1986).

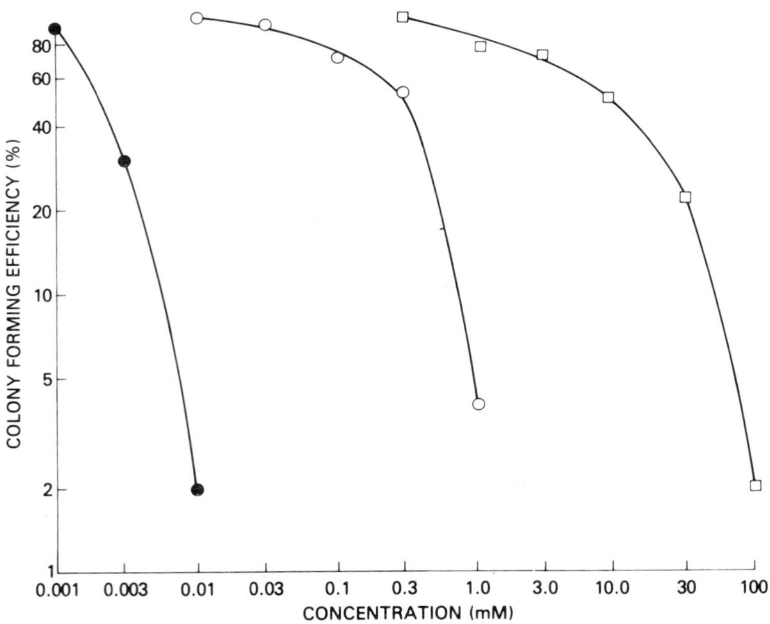

Figure 9.1. Effects of acrolein (*solid circles*), formaldehyde (*open circles*), and acetaldehyde (*squares*) on colony-forming efficiency of normal human bronchial epithelial cells

Table 9.1. DNA single-strand breaks and DNA protein cross-links caused by tobacco-related aldehydes in human bronchial epithelial cells.

Agents	Dose	DNA single-strand breaks per 10^{10} daltons	DNA-protein cross-links per 10^{10} daltons
Formaldehyde	100 μM	1.4	10.8
Acrolein	30 μM	1.4	2.6
Acetaldehyde	10 mM	< 1.0	2.7

Tobacco Smoke-Related Aldehydes

Tobacco smoke contains substantial amounts of formaldehyde, acrolein, and acetaldehyde that are formed by pyrolysis (Wynder and Hoffmann 1979). The concentration of these aldehydes reaching the lower respiratory tract is difficult to assess because, in addition to being present in the gaseous phase, they are also adsorbed to the particulates in the smoke and may thus be carried to the alveoli.

We are systematically investigating the pathobiological effects of aldehydes and peroxides on human lung cells. Using the highly sensitive colony-forming efficiency assay to measure cytotoxicity, acrolein is about 100- and 10000-fold more cytotoxic than formaldehyde and acetaldehyde, respectively (Fig. 9.1). Unsaturated aldehydes such as acrolein and 4-hydroxynonenal are highly reactive and form more stable adducts with macromolecules and deplete intracellular glutathione to a greater extent than saturated aldehydes (Krokan et al. 1985a).

The potency of formaldehyde, acrolein, and acetaldehyde to cause DNA damage has been measured by the alkaline elution technique (Table 9.1). Exposure of cells for 1 h to either formaldehyde or acrolein causes DNA single-strand breaks (SSB) and DNA protein cross-links, whereas acetaldehyde causes DNA-protein cross-links (Grafstrom et al. 1988, 1989; Saladino et al. 1985). Moreover, acetaldehyde causes formation of DNA interstrand breaks (R.C. Grafstrom et al., unpublished results).

Although formaldehyde-induced DNA damage in human cells appears to be partly removed by DNA excision repair (Grafstrom et al. 1984), several different enzymes are involved in the repair of alkyldiazonium ion-induced DNA damage, e.g., O^6-methylguanine is repaired by an O^6-alkylguanine DNA alkyltransferase (Pegg 1986; Yarosh 1985), N^3-methyladenine and N^7-methylguanine are repaired by a specific DNA glycosylase (Male et al. 1981), and O^4-methylthymidine is repaired by an as yet undefined enzyme (Becker and Montesano 1985). O^6-methylguanine-DNA alkyltransferase activity increases before the initiation of DNA synthesis in the S-phase of normal human cells (Kim et al. 1986) and varies widely among different human tissue types and among animal species (Krokan et al. 1986; Grafstrom et al. 1984a; Wild et al. 1986). This DNA repair enzyme has a cysteine at its active site and can be inactivated by alkylating agents (Pegg et al. 1983; Brent 1986) and by aldehydes, including those found in tobacco smoke and

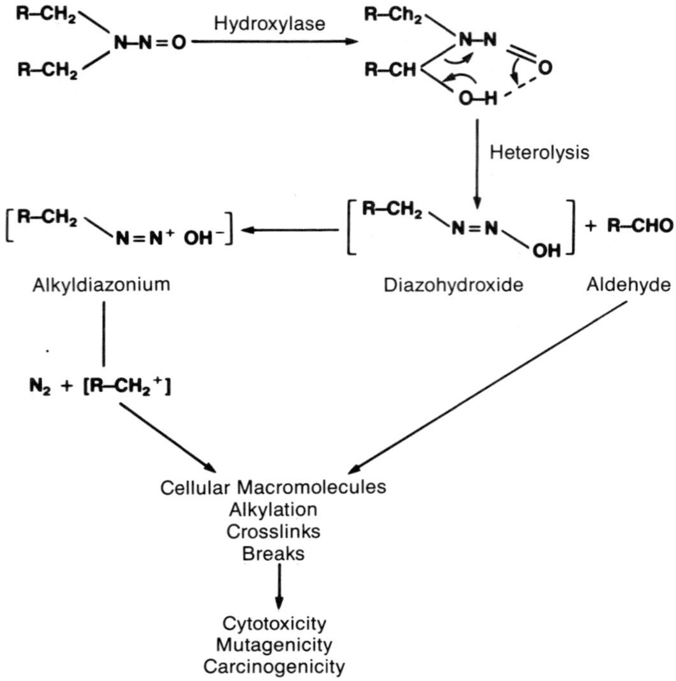

Figure 9.2. Metabolic pathway of *N*-nitrosamines leading to cytotoxic and genotoxic metabolites

formed by the metabolism of *N*-nitrosamines (Fig. 9.2) (Krokan et al. 1985a; Grafstrom et al. 1985a). Normal human bronchial epithelial cells in vitro do not show an adaptive response after treatment with *N*-methyl-*N'*-nitro-*N*-nitrosoguanidine (Krokan et al. 1985b).

To study further the pathobiological consequences of exposure to formaldehyde and inhibition of O^6-methylguanine repair, we investigated the cytotoxic and mutagenic effects of formaldehyde and *N*-methyl-*N*-nitrosourea (NMU) separately and in combination (Grafstrom et al. 1985a). Although the exposure times differed, formaldehyde is three times more mutagenic than NMU on a molar basis in human fibroblasts (Fig. 9.3). NMU and formaldehyde are each weak mutagens at the concentrations tested; however, addition of 50 or 75 μM formaldehyde to 200 μM NMU-treated cells results in a mutation frequency that is significantly greater than that found with either agent alone.

Formaldehyde may increase the mutagenicity of NMU by inhibiting O^6-methylguanine repair. This hypothesis is strengthened by analyzing the frequency of 6-thioguanine-resistant (6-TG^r)mutants as a function of survival. The curves for NMU and for formaldehyde are of similar magnitude and shape (Grafstrom et al. 1985a); however, a much steeper curve is found when formaldehyde and NMU are combined under conditions known to inhibit the repair of O^6-

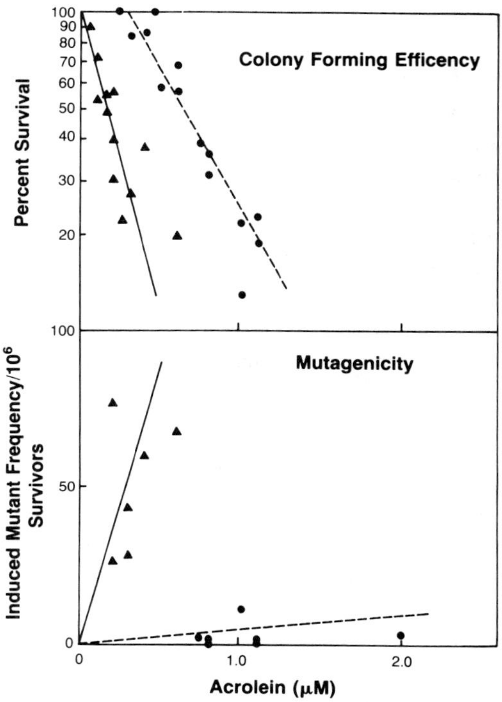

Figure 9.3. Effects of formaldehyde and/or *N*-methyl-*N*-nitrosourea on 6-thioguanine resistance mutation frequency in normal human bronchial fibroblasts (*circles*). *Triangles*, xeroderma pigmentosum cells

methylguanine. The data show that the combination of NMU and formaldehyde is more mutagenic than either agent alone and also indicate that the persistence of the O^6-methylguanine lesion has a more significant effect on induced mutation than it does on cell killing. Thus, there is a good correlation between the inhibition of O^6-methylguanine removal and the synergistic increase in mutation frequency in the presence of formaldehyde. This is also further indirect evidence that O^6-methylguanine is a promutagenic DNA lesion. In addition, NMU may inhibit the repair of promutagenic lesions caused by formaldehyde.

Because acrolein causes DNA SSB (Crook et al. 1986; Grafstrom et al. 1986, 1988), DNA-protein cross-links (Grafstrom et al. 1988), and cyclic deoxyguanosine adducts (Chung et al. 1984), the genotoxic consequences of acrolein have also been recently investigated (Curren et al. 1988). These data show clearly that acrolein is mutagenic to xeroderma pigmentosum (XP) cells at extremely low doses. Concentrations of acrolein as low as 0.2 μM induce approximately 3×10^{-5} 6-TG^r mutants per clonable XP cell. This frequency increases with the dose of acrolein up to 0.6 μM. Above this level the cytotoxicity is too great (<10% survival) to measure mutagenesis reliably.

Tobacco Smoke Condensate

Condensate and its fractions contain compounds that act as tumor initiators and promoters in the mouse skin carcinogenesis model (US Department of Health, Education and Welfare 1988). The B_{Ia} fraction of condensate contains the largest concentration of tumor initiators including several four- and five-ring aromatic hydrocarbons (Wynder et al. 1953). Condensate-derived tumor promoters for mouse skin reside primarily in the strongly polaric neutral subfraction (N_{meoh}) (Bock et al. 1970), the weakly acidic fraction (WA_e) and the phenolic fraction (Hoffmann et al. 1983a).

Because it is not known what compounds may serve as tumor promoters for NHBE cells or what their effects might be, we investigated the effects of representative compounds from three different chemical classes of tumor promoters that are active in the mouse skin carcinogenesis model (Willey et al. 1984, 1985). 12-*O*-tetradecanoylphorbol-13-acetate (TPA), teleocidin B, aplysiatoxin, and debromoaplysiatoxin each induce terminal squamous differentiation in NHBE cells in culture. In order to determine if there might be compounds in condensate that would have similar effects on NHBE cells, we studied the effects of condensate, N_{meoh}, and WA_e on the clonal growth rates, morphology, plasminogen activator and ornithine decarboxylase activities, and cross-linked envelope formation. We also observed the effects of the fractions B_{Ia} and B_{Ib} which are relatively less active in mouse skin tumor promotion studies.

One of the early effects of TPA and other tumor promoters in many epithelial cell culture systems is inhibition of epidermal growth factor (EGF) binding due to a decrease in the affinity of the EGF receptor for its ligand (Blumberg 1981; Horowitz et al. 1983). Therefore, we observed the effect of condensate and its fractions on EGF binding in NHBE cells. Measurement of intracellular thiol levels provides a sensitive indication of the effects on cells of electrophiles such as reactive aldehydes and peroxides. From previous investigations it is known that condensate decreases thiol levels in isolated suspensions of rat liver or lung cells (Moldeus et al. 1985). In addition, the clonal growth inhibitory effects of condensate on cultured human bronchial cells are partially prevented when cells are coincubated with *N*-acetylcysteine, a compound with nucleophile properties similar to glutathione (Grafstrom et al., unpublished results). For these reasons, we measured the effects of condensate and its fractions on thiol levels in bronchial cells. Finally, because condensate is mutagenic (DeMarini 1983) and clastogenic (Leuchtenberger et al. 1973) in other systems, we measured the condensate effects on DNA SSB formation.

In these experiments we determined whether condensate or its fractions could have effects on NHBE cells similar to those observed with TPA, teleocidin B, or the polyacetates (Willey et al. 1987). We found that condensate inhibits clonal growth, induces some morphological changes similar to those seen with TPA, induces the formation of cross-linked envelopes, and inhibits the binding of EGF. The N_{meoh} is a much more potent inhibitor of cell growth, induces a squamous morphological change very similar to that observed after incubation with TPA, increases cross-linked envelopes more than the other fractions, and increases plasminogen activator activity. In addition, N_{meoh} is the most potent inhibitor of EGF binding. Furthermore, this inhibition is due primarily to a decrease in the affinity of EGF receptors for the ligand, rather than down-regulation of, or direct competition for, EGF receptors. This mechanism is also the one by which TPA and other known tumor promoters inhibit EGF binding (Lee and Weinstein 1979). Based on the above findings, we conclude that the tumor-promoting activity in condensate has partitioned primarily into the N_{meoh} fraction. It is unclear why a prolonged incubation in medium containing condensate is necessary before effects on EGF binding are observed. It is possible that a TPA-like product is produced through metabolism of a compound in condensate.

Because the inhibition of growth caused by condensate and its fractions may involve other mechanisms in addition to induction of terminal squamous differentiation, we measured intracellular thiol levels in an effort to determine whether condensate would pose an electrophilic challenge to the cells. Indeed, the most growth inhibitory fraction (N_{meoh}) decreased total thiols, indicating that this fraction may contain small amounts of highly reactive compounds with affinity for sulfhydryls in protein and low molecular weight peptides such as glutathione. Because thiol depletion became statistically significant only at a very high concentration of N_{meoh}, it is likely that this effect plays only a minor role in condensate inhibition of NHBE cell proliferation. Other mechanisms of toxicity, such as inhibition of protein or nucleic acid synthesis, and effects on electrolyte homeostasis, may play a role and remain to be investigated.

There are reports that the irreversible phase of tumor promotion may involve clastogenic events mediated by free radical-induced DNA strand scissions. In support of this speculation, TPA causes DNA SSB (Birnboim 1982) and chromosome aberrations (Emerit and Cerutti 1981) in human leukocytes. Furthermore, mutagenic events that are likely to involve DNA damage have recently been suggested as necessary during later as well as initial stages of carcinogenesis. We find that a significant increase in DNA damage, in the form of SSB, is observed after incubation of the cells with medium containing the 40% growth inhibitory concentration (IC_{40}) of the N_{meoh} fraction (1.9 DNA SSB per 10^{10} daltons). Condensate causes a lower but still significant effect on DNA SSB (1.4 DNA SSB per 10^{10} daltons) at the IC_{40}. Moreover, incubation of N_{meoh}-exposed cells with DNA polymerase inhibitors also resulted in an accumulation of SSB indicating that the neutral fraction causes DNA damage which involves the DNA excision repair pathway. Other investigators report that phosphate-buffered saline exposed to cigarette smoke caused a large number of DNA SSB in cultures of A549 lung carcinoma cells (Nakayama et al. 1985). It is likely that reactive gaseous phase aldehydes, such as acrolein or formaldehyde (Grafstrom et al. 1985b), or labile DNA-damaging agents, e.g., active oxygen generated from cigarette smoke (Grafstrom et al. 1985), are responsible for the differences found between condensate and tobacco smoke.

The differential effects of cigarette smoke condensate (CSC) and its fractions (neutral, basic, and acidic fractions) on proliferation and squamous differentiation of NHBE cells versus human lung carcinoma cells have also been investigated (Miyashita et al. 1989). Condensate, and the neutral and acidic fractions, inhibited cellular proliferation more than the basic fraction. When compared to the acidic and basic fractions, condensate and the neutral fraction were more effective in causing squamous differentiation of NHBE cells and inhibiting specific binding of phorbol dibutyrate. In contrast to the normal epithelial cells, neither HUT-292 nor the three other carcinoma cell lines examined showed marked squamous morphological changes when exposed to either condensate or its fractions, and the carcinoma cells were more resistant to their inhibiting effects on cellular proliferation. These results are consistent with the hypothesis that differential effects of tobacco smoke components on cellular proliferation may allow clonal expansion of preneoplastic and neoplastic human bronchial epithelial cells during the tumor promotion phase of lung carcinogenesis.

As discussed above, DNA damage in NHBE cells resulting from exposure to CSC has been observed through alkaline elution assays. Mutagenicity of tobacco products has been evaluated in the past by the Ames salmonella bacterial assay (Kier et al. 1974). CSC is mutagenic in this system when it is incubated with liver microsomal enzymes. However, although CSC contains agents that are known to be mutagenic without need for activation, CSC is not detectably mutagenic in the absence of metabolic activation. This lack of observed direct mutagenicity may be due to general toxicity in relation to sensitivity of the mutagenesis assay. The A_L CHO human cell hybrid mutagenesis assay (see Waldren et al. 1986) may provide a useful system for testing the direct mutagenicity of CSC, as it is much more sensitive than the other mammalian mutagenesis systems. The A_L cell is a CHO cell that contains only one human chromosome, a single copy of chromosome 11. Three genes have been identified on human chromosome 11 that code for cell surface antigens a_1, a_2, and a_3. a_1 and a_3 are on 11p, while a_2 is on 11q. When the genes coding for these antigens are intact, cells are lysed in the presence of complement and antibody to the appropriate antigen. In contrast, when these genes are mutated and lose function, they are not lysed. The increased sensitivity of this assay results from the fact that the human chromosome 11 is not required for normal function by the CHO cells, and large chromosomal abnormalities such as large deletions and translocations will be detected along with point mutations. We are currently examining the cytotoxic and mutagenic effects of CSC and the neutral, acidic, and basic fractions of CSC in this system without metabolic activation. Preliminary data suggest that CSC is directly mutagenic at high concentrations (unpublished data).

Asbestos

Two major hypotheses for the carcinogenic mechanism of asbestos fibers are: (a) fibers generate oxygen radicals that cause DNA damage; and (b) fibers physically disrupt the mitotic spindle apparatus and cause chromosomal aberrations. We have been testing both of these hypotheses by investigating the effects of asbestos and glass fibers in human lung cells.

Table 9.2. Cytotoxicity of asbestos and glass fibers for human lung cells.

Cell type	Chrysotile	Amosite	Crocidolite	Glass fiber (code 100)
Bronchial fibroblastic	95[a]	225	250	> 100
Bronchial epithelial	5	15	30	100
Mesothelial	0.3	0.5	0.3	5

*Concentration of fiber (μg/ml; 2 ml medium in 60-mm culture dish) required to reduce colony-forming efficiency by 50%.

Human pleural mesothelial cells from noncancerous adult donors with pleural effusions are utilized in these studies. The cultured cells exhibit normal mesothelial cell characteristics, including expression of keratin, secretion of hyaluronic acid-mucin, long, branched microvilli, and retention of the normal human keratinocyte until senescence. The mesothelial cells are ten and 100 times more sensitive to the cytotoxic effects of asbestos fibers than are normal human bronchial epithelial or fibroblastic cells, respectively (Table 9.2). In addition, cultures of mesothelial cells that survived two cytotoxic exposures of amosite fibers are aneuploid with consistent specific chromosomal losses indicative of clonal origin. These aneuploid cells exhibit both altered growth control properties and a population-doubling potential of more than 50 divisions beyond the culture life span (30–40 population doublings) of the control cells.

We have also studied the effects of asbestos fibers on DNA structure. Human mesothelial cells, XP skin fibroblasts, bronchial epithelial cells, and fibroblasts were exposed to doses of 1 and 100 μg/ml of amosite, chrysotile, or crocidolite for between 4 and 24 h and assayed for DNA SSB by the sensitive alkaline elution technique. A dose of 100 μg/ml chrysotile for 4 h caused approximately 0.5 SSB/10^{10} daltons (corresponding to SSB from 18 rads of ionizing radiation) (Fornace et al. 1982) in mesothelial cells. Exposure of mesothelial cells to amosite at doses of 10 or 100 μg/ml for 24 h also caused a similar increase in the elution of DNA. Although the elution of DNA was slightly increased in three different experiments, the calculated amount of damage was always less than 1 SSB/10^{10} daltons, which would be regarded as an insignificant level of induction of SSB with the alkaline elution technique (Fornace et al. 1982). In the other cell types, DNA from asbestos-exposed cells eluted at nondistinguishable rates when compared to DNA from control cells. The presence of the potent DNA polymerase inhibitor combination of 1-β-D-arabinofuranosylcytosine (ara-C) and hydroxyurea (HU) added in combination with asbestos fibers for 4 h did not result in an increased accumulation of SSB in any of the cell types. No enhancement was seen when cells were exposed to ara-C/HU for 4 h either simultaneously with the addition of fibers or after 15- or 24-h preexposure to fibers. These results indicate that neither the uptake nor the presence of asbestos fibers in the cells causes significant levels of SSB. Therefore, we were encouraged to investigate other mechanisms of cellular transformation and chromosome rearrangements.

Normal human cells are characterized by chromosomal stability (DiPaolo 1983) and chromosome rearrangements have rarely been observed in replicative cultures of human cells exposed to asbestos fibers (Craighead and Mossman 1982). In contrast, human mesothelial cells rapidly acquired extensive chromosomal rearrangements, particularly dicentrics, after exposure to amosite (Lechner et al. 1985). Puck (1979) and Tsutsui et al. (1984) have concluded from experimental and epidemiological data that interference with cytoskeletal functions can cause karyotypic instability. In addition, Barrett et al. (1983) and Hesterberg and Barrett (1985) have observed bizarre mitoses in hamster cells exposed to asbestos. They also noted a significant increase in lagging chromosomes at anaphase and telephase in Syrian hamster cells after exposure to asbestos. Results obtained in our laboratory (Lechner et al., unpublished results) support the observations that asbestos disturbs the fidelity of mitosis leading to the induction of aneuploid human mesothelial cells. However, in contrast to the results of Hesterberg and Barrett (1985) our data suggest that asbestos inhibits mitotic tubular polymerization and/or alters centriole separation. These cytotoxicity and cytogenetic results suggest that the uniquely fluid cytoskeleton of the mesothelial cell may be very easily perturbed by penetrating asbestos fibers. This in turn would cause chromosomal instability which could result in activation of proto-oncogenes and inactivation of tumor suppressor genes and eventually neoplastic transformation (Klein and Klein 1984).

There is considerable interest in the possible involvement of free radicals in carcinogenesis. Chemical and physical carcinogens, e.g., ionizing irradiation, as well as tumor promoters have all been linked to the generation of free radicals (Borek and Troll 1983; Kensler and Trush 1984; Cerutti 1985), because the pathobiological consequences of such agents are counteracted by scavengers such as antioxidants, thiols, or enzymes that metabolize radical species (e.g., superoxide dismutase and catalase).

Several observations suggest that free radicals may also have a role in asbestos-related cytotoxicity and carcinogenicity. First, macrophages have been observed to generate oxygen radicals, as detected by chemoluminescence, during phagocytosis of asbestos or a variety of particulates (Hatch et al. 1980). In addition, morphological studies have demonstrated that phagocytosis of asbestos fibers by rat mesothelial cells is followed by the degranulation of lysosomal contents into the phagocytic vacuole (Jaurand et al. 1979), suggesting the generation of oxygen radicals by these cells in response to the fibers (Weiss and LoBuglio 1982). Finally, it has been reported that superoxide dismutase added to culture media reduces asbestos-induced hamster epithelial cell injury, as assayed by the release of intracellular ^{75}Se (Mossman and Landesman 1983).

Our experiments assessed the importance of oxygen radicals in the cytopathology of asbestos in this important target cell by: (a) measuring the possible production of oxygen radicals using the electron paramagnetic resonance assay and the spin trapping agent 5,5-dimethyl-1-pyrroline-1-oxide (DMPO) in mesothelial cells exposed to asbestos; (b) measuring cellular thiol content following asbestos exposure; (c) examining the effects of free radical scavengers on asbestos cytotoxicity; and (d) measuring DNA damage in mesothelial cell cultures exposed to asbestos (Gabrielson et al. 1986).

Electron paramagnetic resonance and spin trapping is a sensitive method (Janzen 1980) capable of detecting oxygen radicals generated in mesothelial cells after a nontoxic exposure to menadione. It is most likely that the DMPO hydroxyl radical detected is a decomposition product of 2,2-dimethyl-5-hydroperoxy-1-pyrrolidinyloxyl, the species directly obtained by the reaction of superoxide with DMPO (Finkelstein et al. 1980). In contrast, no oxygen radicals were detected in mesothelial cells after exposure to asbestos.

It has been suggested that glutathione depletion may be useful for the early detection of toxic metabolites (Welch 1978; Reed and Fariss 1984), especially those that generate oxidative stress (Kappus and Sies 1981). Parallel studies have argued that free radical-mediated cell injury can be inhibited in vitro by the addition of free radical scavengers to the culture medium (Borek and Troll 1983). Compounds such as glutathione, cysteine, and D-α-tocopherol react with free radicals, making the radicals unavailable for reaction with critical cellular macromolecules, and the enzyme superoxide dismutase catalyzes the dismutation of superoxide to hydrogen peroxide and oxygen. In our experiments (Gabrielson et al. 1986), free radical scavengers did not modify asbestos-induced cytotoxicity, and cellular thiol levels did not decrease in mesothelial cells exposed to asbestos, although a marked decrease in cellular thiols was noted in cells exposed to menadione.

Free radical-mediated DNA damage occurring after ionizing irradiation or exposure of cells to hydrogen peroxide can be detected by measurement of DNA SSB with the alkaline elution technique (Saladino et al. 1985). In contrast, neither the uptake nor the presence of asbestos fibers in mesothelial cells causes the induction of significant levels of DNA SSB (Gabrielson et al. 1986).

The results of our studies therefore suggest that oxygen free radicals are not generated in human mesothelial cells by fibers and do not play an important role in mediating asbestos-related mesothelial cell injury. However, oxy-radicals released by macrophages exposed to fibers could play a contributing role in mesothelial carcinogenesis.

Acknowledgment. The editorial aid of Bob Julia is appreciated.

References

Barrett JC, Thomassen DG, Hesterberg TW (1983) Role of gene and chromosomal mutations in cell transformation. Ann NY Acad Sci 407:291–300

Becker RA, Montesano R (1985) Repair of O^6-methyldeoxythymidine residues in DNA by mammalian liver extracts. Carcinogenesis 6:313–317

Birnboim HC (1982) DNA strand breakage in human leukocytes exposed to a tumor promoter, phorbol myristate acetate. Science 215:1247–1249

Blumberg PM (1981) In vitro studies on the mode of action of the phorbol esters, potent tumor promoters. Part 2. CRC Crit Rev Toxicol 8:199–234

Bock FG, Swain AP, Stedman RL (1970) Composition studies on tobacco. XLI. Carcinogenesis assay of subfractions of the neutral fraction of cigarette smoke condensate. J Natl Cancer Inst 44:1305–1314

Borek C, Troll W (1983) Modifiers of free radicals inhibit in vitro the oncogenic actions of x-rays, bleomycin, and the tumor promoter 12-O-tetradecanoylphorbol 13-acetate. Proc Natl Acad Sci USA 80:1304–1307

Brent TP (1986) Inactivation of purified human O^6-alkylguanine-DNA alkyltransferase by alkylating agents or alkylated DNA. Cancer Res 46:2320–2323

Cerutti PA (1985) Prooxidant states and tumor promotion. Science 227:375–381

Chung FL, Young R, Hecht SS (1984) Formation of cyclic 1, N^2-propanodeoxyguanosine adducts in DNA upon reaction with acrolein or crotonaldehyde. Cancer Res 44:990–995

Craighead JE, Mossman BT (1982) The pathogenesis of asbestos-associated diseases. N Engl J Med 306:1446–1455

Crook TR, Souhami RL, McLean AE (1986) Cytotoxicity, DNA cross-linking, and single strand breaks induced by activated cyclophosphamide and acrolein in human leukemia cells. Cancer Res 46:5029–5034

Curren RD, Yang LL, Conklin PM, Grafstrom RC, Harris CC (1988) Mutagenesis of xeroderma pigmentosum fibroblasts by acrolein. Mutat Res 209:17–22

DeMarini DM (1983) Genotoxicity of tobacco smoke and tobacco smoke condensate. Mutat Res 114:59–89

DiPaolo JA (1983) Relative difficulties in transforming human and animal cells in vitro. J Natl Cancer Inst 70:3–8

Doll R, Peto R (1978) Cigarette smoking and bronchial carcinoma: dose and time relationships among regular smokers and lifelong non-smokers. J Epidemiol Community Health 32:303–313

Emerit I, Cerutti PA (1981) Tumour promoter phorbol-12-myristate-13-acetate induces chromosomal damage via indirect action. Nature 293:144–146

Finkelstein E, Rosen GM, Rauckman EJ (1980) Spin trapping of superoxide and hydroxyl radical: practical aspects. Arch Biochem Biophys 200:1–16

Fornace AJ, Lechner JF, Grafstrom RC, Harris CC (1982) DNA repair in human bronchial epithelial cells. Carcinogenesis 3:1373–1377

Gabrielson EW, Rosen GM, Grafstrom RC, Strauss KE, Harris CC (1986) Studies on the role of oxygen radicals in asbestos-induced cytopathology of cultured human lung mesothelial cells. Carcinogenesis 7:1161–1164

Grafstrom RC, Fornace AJ, Harris CC (1984a) Repair of DNA damage caused by formaldehyde in human cells. Cancer Res 44:4323–4327

Grafstrom RC, Pegg AE, Trump BF, Harris CC (1984b) O^6-alkylguanine-DNA alkyltransferase activity in normal human tissues and cells. Cancer Res 44:2855–2857

Grafstrom RC, Curren RD, Yang LL, Harris CC (1985a) Genotoxicity of formaldehyde in cultured human bronchial fibroblasts. Science 228:89–91

Grafstrom RC, Willey JC, Sundqvist K, Harris CC (1985b) Toxicity of tobacco-related aldehydes in cultured human bronchial epithelial cells. US Airforce technical report 84-002, p 255

Grafstrom RC, Curren RD, Yang LL, Harris CC (1986) Aldehyde-induced inhibition of DNA repair and potentiation of N-nitrosocompound-induced mutagenesis in cultured human cells. Prog Clin Biol Res 209:255–264

Grafstrom RC, Dypbukt JM, Willey JC, Sundqvist K, Edman C, Atzori L, Harris CC (1988) Pathobiological effects of acrolein in cultured human bronchial epithelial cells. Cancer Res 48:1717–1721

Grafstrom RC, Edman C, Sundqvist K, Liu Y, Hybbinette SS, Nicoleran P, Dypbukt JM (1989) Cultured human bronchial cells as a model system in lung toxicology and carcinogenesis: implications from studies with acrolein. ATLA (in press)

Harris CC, Yoakum GH, Lechner JF, Willey JC, Gerwin BI, Banks-Schlegel SP, Masui T, Mark GE (1986) Growth, differentiation, and neoplastic transformation of human bronchial epithelial cells. In: Harris CC (ed) Biochemical and molecular epidemiology of cancer. Liss, New York, p 213

Hatch GE, Gardner DE, Menzel DB (1980) Stimulation of oxidant production in alveolar macrophages by pollutant and latex particles. Environ Res 23:121–136

Hecht SS, Carmella S, Mori H, Hoffmann D (1981) A study of tobacco carcinogenesis. XX. Role of catechol as a major cocarcinogen in the weakly acidic fraction of smoke condensate. J Natl Cancer Inst 66:163–169

Hesterberg TW, Barrett JC (1985) Induction by asbestos fibers of anaphase abnormalities: mechanism for aneuploidy induction and possibly carcinogenesis. Carcinogenesis 6:473–475

Hoffmann D, Adams JD (1981) Carcinogenic tobacco-specific N-nitrosamines in snuff and in the saliva of snuff dippers. Cancer Res 41:4305–4308

Hoffmann D, Adams JD, Brunnemann KD, Rivenson A, Hecht SS (1982) Tobacco specific N-nitrosamines: occurrence and bioassays. IARC Sci Publ 4:309–318

Hoffmann D, Hecht SS, Wynder EL (1983a) Tumor promoters and cocarcinogens in tobacco carcinogenesis. Environ Health Perspect 50:247–257

Hoffmann D, Wynder EL, Rivenson A, LaVoie EJ, Hecht SS (1983b) Skin bioassays in tobacco carcinogenesis. Prog Exp Tumor Res 26:43–67

Hoffmann D, Melikian A, Adams JD, Brunnemann KD, Haley NJ (1985) New aspects of tobacco carcinogenesis. Carcinog Compr Surv 8:239–256

Horowitz AD, Fujiki H, Weinstein IB, Jeffrey A, Okin E, Moore RE, Sugimura T (1983) Comparative effects of aplysiatoxin, debromoaplysiatoxin, and teleocidin on receptor binding and phospholipid metabolism. Cancer Res 43:1529–1535

Janzen EG (1980) In: Pryor W (ed) Free radicals in biology, vol 4. Academic, New York, p 116

Jaurand MC, Kaplan H, Thiollet J, Pinchon MC, Bernaudin JF, Bignon J (1979) Phagocytosis of chrysotile fibers by pleural mesothelial cells in culture. Am J Pathol 94:529–538

Kappus H, Sies H (1981) Toxic drug effects associated with oxygen metabolism: redox cycling and lipid peroxidation. Experientia 37:1233–1241

Kensler TW, Trush MA (1984) Oxygen free radicals in chemical carcinogenesis. In: Oberley LW (ed) Superoxide dismutase, vol III. Pathological states. CRC, Boca Raton, p 192

Kier LD, Yamasaki E, Ames BN (1974) Detection of mutagenic activity in cigarette smoke condensates. Proc Natl Acad Sci USA 71:4159–4163

Kim S, Vollberg TM, Ro JY, Kim M, Sirover MA (1986) O^6-methylguanine methyltransferase increases before S phase in normal human cells but does not increase in hypermutable Bloom's syndrome cells. Mutat Res 173:141–145

Klein G, Klein E (1984) Oncogene activation and tumor progression. Carcinogenesis 5:429–435

Krokan H, Grafstrom RC, Sundqvist K, Esterbauer H, Harris CC (1985a) Cytotoxicity, thiol depletion and inhibition of O^6-methylguanine-DNA methyltransferase by various aldehydes in cultured human bronchial fibroblasts. Carcinogenesis 6:1755–1759

Krokan H, Lechner JF, Krokan RH, Harris CC (1985b) Normal human bronchial epithelial cells do not show an adaptive response after treatment with N-methyl-N′-nitro-N-nitrosoguanidine. Mutat Res 146:205–209

Krokan H, Haugen A, Giercksky KE, Nilsen IW, Yoakum GH, Harris CC, Myrnes B (1986) O^6-methylguanine-DNA methyltransferase activity in human tissues and the possible suppressive effect of some activated ras-oncogenes in cultured cells. In: Myrnes B, Krokan H (eds) Repair of DNA lesions introduced by N-nitroso compounds. Norwegian University Press, Oslo, p 199

Lechner JF, Tokiwa T, LaVeck MA, Benedict WF, Banks-Schlegel SP, Yeager H, Banerjee A, Harris CC (1985) Asbestos-associated chromosomal changes in human mesothelial cells. Proc Natl Acad Sci USA 82:3884–3888

Lee LS, Weinstein IB (1979) Mechanism of tumor promoter inhibition of cellular binding of epidermal growth factor. Proc Natl Acad Sci USA 76:5168–5172

Leuchtenberger C, Leuchtenberger R, Schneider A (1973) Effects of marijuana and tobacco smoke on human lung physiology. Nature 241:137–139

Male R, Nes IF, Kleppe K (1981) Purification and properties of 3-methyladenine-DNA glycosylase from L-cells. Eur J Biochem 121:243–248

Miyashita M, Willey JC, Sasajima K, Lechner JF, LaVoie EJ, Hoffmann D, Smith M, Trump BF, Harris CC (1989) Differential effects of cigarette smoke condensate and its fractions on cultured normal and malignant human bronchial epithelial cells. Cancer Res (in press)

Moldeus P, Berggren M, Grafstrom RC (1985) N-acetylcysteine protection against the toxicity of cigarette smoke and cigarette smoke condensates in various tissues and cells in vitro. Eur J Respir Dis [Suppl]139:123–129

Mossman BT, Landesman JM (1983) Importance of oxygen free radicals in asbestos-induced injury to airway epithelial cells. Chest 83:50S–51S

Nakayama T, Kodama M, Nagata C (1984) Generation of hydrogen peroxide and superoxide anion radical from cigarette smoke. Gann 75:95–98

Nakayama T, Kaneko M, Kodama M, Nagata C (1985) Cigarette smoke induces DNA single-strand breaks in human cells. Nature 314:462–464

Pegg AE (1986) Factors affecting O^6-alkylguanine-DNA-alkyltransferase activity. In: Hoffman D, Harris CC (eds) Mechanisms in tobacco carcinogenesis, Banbury report (23). Cold Spring Harbor Press, New York, p 287

Pegg AE, Wiest L, Foote RS, Mitra S, Perry W (1983) Purification and properties of O^6-methylguanine-DNA transmethylase from rat liver. J Biol Chem 258:2327–2333

Puck TT (1979) Studies on cell transformation. Somatic Cell Genet 5:973–990

Reed DJ, Fariss MW (1984) Glutathione depletion and susceptibility. Pharmacol Rev 36:25S–33S

Saladino AJ, Willey JC, Lechner JF, Grafstrom RC, LaVeck MA, Harris CC (1985) Effects of formaldehyde, acetaldehyde, benzoyl peroxide, and hydrogen peroxide on cultured normal human bronchial epithelial cells. Cancer Res 45:2522–2526

Tsutsui T, Maizumi H, Barrett JC (1984) Colcemid-induced neoplastic transformation and aneuploidy in Syrian hamster embryo cells. Carcinogenesis 5:89–93

US Department of Health Education and Welfare (1988) Smoking and health. In: US Department of Health Education and Welfare (ed) A report of the Surgeon General. DHEW publication no (PHS) 79-50066, p 5

Waldren C, Correll L, Sognier MA, Puck TT (1986) Measurement of low levels of x-ray mutagenesis in relation to human disease. Proc Natl Acad Sci USA 83:4839–4843 [published erratum appears in Proc Natl Acad Sci USA 1987, 84(10):3364]

Weiss SJ, LoBuglio AF (1982) Phagocyte-generated oxygen metabolites and cellular injury. Lab Invest 47:5–18

Welch RM (1978) Toxicological implications of drug metabolism. Pharmacol Rev 30:457–467
Wild CP, Umbenhauer D, Chapot B, Montesano R (1986) Monitoring of individual human exposure to aflatoxins (AF) and N-nitrosamines (NNO) by immunoassays. J Cell Biochem 30:171–179
Willey JC, Saladino AJ, Ozanne C, Lechner JF, Harris CC (1984) Acute effects of 12-O-tetradecanoylphorbol-13-acetate, teleocidin B, or 2,3,7,8-tetrachlorodibenzo-p-dioxin on cultured normal human bronchial epithelial cells. Carcinogenesis 5:209–215
Willey JC, Moser CE, Harris CC (1985) Effects of aplysiatoxin and debromoaplysiatoxin on growth and differentiation of normal human bronchial epithelial cells. J Cell Biol 1:145–154
Willey JC, Grafstrom RC, Moser CE, Ozanne C, Sundqvist K, Harris CC (1987) Biochemical and morphological effects of cigarette smoke condensate and its fractions on normal human bronchial epithelial cells in vitro. Cancer Res 47:2045–2049
Winn DM, Blot WJ, Shy CM, Pickle LW, Toledo A, Fraumeni JF (1981) Snuff dipping and oral cancer among women in the southern United States. N Engl J Med 304:745–749
Wynder EL, Grahm EA, Croninger AB (1953) Experimental production of carcinoma with cigarette tar. Cancer Res 13:855–864
Wynder EL, Hoffmann D (1979) Tobacco and health: a societal challenge. N Engl J Med 300:894–903
Yarosh DB (1985) The role of O^6-methylguanine-DNA methyltransferase in cell survival, mutagenesis and carcinogenesis. Mutat Res 145:1–16

10. A Mechanistic Approach to Assess the Inhalation Toxicity and Hazard of Methylisocyanate and Related Aliphatic Monoisocyanates

J. Pauluhn[1]

Introduction

The methylisocyanate (MIC) disaster in Bhopal, India, gave rise to many speculations about the toxic principle of isocyanates in general and MIC in particular (Varma 1987). In an attempt to clarify the mechanisms of the acute inhalation toxicity of MIC, numerous animal inhalation exposure studies were performed and reviewed (Bucher 1987). Thus, MIC can now be regarded as a "standard compound" for short-chained aliphatic monoisocyanates.

The quantitative and qualitative aspects of the acute inhalation toxicity of *n*-propylisocyanate (*n*-PIC), isopropylisocyanate (i-PIC), *n*-butylisocyanate (*n*-BIC), isobutylisocyanate (i-BIC), and tertiary butylisocyanate (t-BIC) were investigated and compared with corresponding MIC data to establish the structure-toxicity relationship of structurally related aliphatic monoisocyanates.

Mechanistic studies with MIC demonstrated that MIC is capable of alterating certain proteins in vitro, e.g., increase of oxygen affinity of hemoglobin (Hb) by carbamylation or inhibition of cholinesterases (ChE) (Troup et al. 1987; Brown et al. 1987). Experimental data did not support the hypothesis that MIC induced alterations to the Hb molecule or that an inhibition of ChE activity was an important factor in the death of animals exposed to MIC (Troup et al. 1987). MIC was described as a sensory and pulmonary irritant inducing a decrease in respiratory rate by lengthening expiration and by apnoeic pauses between the breaths (Nemery et al. 1985; Ferguson et al. 1986). This effect probably contributes to the decline in blood oxygen concentration and pH, and increases pCO_2. Evidence was presented for pulmonary blood shunting and/or severe ventilation perfusion abnormalities as a sequela of obstruction of small airways. Sloughing of epithelium and build-up of mucus and fibrin were recognized as the primary causes for airway obstruction. In short-term inhalation studies, both guinea pigs and rats demonstrated clinical signs that were characteristic for respiratory irritants. Animals were prostrate and were observed to be gasping prior to death.

[1]Department of Toxicology, Friedrich-Ebert-Str. 217-319, D-5600 Wuppertal 1, FRG.

Instantaneous death was observed after high exposure concentrations and was attributed to a bradypnea reflex resulting in respiratory arrest. The major findings in animals that died during the first follow-up days were lung edema, hemorrhages, pleural effusions, congestion, and consolidation of the lungs. Animals subjected to intermediate and low concentrations died 2–3 weeks after exposure.

Chemistry/Physico-chemistry of Alkylmonoisocynates

The isocyanates investigated have a relatively high vapor pressure at ambient temperature (Table 10.1). Accordingly, laboratory experiments via the inhalation route are the most appropriate for comparative hazard evaluations. The value for the heat of vaporization per mole ($\Delta_v H$) was calculated from the slope of the curve of the logarithm of the vapor pressure against 1/T (T = absolute temperature in K). If the slope is multiplied by −R (negative gas constant) the value for $\Delta_v H$ is obtained (Moore 1972). Figure 10.1 shows that Trouton's rule for nonpolar liquids is fulfilled. These calculations demonstrate that under the conditions of animal inhalation studies, the isocyanate vapors should be present in a nonaggregated, i.e., monomer form.

Excellent reviews of the reactions of isocyanates are available (Brown et al. 1987). In chemical reactions isocyanates may react as might be expected of one of the following resonance forms:

$$[\, R-\overset{\ominus}{\underline{\overline{N}}}-\overset{\oplus}{C}=\underline{\overline{O}} \leftrightarrow R-\overline{N}=C=\underline{\overline{O}} \leftrightarrow R-\overline{N}=\overset{\oplus}{C}-\overset{\ominus}{\underline{\overline{O}}}| \,]$$

Such resonance structures suggest the possibility of ionic reactions with electron donors attacking the carbonyl carbon. Potential sites of reaction under exposure conditions are the hydroxyl group (water), the sulfhydryl group, and the amine groups. Although other reactions are possible, they are less likely because of significantly reduced levels of the appropriate form of the functional group under physiological conditions. For example, the amine reaction is optimal with a non-

Table 10.1. Physico-chemical data.

Compound	Boiling point (°C)	Vapor pressure (kPa) at 20°C	C_{sat} (20°C) (mg/m³ air)
MIC	39	51.3	$1119 \cdot 10^3$
n-PIC	88	6.4	$223 \cdot 10^3$
i-PIC	75	12.0	$417 \cdot 10^3$
n-BIC	116	2.2	$90 \cdot 10^3$
i-BIC	102	3.2	$129 \cdot 10^3$
t-BIC	85	8.1	$344 \cdot 10^3$

C_{sat}, vapor saturation concentration.

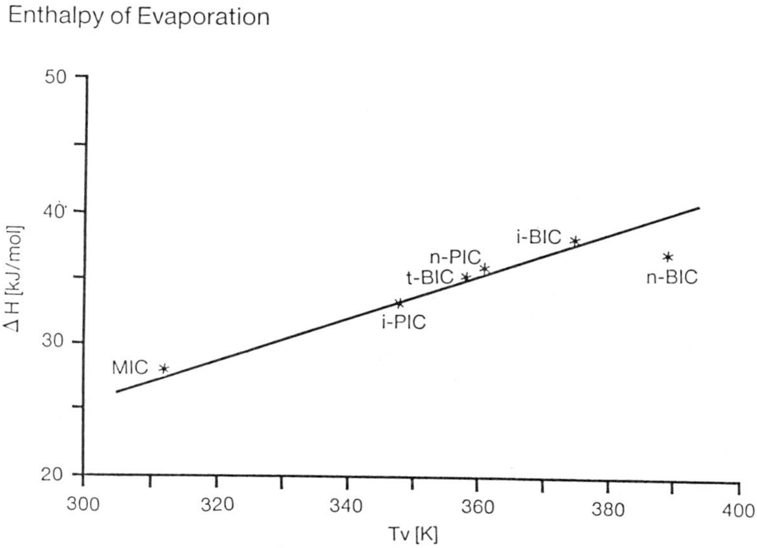

Figure 10.1. Test for validity of Trouton's rule

protonated amine. Under physiological conditions less than 5% of the protein amines are nonprotonated (Brown et al. 1987).

Numerical data of the "chemical reactivity" under conditions prevailing in the respiratory tract are necessary to study the structure-toxicity relationship. Owing to the lack of such data, semiempirical ab initio calculations of the polarizability of electrons in the vicinity of the nitrogen atom of the isocyanate group were performed vs. NH_3 (method: IGLO, computations were performed by M. Schindler, BAYER AG). The resonance structure shows that the electron density of the nitrogen can influence the electrophilic character of the carbonyl carbon of the R−N=C=O group.

Where the destructive nature of isocyanate vapor is a function of the nonspecific reactivity of the R−N=C=O carbonyl group with nucleophilic groups located on the surfaces of the respiratory tract, the electron shifting effect of the side chain (R) to the carbonyl carbon and the acute inhalation toxicity (LC_{50}) should be correlated. Thus, the toxicity of MIC (R = H) should be highest, that of t-BIC (R = tertiary butyl) should be lowest because of the positive inductive effect of the tertiary butyl group. The ensuing animal inhalation toxicity studies were performed to test this hypothesis.

Methodology

Because of the high reactivity of isocyanate vapors with moisture contained in air expired by test animals, the possibility of secondary changes of the test substance in the inhalation chamber is possible. Therefore, the experimental design of an inhalation chamber for animal exposure studies with aliphatic monoisocyanates should prevent mixing of primary test atmosphere and exhaled air (Pauluhn 1988).

It should be noted that accurate predictions of evaporation and condensation rates in test atmospheres are very difficult or impossible as data are not readily available. Therefore, it is far better to realize the possible existence of a generating or sampling problem and avoid it rather than to try to produce a credible correction later based on calculations. If the saturation vapor concentration is much larger than the airborne concentration, the test material will be expected to be found predominantly in the vapor phase, but it may be found in the particle phase if the test atmosphere contains a sufficiently high concentration of airborne particles in which the material is soluble, even if only slightly. Airborne concentrations of water droplets could be relatively massive in many high humidity atmospheres, e.g., as they may occur in whole-body inhalation chambers. Therefore, the nose-only exposure technique is regarded as the exposure technique of choice.

Generation of Test Atmosphere. With regard to the vapor pressure, the test atmosphere can be generated by passing air or nitrogen through a test compound containing reservoir or by taking defined volumes of air from the saturated atmosphere of the reservoir. This atmosphere can be subsequently diluted with dry air to generate lower test concentrations. In all experiments dry air (approximately 5% relative humidity) was used. The advantages of these provisions are that droplet formation can be detected easily and, from the air flow through the vapor generator and the weight loss during the test, the theoretical vapor saturation concentration under the conditions of the experiment can be calculated and compared with exact physical data.

Analytical Characterization of the Test Atmosphere. The air for analytical measurements was sampled in the vicinity of the breathing zone of the animals. The test atmosphere was collected by impingers filled with a nitroreagent containing solvent (Keller and Sandridge 1974). There the isocyanate vapor reacts

Table 10.2. Results of LC_{50} determinations (rats, head-nose only exposure for 4 h, observation period 4 weeks).

Compound	LC_{50} Male (mg/m^3 air)	LC_{50} Female (mg/m^3 air)
MIC	25	25
n-PIC	155	201
i-PIC	499	613
n-BIC	60	55
i-BIC	235	246
t-BIC	1513	1695

into urea derivatives which are characterized by high-performance liquid chromatography (HPLC).

Technique of RD_{50} Determination. Four mice per group were exposed to different concentrations in flow plethysmograph nose-only exposure tubes. The adaptation period was about 15 min, the exposure period was 45 min. During a recovery period of at least 15 min, the reversibility of the observed effects was studied. The measurement and evaluation of data were performed according to published procedures (Ferguson et al. 1986).

Pulmonary Function Measurements. Pulmonary function measurements were performed at the end of observation week 3. The rats were anesthetized using approximately 250 mg hexobarbitone per kilogram body weight (i.p. administration). Measurements of lung mechanics and CO-diffusing capacity were performed in a whole-body flow plethysmograph. Transpulmonary pressure was measured using an orotracheal tube and a water-filled esophageal catheter connected to a differential pressure transducer (Pauluhn et al. 1987).

Animals. In all exposure studies performed, five male and five female young adult inbred Wistar rats (aged about 3 months, mean body weight at the beginning of study about 190 g) were used. Outbred male mice (OF4) were used for sensory irritation tests. At the beginning of these studies the body weights of the mice were about 25 g. The observation period was 4 weeks for rats and 1 week for mice.

Experimental Results

The results of the comparative determination of the acute inhalation toxicity of *n*-, i-PIC, and *n*-, i-, t-BIC in comparison with MIC are summarized in Table 10.2. As observed by other authors (Bucher et al. 1987), male rats were slightly more sensitive than females. This effect might be due to a higher ventilation of male rats (male rat: 1.4 l/ kg body weight; female rat: 1.1 l/kg body weight; all rats exposed to air only; mean body weight 190 g). In all studies a steep concentration-mortality response curve was observed. The MIC LC_{50} value (exposure of rats 1×4 h) of 11 ppm was taken from the most recent publication (Dodd et al. 1986). The lowest published LC_{50} value for the rat (exposure 1×4 h) was 5 ppm (Kimmerle and Eben 1964).

Figure 10.2 demonstrates that the inhalation toxicity (logarithm of LC_{50}) and the chemical shift are correlated. The comparison of $\Delta_v H$ data with LC_{50} data revealed no correlation. Thus, these data support the conclusion that the chemical reactivity of the carbonyl carbon with nucleophils determines the acute inhalation toxicity of these isocyanates. From Fig. 10.2 the conclusion may be drawn that delayed mortality is related to the chemical reactivity. In inhalation studies with i-PIC and t-BIC there was no evidence of delayed mortality,

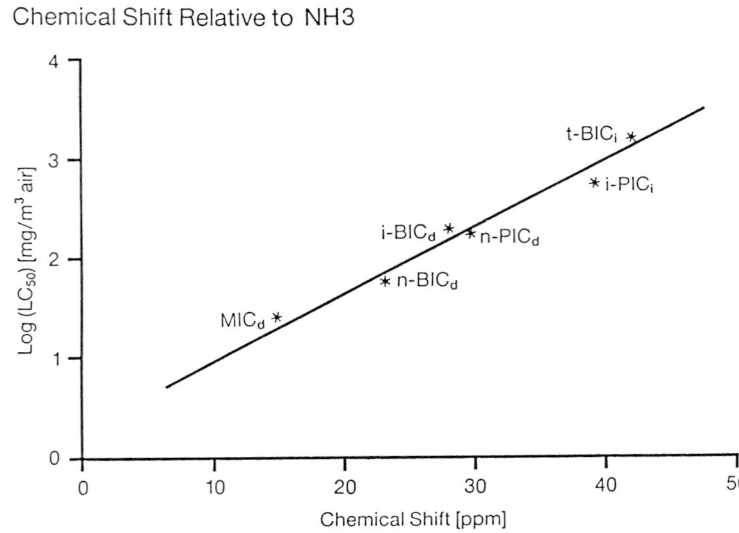

Figure 10.2. Correlation of the positive inductive effect and the logarithm of LC_{50} value. Index: d: delayed mortality, i: immediate mortality

defined as mortality starting with follow-up day 8. Early mortality occurred in rats subjected to high concentrations of *n*-PIC, *n*- and i-BIC. Concentrations lower than the LC_{50} induced delayed mortality.

In an attempt to ascertain the functional sequelae of a single *n*-PIC exposure, a battery of pulmonary function tests were performed. Some essential results are summarized in Fig. 10.3.

The static compliance (C_{stat}) was depressed. Dynamic compliance (C_{dyn}), however, was increased. There was a considerable increase of the residual volume (RV) and a decrease of carbon monoxide diffusing capacity. Mean mid-expiratory flow (MMEF) was decreased as evidence of smaller airway obstruction. These data may be the result of an inefficient communication between lung compartments and subsequent inhomogeneous distribution of ventilated air. The same profile of functional evidence of airway obstruction was found in MIC-exposed rats (Stevens et al. 1987). Thus, the pathophysiological mechanism of delayed effects might be due to obstructive processes in smaller airways. This finding corresponds with proliferative responses in the smaller airways observed in MIC-exposed rats (Bucher et al. 1987).

The clinical signs were dominated by perinasal wetness, lacrimation, laboured and slow breathing, irritation of mucous membranes, and rales. These were the major clinical signs in rats exposed to all isocyanates investigated. The degree of sensory irritation and of acute inhalation toxicity (LC_{50}) was correlated.

Animals exposed to isocyanates reduce their ventilation in a concentration-dependent manner. This is accompanied by secondary effects, e.g., decreased body temperature (Thorne et al. 1987), or change in blood gases or pH (Nemery

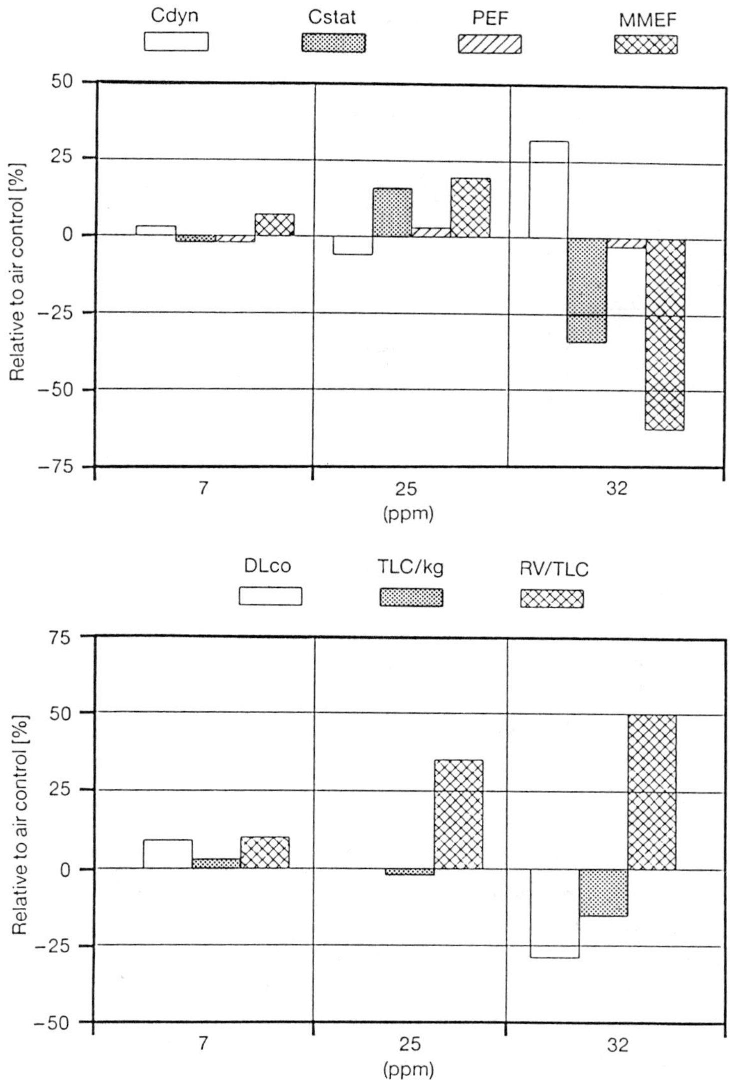

Figure 10.3. *n*-PIC–pulmonary function measurements (rat, exposure 1 × 4 h; measurement 3 weeks after exposure 1 ppm = 4.1 mg/m³ air)

et al. 1985). Body temperature and pH have a great influence on Hb-oxygen affinity. Rats exposed to *n*-BIC concentrations which elicited a decrease in respiration showed a tremendous change in body temperature (Fig. 10.4). Hypothermia increases the Hb-oxygen affinity (Jacquez 1979).

The resulting hypoxia may be a pathophysiological mechanism of a potential reproductive toxicity. In MIC-exposed mice (1×3 h on gestation day 8) an effect

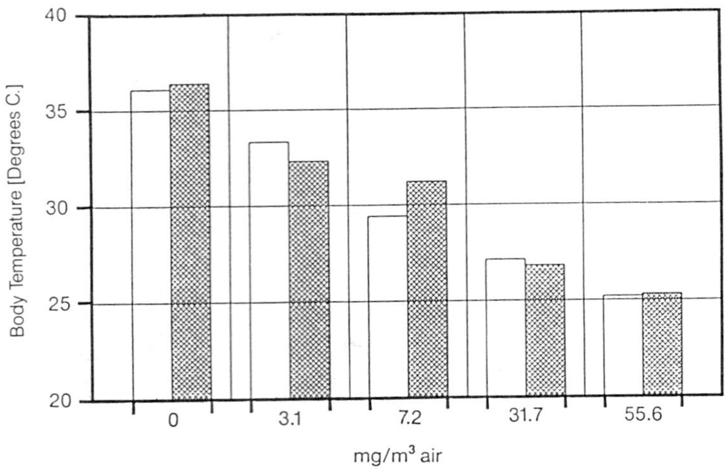

Figure 10.4. Rectal temperature immediately after ceasing a head-nose only exposure of rats; duration, 4 h; test compound, *n*-BIC. *Open columns*, male; *hatched columns*, female

on fetuses and placentae was noticed (Varma et al. 1987). Thus, severe hypothermia and concomitant hypoxia may be a possible mechanism for the reproductive effects induced by MIC.

Conclusions

According to qualitative data, the toxic principle of all isocyanates investigated is comparable with that of MIC. Quantitative data (RD_{50}, LC_{50}, no-observable effect level) revealed that the acute inhalation toxicity of MIC and t-BIC covers a range from 25 to 1695 mg/m³ air. Thus, each aliphatic monoisocyanate has to be considered as an individual chemical compound.

Table 10.3. Hazard index.

Compound	C_{sat}/LC_{50} (rat, female) (based on mg/m³ air values)
MIC	44 760
n-PIC	1109
i-PIC	680
n-BIC	1636
i-BIC	524
t-BIC	203

C_{sat}, vapour saturation concentration, 20°C.

The inhalation toxicity and the nucleophilic chemical reactivity are correlated. The chemical/physico-chemical nature of the nonfunctional part of the molecule is a major determinant of the inhalation toxicity and inhalation hazard (Table 10.3). It would appear that the high reactivity of these chemicals is their major mechanism of toxic effects to the respiratory tract. This might be supported by the fact that injury was localized to the point of initial contact, which in these exposure studies was primarily the respiratory tract.

The disruption of the control of breathing by reflex inhibition is a conceivable mechanism for secondary effects such as hypoxia, hypothermia, or shifts of pH in animals exposed to intermediate and lower concentrations. Thus threshold phenomena (reflex bradypnea-induced hypothermia) have to be taken into account in the assessment of inhalation hazards and extrapolation of experimental data.

References

Brown WE, Green AH, Cedel TE, Cairns J (1987) Biochemistry of protein-isocyanate interactions: a comparison of the effects of aryl vs. alkyl isocyanates. Environ Health Perspect 72:5–11

Bucher JR (1987) Methyl isocyanate: a review of health effects research since Bhopal. Fundam Appl Toxicol 9:367–379

Bucher JR, Gupta BN, Adkins B, Thompson M, Jameson CW, Thigpen JE, Schwetz BA (1987) Toxicity of inhaled methyl isocyanate in F344/N rats and B6C3F1 mice. I. Acute exposure and recovery studies. Environ Health Perspect 72:53–61

Dodd ED, Fowler EH, Snellings WM, Pritts IM, Baron RL (1986) Acute inhalation studies with methyl isocyanate vapor. I. Methodology and LC_{50} determinations in guinea pigs, rats, and mice. Fundam Appl Toxicol 6:747–755

Ferguson JS, Schaper M, Stock MF, Weyel DA, Alarie Y (1986) Sensory and pulmonary irritation with exposure to methyl isocyanate. Toxicol Appl Pharmacol 82:329–335

Jacquez JA (1979) Respiratory physiology. Hemisphere, Washington, p 170

Keller J, Sandridge RL (1974) Determination of isocyanates in the working atmosphere by thin-layer chromatography. Anal Chem 46:1845

Kimmerle G, Eben A (1964) Zur Toxizität von Methylisocyanat und dessen quantitativer Bestimmung in der Luft. Arch Toxicol 20:235–241

Moore WJ (1972) Physical chemistry, 5th ed. Longman, London, pp 213

Nemery B, Dinsdale D, Sparrow S, Ray DE (1985) Effects of methyl isocyanate on the respiratory tract of rats. Br J Ind Med 42:799–805

Pauluhn J (1988) Comparison and interpretation of different methods used in acute and subchronic inhalation studies with potential lung irritants with particular attention to pulmonary function measurements. In: Mohr U (ed) Inhalation toxicology—the design and interpretation of inhalation studies and their use in risk assessment. Springer, Berlin Heidelberg New York, pp 87–102

Pauluhn J, Machemer L, Kimmerle G (1987) Effects of inhaled cholinesterase inhibitors on bronchial tonus and on plasma and erythrocyte acetylcholine esterase in rats. Toxicology 46:177–190

Stevens MA, Fitzgerald S, Menache MG, Costa DL, Bucher JR (1987) Functional evidence of persistent airway obstruction in rats following a two-hour inhalation exposure to methyl isocyanate. Environ Health Perspect 72:89–94
Thorne PS, Yeske CP, Karol MH (1987) Monitoring of guinea pig core temperature by telemetry during inhalation exposures. Fudam Appl Toxicol 9:398–408
Troup CM, Dodd DE, Fowler EH, Frank FR (1987) Biological effects of short-term high concentration exposure of methyl isocyanate. II. Blood chemistry and hematologic evaluations. Environ Health Perspect 72:21–28
Varma DR (1987) Anatomy of the methyl isocyanate leak in Bhopal. Hazard assessment of chemicals. Curr Dev 5:233–289
Varma DR, Ferguson JS, Alarie Y (1987) Reproductive toxicity of methyl isocyanate in mice. J Toxicol Environ Health 21:265–275

PART III

Case Studies

Section 1. Formaldehyde

11. Inhalation Toxicity and Carcinogenicity of Formaldehyde in Animals: Significance for Assessment of Human Health Risk

V.J. Feron,[1] J.W.G.M. Wilmer,[1] R.A. Woutersen,[1] and A. Zwart[1]

Introduction

Formaldehyde is a common chemical which is used mainly in the manufacture of a wide variety of commercial products such as textiles, papers and synthetic resins. Most people will be exposed to formaldehyde during their life, either occupationally or via cigarette smoke, automotive exhaust, photochemical smog, incinerators, or release from urea-formaldehyde products (ECETOC 1981).

Formaldehyde is a highly irritating, directly-acting genotoxic carcinogen capable of inducing malignant nasal tumours in rats, and probably in mice, after inhalation exposure (Swenberg et al. 1980; Albert et al. 1982; Tobe et al. 1985; Feron et al. 1988; Ma and Harris 1988). It also induces papillary hyperplasia (and possibly papillomas) in the forestomach of rats after administration in the drinking water (Takahashi et al. 1986; Til et al. 1989). Nasal tumours have been found only at exposure concentrations (of about 15 ppm) also inducing severe degenerative, hyperplastic and metaplastic changes of the nasal respiratory epithelium. These data indicate an important role of the cytotoxic effects of formaldehyde in its carcinogenic potential. Correspondingly, at non-cytotoxic concentrations, the carcinogenic potential of formaldehyde may be very small or even virtually nonexistent. In this paper an analysis is given of the nasal toxicity and carcinogenicity data of formaldehyde in animals to find out their relevance for assessing the health risk of humans being exposed to formaldehyde vapour.

Nasal Cytotoxicity

The nasal toxicity of formaldehyde is characterised by inhibition of mucociliary function (Morgan et al. 1986a), reaction with albumin and other small proteins occurring in mucus (Bogdanffy et al. 1987), reaction with glutathione and detox-

[1]TNO-CIVO Toxicology and Nutrition Institute, P.O. Box 360, 3700 AJ Zeist, The Netherlands.

ification by formaldehyde dehydrogenase (Heck and Casanova-Schmitz 1984), and, when biotransformation is overwhelmed or even inactivated, rhinitis, degeneration and necrosis followed by regenerative hyperplasia and metaplasia of the respiratory epithelium (Swenberg et al. 1983). The aforementioned and other histopathological nasal changes have been found in rats exposed to formaldehyde concentrations ranging from 2 to 20 ppm for periods varying from a few weeks to 28 months, the "no toxic effect level" being 1 ppm (Kerns et al. 1983; Wilmer et al. 1986, 1987; Woutersen et al. 1987, 1989; Appelman et al. 1988). Occasionally, effects on the nasal respiratory epithelium (protuberances of the cell membrane of cilia, disarrangement of cells, irregular and indented nuclei, variation in cell turnover, increased rhinitis and hyperplasia/metaplasia) have been reported at concentrations of 1 ppm and lower (0.3 and 0.5 ppm) (Tobe et al. 1985; Monteiro-Riviere and Popp 1986; Zwart et al. 1988). However, in each case the effects either could not unequivocally be attributed to formaldehyde or were considered the manifestation of adaptation rather than of toxicity. Nasal effects of formaldehyde vapour have also been reported in mice, hamsters, guinea pigs and monkeys (Mitchell et al. 1979; Dalbey 1982; Marshall et al. 1983; Rusch et al. 1983; BGA 1985; Maronpot et al. 1986). Mice and guinea pigs, and also Syrian golden hamsters, seem to be less sensitive than rats, while monkeys are as sensitive as rats to the cytotoxic properties of formaldehyde. Recently, studies in Rhesus monkeys have demonstrated significant differences in the distribution of formaldehyde-induced respiratory tract lesions between rats and monkeys (Monticello et al. 1989). The above data clearly demonstrate the nasal cytotoxicity of formaldehyde in a number of animal species; 1 ppm appears to be the highest "no observed adverse effect level".

Nasal Carcinogenicity

In a long-term inhalation study with F-344 rats, a high incidence of nasal carcinomas (about 45%) was found at an exposure level of 14 ppm formaldehyde, whereas at 6 ppm a low, not statistically significant incidence of nasal carcinomas (about 1%) occurred (Swenberg et al. 1980). The clear tumour response at 14 ppm was confirmed in two other long-term studies in rats using a similar exposure concentration (Albert et al. 1982; Tobe et al. 1985). In a recent 28-month inhalation study in Wistar rats, no formaldehyde-related nasal tumours were observed at 0.1, 1.0 and 10 ppm formaldehyde (Feron et al. 1987; Woutersen et al. 1989). Moreover, 3-month inhalation exposure of Wistar rats of the same strain to 10 or 20 ppm formaldehyde followed by an observation period of 27 months was associated with a significant tumour response of the nasal respiratory epithelium in the 20-ppm group but not in the 10-ppm group (Feron et al. 1988).

In a well-conducted 24-month inhalation study with formaldehyde in mice, nasal carcinomas were seen only in two males (about 2%) of the high-concentration (14 ppm) group (Kerns et al. 1983). Most probably these tumours are formaldehyde induced, although the response did not reach the level of statistical significance. No nasal tumours or tumours at other sites were observed in golden hamsters exposed to formaldehyde vapour at a concentration of 10 ppm for 5 hours a day, 5 days a week for life (Dalbey 1982). In conclusion, formaldehyde is a nasal carcinogen in rats and probably also in mice; a significant tumour response occurs at exposure concentrations of 14 ppm and higher.

Nasal Carcinogenicity Associated with Nasal Cytotoxicity

The aforementioned data on nasal toxicity and carcinogenicity of formaldehyde clearly demonstrate that exposure levels causing nasal tumours also cause rhinitis, necrosis, and hyper- and metaplasia of the nasal mucosa. Moreover, the dose-response curve for nasal tumours appears to be very steep and extremely non-linear and to correspond with those for inhibition of the mucociliary function, increased cell proliferation, and hyper- and metaplasia of the nasal respiratory epithelium (Heck and Casanova-Schmitz 1984). Obviously, an association exists between the cytotoxic and carcinogenic effects. In other words, the steep, non-linear dose-response curve for nasal tumours – indicating a disproportionate decrease in cancer risk at low concentrations – strongly suggests a high effectiveness of the nasal defence mechanisms (mucociliary clearance, detoxification by dehydrogenase, regenerative hyperplasia/metaplasia) at low concentrations, and overwhelming and inactivation of defence systems at high concentrations. Apparently, the formation of nasal carcinomas in rats by formaldehyde requires exposure to concentrations that cause considerable damage to the nasal epithelium followed by regenerative hyperplasia and metaplasia. The increased cell replication may strongly enhance the likelihood of relevant DNA damage, and, moreover, the progression of initiated or preneoplastic cells to cancer. This also implies that formaldehyde in concentrations not resulting in cell damage most probably cannot act as a complete carcinogen and, consequently, is most likely unable to induce cancer by itself. In conclusion, cytotoxicity is of paramount importance, perhaps even a *conditio sine qua non*, for the formation of nasal tumours by formaldehyde.

Formaldehyde Carcinogenesis: A High-Dose Phenomenon

Does exposure to cytotoxic concentrations of formaldehyde entail a real carcinogenic risk? There is experimental evidence to underpin that most likely only *high* cytotoxic concentrations bring on such a risk:

1. In a study on the more precise localisation of formaldehyde-induced nasal tumours in rats, Morgan et al. (1986b) found a good correlation between the sites of toxic effects and those of tumours, except for one area, viz. the medial maxilloturbinate where clear hyper- and metaplasia were invariably present but where no carcinomas were found. The absence of tumours in this particular area may

Table 11.1. Incidence of nasal tumours in male rats with a damaged or undamaged nasal mucosa exposed to formaldehyde vapour for 3 or 28 months.

Formaldehyde concentration (ppm)	Exposure period and number of rats with a damaged or undamaged nose bearing a nasal tumor/number of rats examined			
	3 months[a]		28 months	
	Damaged	Undamaged	Damaged	Undamaged
0	0/57	0/26	1/54	0/26
0.1	2/57	0/30	1/58	1/26
1.0	2/53	0/29	0/56	1/28
10	2/54	2/26	17/58	1/26

[a] Followed by a 25-month non-exposure observation period.

well be due to a lower local formaldehyde dose in this area than in tumour-bearing areas; the dose delivered to the medial maxilloturbinate may invariably have been high enough for the induction and persistence of hyperplasia/metaplasia but not for the development of neoplasia.

2. The significance of damage to the nasal mucosa for the induction of nasal tumours by formaldehyde at the site of damage was clearly demonstrated by the results of a recently completed inhalation study in which male Wistar rats were exposed to 0 (controls), 0.1, 1 and 10 ppm formaldehyde for 6 h a day, 5 days a week, for either 28 months or 3 months followed by an observation period of 25 months (Woutersen et al. 1989). The damage of the nasal mucosa was induced by bilateral intranasal electrocoagulation (Immel and Feron 1987, 1988). The incidences of nasal tumours in the various groups are given in Table 11.1, which shows that treatment-related tumours only occurred in rats with a damaged nose and exposed to 10 ppm formaldehyde for the whole period of 28 months. Obviously, severe damage to the nasal mucosa in combination with prolonged exposure to a high cytotoxic concentration of formaldehyde leads to tumours; 10 ppm induced extensive and severe hyperplasia and metaplasia in the intact nasal mucosa but no tumours. Clearly, for tumour formation "drastic" conditions are required: severe damage plus a high dose.

3. Another piece of evidence originates from a 30-month study performed to examine the long-term effects on the nasal mucosa of exposure to definitely cytotoxic concentrations of formaldehyde during short exposure periods (Feron et al. 1988). In this experiment male Wistar rats were exposed to 0 (controls), 10 or 20 ppm formaldehyde vapour (6 h a day, 5 days a week) for 1, 2 or 3 months, followed by observation periods of 29, 28 or 27 months, respectively. The incidences of nasal tumours found in this study are presented in Table 11.2, which shows that rats exposed for 3 months to 20 ppm formaldehyde developed six tumours, which is clearly above background. The total score was two nasal tumours in the controls, two nasal tumours in the 10-ppm group and ten nasal tumours (six squamous cell carcinomas, one carcinoma in situ, two polypoid adenomas and one ameloblastoma) in the 20-ppm group. Of course, there were also hyperplastic and metaplastic changes of the nasal respiratory epithelium in the

Table 11.2. Incidence of nasal tumours in male rats exposed to high concentrations of formaldehyde for short periods of time.

Formaldehyde concentration (ppm)	Exposure period and number of rats with a nasal tumour/number of rats examined.			
	1 month[a]	2 months[b]	3 months[c]	Total
0	0/44	2/45	0/45	2/134
10	0/44	1/44	1/44	2/132
20	2/45	2/43	6/44	10/132

[a] Post-treatment period 29 months.
[b] Post-treatment period 28 months.
[c] Post-treatment period 27 months.

10- and 20-ppm groups. The point to be stressed is that these non-neoplastic alterations occurred in many of the exposed rats despite recovery periods of, in many cases, over 2 years. Clearly, only 20 ppm—a very high (cytotoxic) level—and not 10 ppm—still a high (cytotoxic) level—was capable of inducing nasal tumours in rats under these experimental conditions.

Implications for Human Health Risk

The crucial role of tissue damage followed by hyperplasia and metaplasia of the nasal respiratory epithelium in formaldehyde carcinogenesis has now been widely recognised and should be included in human cancer risk assessment. Despite differences in anatomy and physiology of the nose between rats and humans, the respiratory tract defence systems are similar in both species. It is, therefore, reasonable to conclude that the response of the respiratory tract to formaldehyde will be qualitatively similar in rats and humans. If in humans exposure to formaldehyde were to be accompanied by recurrent tissue damage and repair at the site of contact, formaldehyde in cytotoxic concentrations may be assumed to have carcinogenic potential in man. It has been shown that in rats such recurrent tissue damage must be accompanied by exposure to *high* cytotoxic concentrations of formaldehyde in order that nasal tumours be formed (Feron 1988; Woutersen et al. 1989). Correspondingly, if the respiratory tract tissue is not recurrently injured, exposure of humans to relatively low non-cytotoxic levels of formaldehyde can be assumed to represent a negligible cancer risk.

Human exposure to formaldehyde should be minimised for its potency to damage tissue. At the workplace formaldehyde exposure should be controlled to levels below that likely to produce a significant irritant effect. This threshold is about 1 ppm. Thus, an occupational exposure limit of 1 ppm would be likely to virtually eliminate any carcinogenic risk that might have existed for workers exposed to higher concentrations in the past. In the Netherlands a Special Advisory Committee on Formaldehyde (1982) has recommended an atmospheric exposure limit of 0.1 ppm for the general population, which was considered a virtually safe exposure level. Whether this approach is also supported by the molecular (geno)toxicology and the epidemiology of formaldehyde will appear from data presented in the next two papers (Heck and Casanova, this volume; Ulm, this volume).

Conclusions

- Formaldehyde is cytotoxic to the nasal mucosa of a number of animal species including Rhesus monkeys; 1 ppm is the highest "no observed adverse effect level".
- Formaldehyde is a nasal carcinogen in rats and probably also in mice.
- Exposure levels of formaldehyde causing nasal cancer also cause inflammatory, degenerative and proliferative changes in the nose.
- Cytotoxicity is of paramount importance, if not a *conditio sine qua non*, for the formation of nasal tumours by formaldehyde.
- Formaldehyde carcinogenesis is a high-dose phenomenon.
- Exposure of humans to non-cytotoxic concentrations of formaldehyde is considered to represent a negligible cancer risk.

References

Albert RE, Sellakumar AR, Laskin S, Kuschner M, Nelson N, Snyder CA (1982) Gaseous formaldehyde and hydrogen chloride induction of nasal cancer in the rat. J Natl Cancer Inst 68:597–603

Appelman LM, Woutersen RA, Zwart A, Falke HE, Feron VJ (1988) One year inhalation toxicity study of formaldehyde in male rats with a damaged or undamaged nasal mucosa. J Appl Toxicol 8:85–90

BGA (1985) Formaldehyde. A joint report of the Bundesgesundheitsamt, the Bundesanstalt für Arbeitsschutz and the Umweltbundesamt. MMV Medizin, Munich

Bogdanffy MS, Morgan PH, Starr TB, Morgan KT (1987) Binding of formaldehyde to human and rat nasal mucus and bovine serum albumin. Toxicol Lett 38:145–154

Dalbey WE (1982) Formaldehyde and tumors in hamster respiratory tract. Toxicology 24:9–14

ECETOC (1981) Assessment of data on the effects of formaldehyde. Technical report no 1. European Chemical Industry Ecology and Toxicology Centre, Brussels

Feron VJ (1988) Recent inhalation and chronic oral studies with formaldehyde in rats. In: Transcript of the toxicology forum 1988 annual summer meeting. Aspen, Washington, pp 116–139

Feron VJ, Immel HR, Wilmer JWGM, Woutersen RA, Zwart A (1987) Nasal tumours in rats after severe injury to the nasal mucosa and exposure to formaldehyde vapour: preliminary results. In: Tyihák E, Gullner G (eds) The role of formaldehyde in biological systems. Sote, Budapest, pp 73–77

Feron VJ, Bruyntjes JP, Woutersen RA, Immel HR, Appelman LM (1988) Nasal tumours in rats after short-term exposure to a cytotoxic concentration of formaldehyde. Cancer Lett 39:101–111

Heck HA, Casanova-Schmitz M (1984) Biochemical toxicology of formaldehyde. Rev Biochem Toxicol 6:155–189

Higginson J, Hensen OM, Kinlen L, Kirsten WH, MacMahon B, Matanoski GM, Smith TJ, Thomas DC (1987) Epidemiology of chronic occupational exposure to formaldehyde. Report of the ad hoc panel on health effects of formaldehyde. Universities Associated for Research and Education in Pathology, Bethesda

Immel HR, Feron VJ (1987) Intranasal electrocoagulation in rats. Z Versuchstierkd 29:265–269

Immel HR, Feron VJ (1988) Restoration of the rat nasal septum and turbinates after damage by intranasal electrocoagulation. Z Versuchstierkd 31:233–240

Kerns WD, Pavkov KL, Donofrio DJ, Gralla EJ, Swenberg JA (1983) Carcinogenicity of formaldehyde in rats and mice after long-term inhalation exposure. Cancer Res 42:4382–4392

Ma TH, Harris MM (1988) Review of the genotoxicity of formaldehyde. Mutat Res 196:37–59

Maronpot RR, Miller RA, Clarke WJ, Westerberg RB, Decker JR, Moss OR (1986) Toxicity of formaldehyde vapor in $B_6C_3F_1$ mice exposed for 13 weeks. Toxicology 41:253–266

Marshall TC, Hahn FF, Henderson RF, Silbaugh SA, Wolff RK (1983) Subchronic inhalation exposure of guinea pigs to formaldehyde. Annual meeting of the Society of Toxicology, Las Vegas

Mitchell RI, Pavkov KL, Persing RL, Holzworth DA (1979) Final report on a 90-day inhalation toxicity study in rats and mice exposed to formaldehyde. Chemical Industry Institute of Toxicology, Research Triangle Park

Monteiro-Riviere NA, Popp JA (1986) Ultrastructural evaluation of acute nasal toxicity in the rat respiratory epithelium in response to formaldehyde gas. Fundam Appl Toxicol 6:251–262

Monticello TM, Morgan KT, Everitt JI, Popp JA (1989) Pathology and cell proliferation in the respiratory tract of the rhesus monkey exposed to formaldehyde gas. In: Feron VJ, Bosland MC (eds) Nasal carcinogenesis in rodents: relevance to human health risk. PUDOC, Wageningen, p 216

Morgan KT, Gross EA, Patterson DL (1986a) Distribution, progression, and recovery of acute formaldehyde-induced inhibition of nasal mucociliary function in F-344 rats. Toxicol Appl Pharmacol 86:448–456

Morgan KT, Jiang XZ, Starr TB, Kerns WD (1986b) More precise localization of nasal tumors associated with chronic exposure of F-344 rats to formaldehyde gas. Toxicol Appl Pharmacol 82:264–271

Rusch GM, Clary JJ, Rinehart WE, Bolte HF (1983) A 26-week inhalation toxicity study with formaldehyde in the monkey, rat or hamster. Toxicol Appl Pharmacol 68:329–343

Special Advisory Committee on Formaldehyde (1982) Formaldehyde and cancer in humans. Ministry of Public Health and Environment, Staatsuitgeverij, 's-Gravenhage

Swenberg JA, Kerns WD, Mitchell RI, Gralla EJ, Pavkov KL (1980) Induction of squamous cell carcinomas of the rat nasal cavity by inhalation exposure to formaldehyde vapor. Cancer Res 40:3398–3402

Swenberg JA, Barrow CS, Boreiko CJ, Heck HA, Levine RJ, Morgan KT, Starr TB (1983) Non-linear biological responses to formaldehyde and their implications for carcinogenic risk assessment. Carcinogenesis 4:945–952

Takahashi M, Hasegawa R, Furukawa F, Toyoda K, Sato H, Hayashi Y (1986) Effects of ethanol, potassium metabisulfite, formaldehyde and hydrogen peroxide on gastric carcinogenesis in rats after initiation with N-methyl-N′-nitro-N-nitrosoguanidine. Gann 77:118–124

Til HP, Woutersen RA, Feron VJ, Hollanders VHM, Falke HE, Clary JJ (1989) Two-year drinking water study of formaldehyde in rats. Food Chem Toxicol 27:77–87

Tobe M, Kaneko T, Uchida Y, Kamata E, Ogawa Y, Ikeda Y, Saito M (1985) Studies on the inhalation toxicity of formaldehyde. Report of the National Sanitary and Medical Laboratory Service, Toxicity Department of the Organism Safety Research Center, Tokyo

Wilmer JWGM, Appelman LM, Woutersen RA (1986) Subchronic (13-week) inhalation toxicity study of formaldehyde in male rats: 8-hour continuous versus 8-hour intermittent exposures. Report no V 86.361/250283. TNO-CIVO Toxicology and Nutrition Institute, Zeist

Wilmer JWGM, Woutersen RA, Appelman LM, Leeman WR, Feron VJ (1987) Subacute (4-week) inhalation toxicity study of formaldehyde in male rats: 8-hour intermittent versus 8-hour continuous exposures. J Appl Toxicol 7:15–16

Woutersen RA, Appelman LM, Wilmer JWGM, Falke HE, Feron VJ (1987) Subchronic (13-week) inhalation toxicity study of formaldehyde in rats. J Appl Toxicol 7:43–49

Woutersen RA, van Garderen-Hoetmer A, Bruyntjes JP, Zwart A, Feron VJ (1989) Nasal tumours in rats after severe injury to the nasal mucosa and prolonged exposure to 10 ppm formaldehyde. J Appl Toxicol 9:39–46

Zwart A, Woutersen RA, Wilmer JWGM, Spit BJ, Feron VJ (1988) Cytotoxic and adaptive effects in rat nasal epithelium after 3-day and 13-week exposure to low concentrations of formaldehyde vapour. Toxicology 51:87–99

12. Formaldehyde: Evidence of the Carcinogenic Potential from Epidemiological Data

K. Ulm[1]

Introduction

The widespread use of formaldehyde in consumer products and industrial processes, and its natural occurrence makes the question of its effect on human health of considerable importance. Ever since a 1980 report that squamous cell carcinomas were produced in the nasal cavities of rats exposed to high concentrations of formaldehyde vapor (Swenberg et al. 1980), the regulatory and public health concerns have focused primarily on its carcinogenic potential. Although the inhalation findings in rats have been confirmed, evidence of inhalation carcinogenesis in other species is much less compelling.

Estimates of the population in the United States at risk of occupational exposure to free formaldehyde vapor, its particulate polymers, or formaldehyde adsorbed onto ambient airborne particles exceed 1.5 million (Kerns et al. 1983). Large segments of the population may also be exposed via consumer products, construction materials, and insulation. The content of formaldehyde in cigarettes in mainstream smoke (30 μg), and to a far greater extent in sidestream smoke (> 1500 μg) (Jermini et al. 1976) means a potential source for the general population.

The aim of this paper will be to give an overview of the epidemiological studies in relation to formaldehyde exposure, followed by conclusions on the evidence of its carcinogenic effects.

Overview of the Epidemiological Studies

A variety of studies using different designs have been undertaken in order to address the question of the carcinogenic effects of formaldehyde in humans. Many of these epidemiological studies have been published within the last few

---

[1]Institute for Medical Statistics and Epidemiology, Technical University Munich, Ismaninger Strasse 22, D-8000 Munich 80, FRG.

years. Only data from studies using either a cohort or a case-control design were analyzed for this review since they yield the best quality of information to judge causality.

In cohort studies, the mortality of the exposed individuals is compared to the mortality experience among unexposed persons in terms of the standard mortality ratio (SMR). Mostly, the general population is used as a reference group. The observed number of deaths from a specific cause of death within the cohort is divided by the expected number based on the mortality rates in the population and the person years at risk in the cohort (Breslow and Day 1987). Within this design, a variety of causes of death can be considered. Studies of this type are probably the best ones; however, even the very large ones can include only a few cases if rare cancer sites are being considered.

In some studies, proportional mortality rates (PMR) are calculated. In this type of study, only the deaths over a certain period of time are analyzed. The cause-specific proportions of mortality among the exposed (observed deaths) are compared to the expected proportion of deaths among unexposed individuals. Again, the general population is mostly used as a reference group. In a PMR study, large excesses of deaths due to one cause can deflate the remaining proportions and can, thus, bias comparisons of the other causes of death. In this type of study, a variety of different causes of death can be considered. In the third type of study analyzed within this report, the case-control design, one cause of death is investigated. The number of deaths due to a specific cause is compared to controls with respect to formaldehyde exposure. The odds is the ratio between the exposed and the unexposed cases or controls. The odds ratio (OR) is the ratio of both odds. If the disease under study is rare, the OR is numerically very close to its associated relative risk, but the causal inference is not as direct. In addition, an OR obtained from a case-control study nested within a cohort design can be used to support conclusions from the cohort study.

In Table 12.1 all the studies analyzed in this report are summarized together with information regarding the design and a short description of the persons included in the study. Altogether, 19 studies were reviewed. In other reports the number of studies analyzed is sometimes larger, but they include studies which are not directly related to formaldehyde. Among the 19 studies 11, used either an SMR ($n = 7$) or a PMR design ($n = 4$). Six studies were of chemical or industrial workers, and five studies were of medically related professions. For these professions, e.g., morticians, embalmers, anatomists, and pathologists, the exposure was to formalin. This group has diverse chemical exposures, but formalin is one exposure which is common. Eight case-control studies examined occupational etiologies of sinonasal cavity, pharyngeal, or lung cancers.

The studies resulted in different findings. Some studies found a significant increase in the occurrence of tumors of one of the following sites: nasal cavity and sinuses, nasopharynx, buccal cavity, lung, colon, liver, gallbladder, connective tissue, skin, prostate, genitourinary tract, and brain, as well as the occurrence of leukemia. Other studies did not show any effect. There is no consistency in the findings among these studies.

Table 12.1. Summary of the studies related to formaldehyde which are analyzed in this report.

Study design	Reference		Study group
SMR			
	Acheson et al.	1984	Chemical workers
	Bertazzi et al.	1986	Formaldehyde resin workers
	Blair et al.	1986, 1987	Formaldehyde producers or users
	Harrington and Oakes	1984	Pathologists
	Levine et al.	1984	Ontario morticians
	Stayner et al.	1988	Garment workers
	Stroup et al.	1986	Anatomists
PMR			
	Marsh	1982	Chemical workers (male)
	Walrath and Fraumeni	1983	New York embalmers and funeral directors
	Walrath and Fraumeni	1984	Californian embalmers
	Stayner et al.	1985	Garment workers
CC			
	Coggan et al.	1984	Lung and bladder cancer in the UK
	Fayerweather et al.	1982	Chemical workers
	Hayes et al.	1986	Nasal and sinonasal cancer patients
	Olsen et al.	1984	Nasal and sinonasal cancer patients in Denmark
	Partanen et al.	1985	Respiratory cancer patients from a retrospective cohort of male wood-workers
	Roush et al.	1985	Sinonasal and pharyngeal cancer patients in Connecticut
	Vaughan et al.	1986a,b	Sinonasal and pharyngeal cancer patients in Washington

CC, case-control study.

Because of the rapid metabolism of formaldehyde in the human body, it is highly unlikely that formaldehyde can cause tumors in organs remote from the point of entry, which is the upper respiratory tract. Therefore, the main discussion was focused on the association between formaldehyde exposure and cancer of the upper respiratory tract, such as nasal and nasopharyngeal cancers and cancer of the buccal cavity. This review deals with these sites together with lung cancer. The results are summarized in Tables 12.2–12.5 according to the classification of these cancer sites in the 9th revision of the International Classification of Diseases and Death (ICD).

As well as the authors and the design of the studies, these tables also summarize the observed and expected numbers together with the estimated risk and the 95% confidence interval for this risk. This is easily done for SMR and PMR studies. For case-control studies, the "observed" column shows the ratio of the exposed vs. unexposed persons among all cases, and the "expected" column shows the same ratio for controls. The estimated risk is the OR.

Table 12.2. Risk of buccal cavity cancer (ICD 140–149) in relation to formaldehyde exposure.

Study design	Reference		O	E	Risk (O/E 100%)	95% Confidence interval
SMR						
	Acheson et al.	1984[a]	5	4.6	109	35– 278
	Blair et al.	1986[b]	18	19.0	96	57– 152
	Levine et al.	1984	1	2.1	48	1– 267
	Stayner et al.	1988	6	3.9	155	57– 338
	Stroup et al.	1986	1	26.8	15	38– 369
			31	36.4	85	57– 121
PMR						
	Walrath and Fraumeni	1983	8	7.1	113	49– 222
	Walrath and Fraumeni	1984	8	6.1	131	56– 258
	Stayner et al.	1985	3	1.3	229	47– 669
			19	14.5	131	79– 202
CC						
	Fayerweather et al.	1982	1/6	1/6	100	7–1430
	Vaughan et al.	1986a[c]	69/163	171/552	137	98– 190
	Vaughan et al.	1986b[d]	36/196	82/469	105	69– 161
					123	95– 160

[a] ICD 143–149.
[b] White men.
[c] ICD 146–148 – exposure level background vs. rest.
[d] ICD 146–148 – exposure level in mobile home O vs. > 1 year.
O, observed; E, expected; CC, case-control study.

The results for the four sites were summarized separately for each of the three study designs in the following way: the SMR and PMR studies were merely pooled, the case-control studies were added up using the Mantel-Haenszel procedure (see Breslow and Day 1980).

Cancer of the Buccal Cavity

Cancer of the buccal cavity was considered in 11 out of the 19 studies (Table 12.2). None of these studies reported a statistically significant excess. This is also true for the pooled data. The risk varies between 85% (SMR studies) and 131% (PMR studies). It is interesting to note that the risk estimated by the SMR studies is lower compared to that obtained by the other designs.

Nasopharyngeal Cancer

The findings of six studies of nasopharyngeal cancer are summarized in Table 12.3. In the study of Acheson et al. (1984), no death from this site was observed, but the expected numbers are not given. In one study (Blair et al. 1986), how-

Table 12.3. Risk of nasopharyngeal cancer (ICD 147) in relation to formaldehyde exposure.

Study design	Reference		O	E	Risk (O/E 100%)	95% Confidence interval
SMR						
	Acheson et al.	1984	0	Not given	0	–
	Blair et al.	1986	6	2	300	110–654
			6	2	300	110–654
CC						
	Olsen et al.	1984[a]	6/209	68/1563	66	28–153
	Roush et al.	1985	–	–	110	70–190
	Vaughan et al.	1986a[b]	11/16	171/381	153	70–335
	Vaughan et al.	1986 b[c]	8/19	82/469	241	104–555
					124	78–199

[a] Males only.
[b,c] See Table 2.
O, observed; E, expected.

ever, a significant excess was found (6 observed vs. 2.0 expected). But there was no increase in the risk with cumulative exposure (Blair et al. 1986). Later, Blair et al. (1987) reported that of the six exposed cases, five had worked in dusty conditions (four in one plant) and thus had simultaneous exposure to formaldehyde and particulates. In these persons, a positive, though not significant, trend was found with cumulative exposure to formaldehyde. The paper does not mention which agents the four patients with nasopharyngeal cancer, all from one plant, were exposed to. The problem of the dose-response relationship is discussed on p. 147 ff.

In three out of the four studies using a case-control design, which is appropriate for this rare site, an excess was reported. This excess was statistically significant in one study (Vaughan et al. 1986b) in which a residential, as opposed to an occupational, exposure was considered. This point is discussed on p. 146 ff in greater detail in connection with possible confounding factors. Combining the studies of Olsen et al. (1984) and of Vaughan et al. 1986a), both of which consider occupational exposure, the odds ratio is 99%.

Cancers of the Nasal Cavity and Nasal Sinuses (Sinonasal Cancer)

The results of animal studies show that the production of nasal tumors in rats by heavy and prolonged exposure to formaldehyde gives sinonasal cancer a special claim to attention in human studies. The results of nine such studies (four SMR and five case-control studies) are summarized in Table 12.4. The SMR studies do not support the hypothesis that there is an association between formaldehyde exposure and sinonasal cancer. The overall risk is 50%. However, two out of the

Table 12.4. Risk of sinonasal cancer (ICD 160 = cancer of the nasal cavity and nasal sinuses) in relation to formaldehyde exposure.

Study design	Reference		O	E	Risk (O/E 100%)	95% Confidence interval
SMR						
	Acheson et al.	1984	0	1.1	0	0– 273
	Blair et al.	1986	2	2.2	90	11– 324
	Levine et al.	1984	0	0.2	0	0–1500
	Stroup et al.	1986	0	0.5	0	0– 600
			2	4.0	50	6– 181
CC						
	Hayes et al.	1986	15/48	18/143	248	118– 523
	Olsen et al.	1984	60/500	68/1563	276	195– 391
	Roush et al.	1985	–	–	80	50– 113
	Vaughan et al.	1986a	12/41	171/381	65	34– 127
	Vaughan et al.	1986b	5/48	82/469	60	23– 153
					170	131– 222

O, observed; E, expected.

five case-control studies found a significant excess due to occupational exposure. The estimated risk pooled over the five case-control studies is 170% with a 95% confidence interval from 131% to 222%. There is a statistically significant excess. The main problem in interpreting these findings is the mixture of other exposures, mainly to wood dust, a point which is discussed on p. 146 ff.

Lung Cancer

Most of the data available investigate a possible association between formaldehyde and lung cancer. This cause of death is analyzed in 13 out of the 19 studies (Table 12.5). In three big studies (Acheson et al. 1984; Blair et al. 1986; Stayner et al. 1988) the SMR was slightly above 100. As in most retrospective cohort studies, no account could be taken of smoking habits. No correlation was found with intensity or duration of exposure to formaldehyde or with cumulative, average, or peak exposures.

In one cohort study from Italy, a significant increase in lung cancer (SMR = 236) was found in workers of a formaldehyde resin plant (Bertazzi et al. 1986). This excess, however, was similar among those exposed to compounds other than formaldehyde (SMR = 148), as well as among those exposed to formaldehyde (SMR = 136), though the greatest increase was found in those with unknown exposure (SMR = 388). A significant association with formaldehyde exposure was found in one case-control study, but there was no exposure-response relationship nor an increase in risk among men in the highest exposure category (Coggan et al. 1984).

Table 12.5. Lung cancer (ICD 162) in relation to formaldehyde exposure.

Study design	Reference		O	E	Risk (O/E 100%)	95% Confidence interval
SMR						
	Acheson et al.	1984	236	231	102	90–117
	Bertazzi et al.	1986	18	7.6	236	140–373
	Blair et al.	1986	201	182	111	96–127
	Harrington and Oakes	1984	9	22	41	19– 78
	Levine et al.	1984	19	20.2	94	56–147
	Stayner et al.	1988	39	34.1	114	81–155
	Stroup et al.	1986	12	43.1	30	10– 50
			534	540	99	91–108
PMR						
	Stayner et al.	1985	11	11.62	95	47–169
	Walrath and Fraumeni	1983	72	66.8	108	85–137
	Walrath and Fraumeni	1984	41	42.9	96	68–130
			124	121	102	85–122
CC						
	Coggan et al.	1984	296/302	472/708	147	121–179
	Fayerweather et al.	1982	36/142	39/142	92	55–154
	Partanen et al.	1985	11/41	10/44	118	45–308
					137	115–165

O, observed; E, expected.

The estimated overall risk is 99% (range: 91%–108%) for all SMR studies, 102% (range: 85%–122%) for all PMR studies, and 137% (range: 115%–165%) for all case-control studies. Only in the case-control studies is the risk statistically increased, but this result is based on only one study (Coggan et al. 1984). With respect to the results in the SMR studies with more than 500 observed cases, the overall evidence is not indicative of an association between formaldehyde exposure and lung cancer risk.

Possible Confounding Factors

For all sites, the possibility of confounding factors must be stressed. Can the increased risk be explained only by the exposure to formaldehyde, or are there other explanations as well? The possible confounding factors are discussed in relation to the specific sites.

Nasopharyngeal Cancer

Two studies (Blair et al. 1986; Vaughan et al. 1986b) demonstrate an increased risk. The use of tobacco is discussed as being related to nasopharyngeal and oral cancer (Henderson et al. 1976; Winn et al. 1981). The type of snuff dipping prevalent in the southern United States is particularly mentioned as being a risk factor. Similar findings regarding this smokeless use of tobacco, also in the form of chewing, are reported by the IARC (Zaridze and Peto, 1985). Some of the plants analyzed by Blair et al. (1986) are located in the southern United States, but there is no data on these factors in their paper.

In the study of Vaughan et al. (1986b), a significantly increased risk of nasopharyngeal cancer due to residential exposure was reported. In this study, an association was found with living in a mobile home. When the exposure was defined as living in a mobile home for 1–9 years, the risk (OR = 210%) was increased but not statistically significant. The risk from living in a mobile home for 10 years or longer increased to 549%, which is statistically significant. This result is based on only four exposed cases. The main problem here is that no measurements were taken of the formaldehyde exposure or of any other exposures. Therefore, it remains unclear whether this risk is solely due to formaldehyde.

Sinonasal Cancer

The association between the occupational exposure to wood dust, especially from hard wood, and nasal cancer was established a couple of years ago. Therefore, in all studies related to nasal cancer the exposure to wood dust has to be taken into account, and the estimate of the risk must be adjusted for this additional exposure. When this was done, no increased risk due to formaldehyde could be found. As an example, in the study of Olsen et al. (1984) the risk is reduced from 276% to 160%, which is not statistically significant, but still above 100%. It should be pointed out that in the paper of Olsen et al. (1984), nearly all the patients with sinonasal cancer were either unexposed to formaldehyde and wood dust (80%) or exposed to both factors (14%).

Problems in Detecting Small Risks by Epidemiological Studies

I want to demonstrate the problems of detecting a small risk in epidemiological studies, especially if the incidence of the cause of death in question is also very low. As an example, in the study by Stayner et al. (1988) 11030 workers were followed for approximately 18 years which yields 188025 person years at risk. In considering cancer of the buccal cavity, six cases were observed compared with 3.9 expected. This yields an SMR of 155%, indicating an increased risk of about 55%. The risk is statistically not significant, the 95% confidence interval is from 57% to 338%. This means that, with a probability of 95%, the true risk can range from 57%, indicating a protective effect of the exposure to formaldehyde, to a more than threefold risk.

If the result is statistically not significant apart from the confidence interval, the power of detecting an excess risk within this specific study should also be calculated. The power is the probability of overlooking a certain risk. As an example, if we assume a twofold risk of dying from cancer of the buccal cavity, the power in the study of Stayner et al. (1988) would be 50%. If, indeed, the risk is higher, the power increases, otherwise it decreases.

A low risk also has to be considered in connection with the well-known healthy worker effect. This makes it even more difficult to detect a low risk. Among the criteria used in epidemiology to judge whether an association is caused by a certain exposure, two criteria are of great importance in the situation of detecting a small risk together with low incidence: (a) dose-response relationship; and (b) consistency across the studies.

The main problem in demonstrating a dose-response relationship is the lack of reliable measurements of the concentration in the past. In the study of Blair et al. (1987), an attempt was made to estimate the exposure. In Table 1 of the paper of Blair et al. (1987), the SMR for nasopharyngeal cancer has been classified into four different categories according to the cumulative exposure. One can assume a dose-response relationship only in the group with a simultaneous exposure to formaldehyde and particulates. In the two other combinations (no exposure to particulates and ignoring this exposure), no association can be found. But, owing to the small numbers of observed cases, this relationship is statistically not significant. Using isotonic regression techniques, the log-likelihood ratio R for testing a dose-response relationship based on the Poisson distribution is $R = 2 \Sigma O_i \ln (SMR_i/SMR)$ with O_i the observed number of cases and SMR_i the estimated risk in the different subgroups i ($i = 1, \ldots, I$) with respect to the cumulative exposure. SMR is the overall SMR. The data in the group with exposure to formaldehyde and particulates yields a value of $R = 1.46$. The critical value with $p < 0.05$ is 4.53. Therefore, the sample has to be increased by a factor of about 3 to obtain a significant result. In the study of Blair et al. (1986), about 25000 persons were followed for approximately 24 years. This means that a cohort has to cover about 75000 persons with a follow-up over the same period of time. This seems unrealistic in practice. Consequently, an important factor in demonstrating an association is to have consistency across the studies, which is, of course, also problematic in the situation of low risk. Probably the best way to estimate the risk is by pooling all case-control studies, including those nested within cohort studies.

Conclusions

There are two specific cancer sites which are of special interest: the nasopharynx and the nasal cavity and sinuses. For both sites some studies show a significant excess, but the findings are not consistent. Therefore, based on all the studies available, there is still no convincing evidence that formaldehyde exposure is related to any cancer in humans (see also Universities Associated for Research and Education in Pathology 1988). It should, however, be pointed out that this result does not mean that there is no risk. A certain risk cannot be excluded. With a probability of 95%, the true risk is within the corresponding confidence interval. These findings induced the authorities of the Federal Republic of Germany, in contrast to other regulatory bodies, not to classify formaldehyde as a carcinogenic compound. It remains in the group which includes compounds justifiably suspected of having carcinogenic potential.

References

Acheson ED, Barnes HR, Gardner MJ et al. (1984) Formaldehyde in the British chemical industry. An occupational cohort study. Lancet 1:611–616

Bertazzi PA, Pesatori AC, Radice L et al. (1986) Exposure to formaldehyde and cancer mortality in a cohort of workers producing resins. Scand J Work Environ Health 12:461–468

Blair A, Stewart P, Gaffey W et al. (1986) Mortality among industrial workers exposed to formaldehyde. JNCI 76:1071–1084

Blair A, Stewart PA, Hoover RN et al. (1987) Cancers of the nasopharynx and oropharynx and formaldehyde exposure. JNCI 78:191

Breslow NE, Day NE (1980) Statistical methods in cancer research. I. The analysis of case-control studies. Oxford Press (IARC sci publ no 32)

Breslow NE, Day NE (1987) Statistical methods in cancer research. II. The design and analysis of cohort studies. Oxford Press (IARC sci publ no 82)

Coggan D, Pannett B, Acheson ED (1984) Use of job exposure matrix in an occupational analysis of lung and bladder cancer on the basis of death certificates. JNCI 72:61–65

Fayerweather WE, Pell S, Bender JR (1982) Case control study of cancer death in DuPont workers with potential exposure to formaldehyde. In: Clary JJ, Gibson JE, Waritz RS (eds) Formaldehyde–toxicology, epidemiology, mechanisms. Dekker, New York, pp 47–125

Harrington JM, Oakes D (1984) Mortality study of British pathologists 1974–1980. Br J Ind Med 41:188–191

Hayes RB, Raatgever JW, Debruyn A et al. (1986) Cancer of the nasal cavity and paranasal sinuses and formaldehyde exposure. Int J Cancer 37:487–492

Henderson BE, Louie E, Soohoo J et al. (1976) Risk factors associated with nasopharyngeal carcinoma. N Engl J Med 11:1101–1106

Levine RJ, Jelkovich DA, Shaw LK (1984) The mortality of Ontario undertakers and a review of the formaldehyde-related mortality studies. J Occup Med 26:740–746

Kerns WD, Pavkov KL, Donofrio DJ et al. (1983) Carcinogenicity of formaldehyde in rats and mice after long-term inhalation exposure. Cancer Res 43:4382–4392

Jermini C, Weber A, Grandjean E (1976) Quantitative determination of various gas phase components of the side stream smoke of cigarettes in the room air as a contribution to the problem of passive smoking. Int Arch Occup Environ Health 36:169–181

Marsh GM (1982) Proportional mortality patterns among chemical plant workers exposed to formaldehyde. Br J Ind Med 39:313–322

Olsen H, Jensen SP, Hink M et al. (1984) Occupational formaldehyde exposure and increased nasal cancer risk in man. Int J Cancer 34:639–644

Partanen T, Kauppinen T, Nurminen M et al. (1985) Formaldehyde exposure and respiratory and related cancers. A case-referent study among woodworkers. Scand J Work Environ Health 11:409–415

Roush G, Walrath J, Stayner L et al. (1985) Formaldehyde-related occupations and cancers of the nasopharynx and sinonasal region. Am J Epidemiol 122:518 (abstr)

Stayner L, Smith AB, Reeve G et al. (1985) Proportionate mortality study of workers in the garment industry exposed to formaldehyde. Am J Ind Med 7:229–240

Stayner L, Elliott L, Blade L et al. (1988) A retrospective cohort mortality study of workers exposed to formaldehyde in the garment industry. Am J Ind Med 13:667–681

Swenberg JA, Kerns WD, Mitchell RE et al. (1980) Induction of squamous cell carcinomas of the rat nasal cavity by inhalation exposure to formaldehyde vapor. Cancer Res 40:3398–3402

Stroup NE, Blair A, Erickson GE (1986) Brain cancer and other causes of death in anatomists. J Natl Cancer Inst 77:1217–1224

Universities Associated for Research and Education in Pathology (1988) Epidemiology of chronic occupational exposure for formaldehyde: report of the ad hoc panel on health aspects of formaldehyde. Toxicol Ind Health 4:77–90

Vaughan TL, Strader C, Davis S et al. (1986a) Formaldehyde and cancers of the pharynx sinus and nasal cavity. I. Occupational exposures. Int J Cancer 38:677–683

Vaughan TL, Strader C, Davis S et al. (1986b) Formaldehyde and cancers of the pharynx sinus and nasal cavity. II. Residential exposures. Int J Cancer 38:685–688

Walrath J, Fraumeni JF Jr (1983) Mortality patterns among embalmers. Int J Cancer 31:407–411

Walrath J, Fraumeni JF Jr (1984) Cancer and other causes of death among embalmers. Int J Cancer 44:4638–4641

Winn DM, Blot WJ, Shy CM et al. (1981) Snuff dipping and oral cancer among women in the southern United States. N Engl J Med 13:747–748

Zaridze DG, Peto R (1985) Tobacco: a major international health hazard. (IARC sci publ no 74)

13. Covalent Binding of Formaldehyde to DNA: Characterization of Reaction Products and Quantitation Following Inhalation Exposure

H.d'A. Heck[1] and M. Casanova[1]

Introduction

The induction of nasal cancer by formaldehyde in rats is a highly *sub*linear function of concentration, i.e., the carcinogenic potency of formaldehyde increases in more-than-direct proportion to the airborne concentration as the concentration increases above 6 ppm (Kerns et al. 1983). This sublinear concentration response for nasal cancer is consistent with the fact that cell turnover in the nasal mucosa (Swenberg et al. 1983; Wilmer et al. 1987), squamous metaplasia (Rusch et al. 1983), the binding of formaldehyde to nasal mucosal DNA (Casanova et al. 1989a,b), and the mutagenic efficiency of formaldehyde (Craft et al. 1987) are either sublinear functions of concentration or have been shown to increase with concentration at a constant total dose. These findings strongly suggest that the carcinogenic potency of formaldehyde is modulated by defense mechanisms that are saturable at high concentrations.

This article summarizes the results of recent experiments which have led to an improved understanding of the interactions of formaldehyde with cellular DNA, and which have elucidated the concentration dependence of the delivered dose in selected tissues of the rat and monkey. Formaldehyde is mutagenic and clastogenic in vitro (Goldmacher and Thilly 1983; Natarajan et al. 1983), it is an initiator of cell transformation in vitro (Ragan and Boreiko 1981), and it can form DNA-protein cross-links in vitro and in vivo (Ross and Shipley 1980; Grafstrom et al. 1983; Casanova-Schmitz and Heck 1983; Casanova-Schmitz et al. 1984). Thus, the binding of formaldehyde to DNA may play an important role in formaldehyde carcinogenesis. The studies described in this article suggest that quantitative measurements of DNA binding interpreted in terms of a suitable pharmacodynamic model can be a very important component of the risk assessment process, providing significant insight into the behavior of biological systems exposed to low concentrations of toxicants.

---

[1]Chemical Industry Institute of Toxicology, Research Triangle Park, North Carolina 27709, USA.

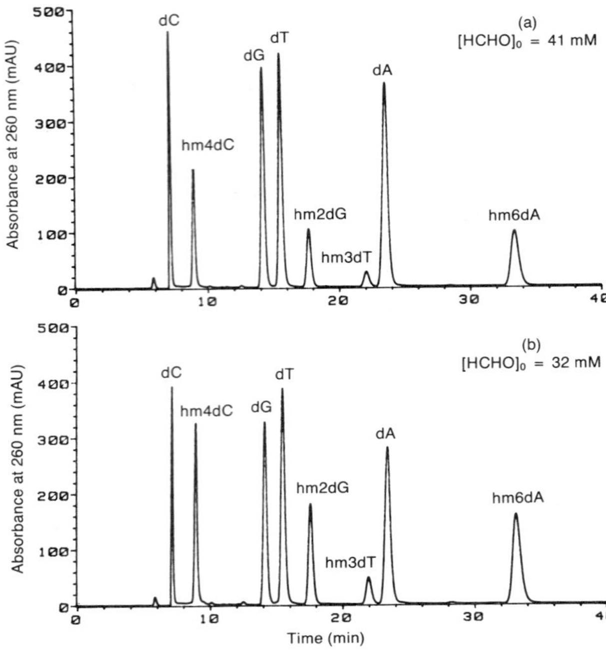

Figure 13.1. Reversed-phase HPLC of a mixture of deoxyribonucleosides incubated with formaldehyde (pH 7.1, 23°C, 12–16 h); initial concentrations of formaldehyde (HCHO) are shown in each figure. (From Casanova et al. 1989a, with permission)

Covalent Reactions of Formaldehyde with DNA

Formaldehyde can form hydroxymethyl adducts with the four major nucleic acid bases (McGhee and von Hippel 1975a,b). These reactions have been extensively investigated using nucleosides, nucleotides, and protein-free DNA, and the rates and equilibrium constants are known. The adducts appear to form readily in single-stranded regions of protein-free DNA, but they do not form in double-stranded DNA (McGhee and von Hippel 1977).

The adducts of formaldehyde with deoxyribonucleosides are unstable and dissociate readily in aqueous solution. Despite their instability, all of the adducts can be chromatographed and analyzed by reversed-phase high-performance liquid chromatography (HPLC) with negligible dissociation of the adducts having occurred (Fig. 13.1) (Casanova et al. 1989a). Experimental (Funderburk et al. 1978) and theoretical (Williams 1987) studies suggest that water plays an important *catalytic* role in the formation and dissociation of formaldehyde adducts with oxygen- and nitrogen-containing nucleophiles. The remarkable stability of the hydroxymethyl adducts in HPLC may, therefore, be due to partitioning of the adducts into the bonded stationary phase of the reversed-phase column with consequent removal of the adducts from direct contact with water.

Although the existence of hydroxymethyl adducts in protein-free DNA is unquestioned, there is no compelling evidence to indicate that these adducts are formed in *cellular* DNA, which is in intimate contact with basic histones and other nuclear proteins. A very small quantity of an adduct, N^6-hydroxymethyl-deoxyadenosine (hm6dA), was detected by one group of investigators (Beland et al. 1984) following the extraction of DNA from mammalian cells incubated with [^{3}H]formaldehyde, enzymatic hydrolysis of the DNA, and analysis of the hydrolyzate by HPLC with liquid scintillation counting (LSC). However, the investigators also showed that the great majority ($>$ 99%) of the [^{3}H]formaldehyde associated with the DNA was not bound to deoxyribonucleosides but was released by hydrolysis of the DNA. (The released [^{3}H]formaldehyde was denoted as "nonreacted" or "noncovalently bound," although it was not removed from the DNA by gel permeation chromatography.) The quantity of [^{3}H]formaldehyde detected as the adduct, hm6dA, represented approximately 0.3% of the quantity released.

Subsequent experiments were carried out in our laboratory to check these results, and a different conclusion was reached (Casanova et al. 1989a). [^{14}C]Formaldehyde [corresponding to the released (or noncovalently bound) formaldehyde in the preceding experiments (Beland et al. 1984)] was incubated with a mixture of the four major deoxyribonucleosides under conditions very similar to those used to hydrolyze the DNA from mammalian cells, and the solutions were analyzed by HPLC-LSC. After correcting for minor differences in the concentrations of the reactants, the yield of hm6dA was indistinguishable from that reported in DNA hydrolyzates (Beland et al. 1984), indicating that the hm6dA detected previously was probably formed *after* the exposure to [^{3}H]formaldehyde by a reaction between the formaldehyde released from the DNA by hydrolysis and deoxyadenosine.

In addition to demonstrating that hm6dA can be formed during the hydrolysis of DNA, these experiments also showed that the types and quantities of adducts produced are dependent on the buffer used for incubation. In the particular combination of buffers that was used previously (Beland et al. 1984), only hm6dA is produced. However, under slightly different conditions, it is possible to produce either all four hydroxymethyl adducts or no adducts at all (Casanova et al. 1989a). The fundamental determinant of adduct formation is the nature of the amino group (primary or tertiary) present in the buffer.

An important question, however, concerned the origin of the released formaldehyde in the DNA of mammalian cells incubated with [^{3}H]formaldehyde (Beland et al. 1984). Insight into this question was obtained in experiments in which rats were exposed to [^{3}H]- and [^{14}C]formaldehyde, and the DNA was extracted from the nasal respiratory mucosa. Extraction of the DNA separated it into aqueous (AQ) and interfacial (IF) fractions. Previous experiments had shown that the IF DNA but not the AQ DNA is cross-linked by formaldehyde to proteins (Casanova-Schmitz and Heck 1983; Casanova-Schmitz et al. 1984). The IF DNA was solubilized by digesting the interfacial proteins with proteinase K, and both fractions were purified using hydroxyapatite chromatography, followed by incubation with ribonuclease and ultrafiltration.

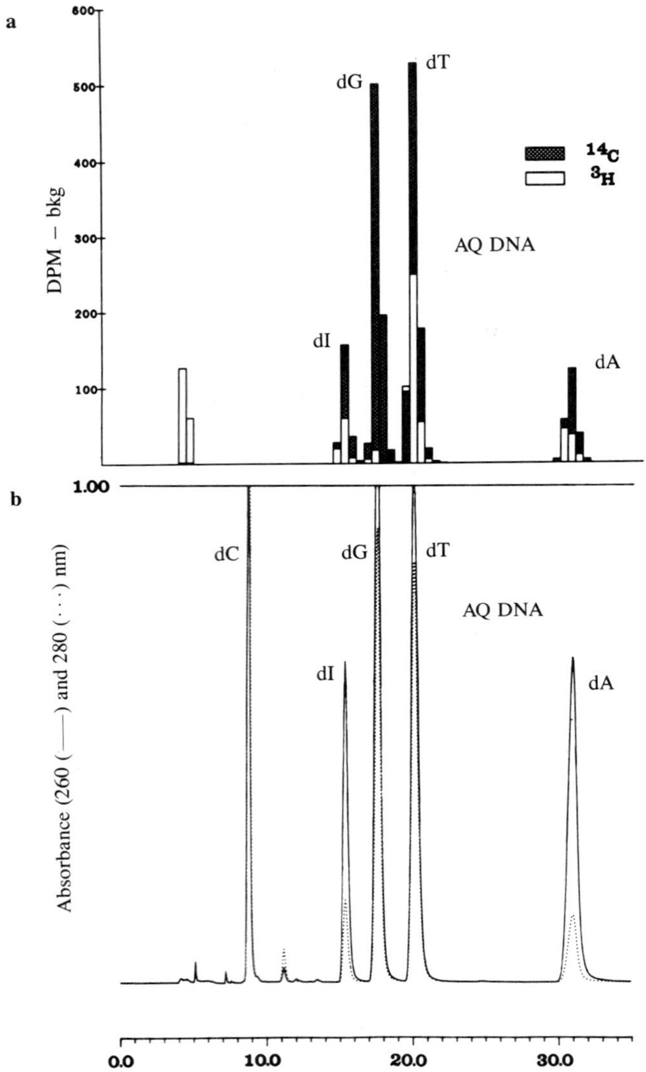

Figure 13.2. Reversed-phase HPLC of aqueous (AQ) DNA isolated from the nasal mucosa of Fischer 344 rats exposed to [^{3}H]- and [^{14}C]formaldehyde (6 ppm, 6 h). (From Casanova et al. 1989a, with permission)

The two DNA fractions were hydrolyzed enzymatically and were analyzed by HPLC-LSC (Casanova et al. 1989a). The results of these experiments are shown in Fig. 13.2 for the AQ DNA and in Fig. 13.3 for the IF DNA, respectively. In the AQ DNA, all of the ^{14}C and most of the ^{3}H was due to normal metabolic incorporation into purines and thymine (Fig. 13.2). Note that the deoxyribonucleosides contained more ^{14}C than ^{3}H; this indicates that formaldehyde was

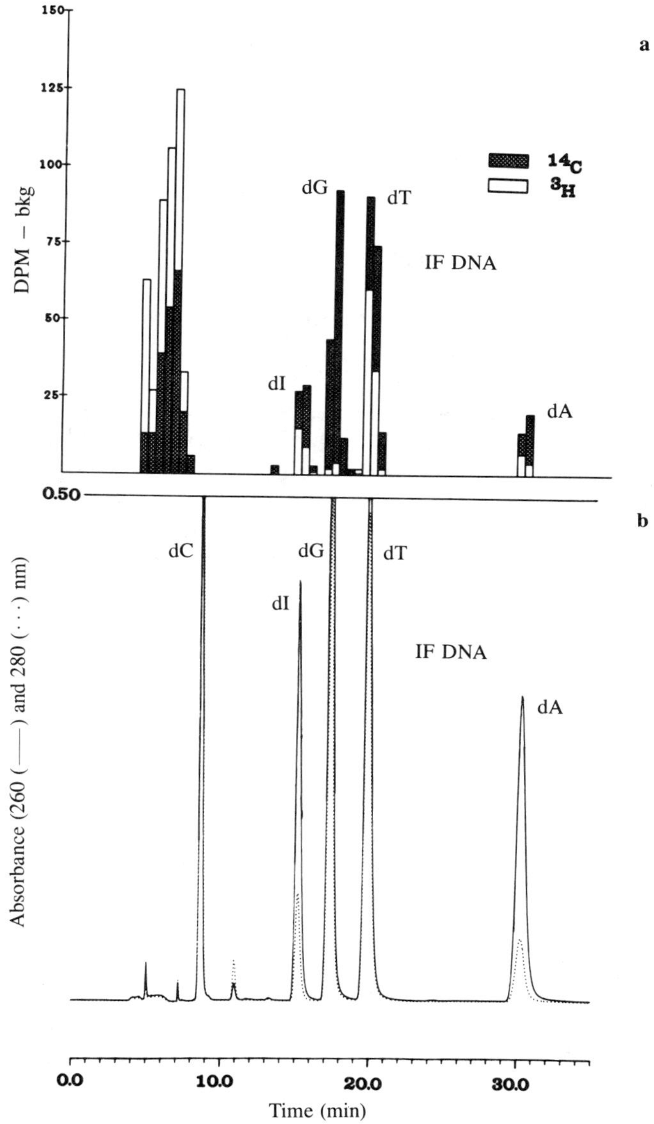

Figure 13.3. Reversed-phase HPLC of interfacial (IF) DNA isolated from the nasal mucosa of Fischer 344 rats exposed to [^{3}H]- and [^{14}C]formaldehyde (6 ppm, 6 h). (From Casanova et al. 1989a, with permission)

oxidized to the level of formate prior to or during its incorporation into the deoxyribonucleosides. ^{3}H (but not ^{14}C) was also eluted very early in the chromatographic profile in a region with little UV absorbance (Fig. 13.2). Since this early-eluting peak did not contain ^{14}C, it was probably caused by isotopic exchange of [^{3}H]H_2O with the DNA.

In the IF DNA, incorporation of ^{3}H and ^{14}C into deoxyribonucleosides was also observed, similar to that which occurred in the AQ DNA (Fig. 13.3). However, unlike the AQ DNA, both ^{3}H *and* ^{14}C were eluted near the void volume of the column before any of the nucleosides had emerged (Fig. 13.3). The ^{3}H/^{14}C ratio of the early-eluting peak was substantially higher than those of the nucleoside peaks, suggesting that the early-eluting peak was due to unoxidized formaldehyde. (Subsequent experiments identified the early-eluting radioactivity as [^{14}C]- and [^{3}H]formaldehyde. This peak was also found to contain a small amount of [^{3}H]H_2O, similar to the AQ DNA).

These experiments demonstrated that formaldehyde present as DNA-protein cross-links is quantitatively released from DNA by hydrolysis. Thus, cross-links appear to be stable only in unhydrolyzed DNA; they are quite labile in hydrolyzed DNA and rapidly dissociate, releasing free formaldehyde. On the basis of this finding and the fact that [^{3}H]- and [^{14}C]formaldehyde were released *only* from the IF DNA, we have suggested that the so-called "noncovalently bound" formaldehyde reported previously (Beland et al. 1984) was in fact due to the formation of DNA-protein cross-links, and that these cross-links were liberated by hydrolysis of the DNA (Casanova et al. 1989a). This interpretation is supported by several experimental studies showing that N-CH_2-N cross-links between nucleosides and amino acids are unstable in aqueous solution (Sloan and Siver 1984; Volkov et al. 1985; Fennell et al. 1987), and by theoretical studies indicating that water catalyzes the decomposition of formaldehyde adducts (Williams 1987).

Dosimetry Studies

Following the above exploratory experiments, an improved method to investigate the covalent reactions of formaldehyde with cellular DNA was developed. In this method, it is unnecessary to use [^{3}H]formaldehyde. (Avoidance of ^{3}H is advantageous, due to complications arising from an isotope effect that occurs in the oxidation of [^{3}H]formaldehyde (Heck and Casanova 1987).) It is also unnecessary to separately purify the AQ and IF DNA; both can be purified together by digesting the proteins with proteinase K before extracting the DNA. Using this simpler method, the formation of DNA-protein cross-links by [^{14}C]formaldehyde was determined over a wide range of airborne concentrations (0.3 to 10 ppm) (Casanova et al. 1989a). The results, which are shown in Fig. 13.4, are expressed as a *dose rate*, B/*t*, where B is the amount of [^{14}C]formaldehyde bound per mg DNA and *t* is the exposure time.

►

Figure 13.5. Pharmacodynamic model for disposition of inhaled formaldehyde in the rat nasal mucosa to metabolites (detoxication) or to DNA-protein cross-links. *A*, Extracellular (administered) formaldehyde; *D*, intracellular (delivered) formaldehyde consisting of free (D_f) and glutathione (*GSH*)-bound (D_b) forms. (From Casanova and Heck 1987, with permission)

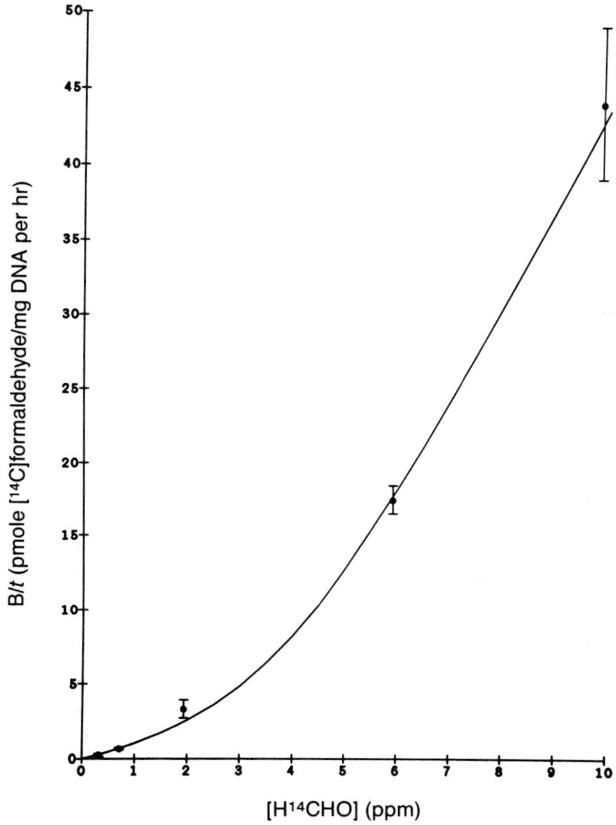

Figure 13.4. Rate of formation of DNA-protein cross-links in the nasal respiratory mucosa of rats exposed for 6 h to 0.3, 0.7, 2, 6, or 10 ppm of [^{14}C]formaldehyde. Values are mean ± SE, $n = 3$ exposures per concentration, four rats per exposure group. (Adapted from Casanova et al. 1989a)

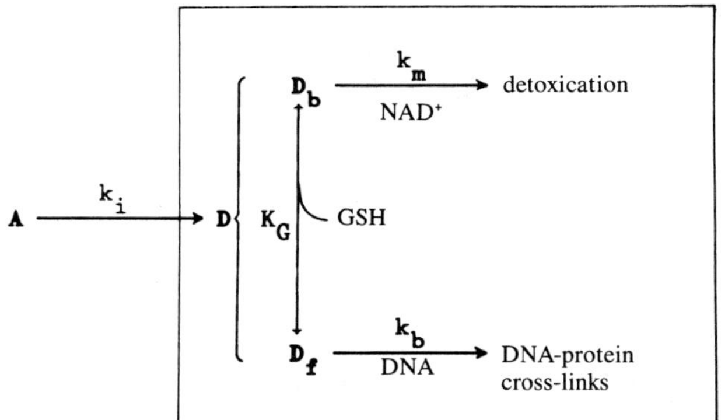

Covalent binding to DNA was expressed as a dose rate owing to a pharmacodynamic analysis, which suggested that the dose rate is proportional to the *intracellular* [^{14}C]formaldehyde concentration (Casanova et al. 1989a). The pharmacodynamic model is shown in Fig. 13.5; the smooth curve shown in the figure was calculated by computer after carrying out a nonlinear least-squares fit of the model to the data points. In terms of model parameters, the specific relationship between the dose rate and the intracellular concentration is defined as:

$$B/t = k_b(1 - \varphi)D$$

where k_b is the first-order rate constant for covalent binding to DNA, $(1 - \varphi)$ is the fraction of the intracellular formaldehyde that is not bound to glutathione, and D is the intracellular (*delivered*) concentration of formaldehyde, i.e., the biologically relevant measure of exposure. Thus, the dose rate is a surrogate for the intracellular concentration. Estimates of cancer risk that were calculated using the multistage model were about 50-fold lower at 0.1 ppm when the surrogate rather than the administered concentration was used (Starr and Buck 1984).

As shown in Fig. 13.4, DNA-protein cross-links were detected in the rat nasal respiratory mucosa even at the lowest airborne concentration examined (0.3 ppm). Thus, there is presently no evidence to suggest that the formation of cross-links involves a low-concentration threshold. However, the reaction of formaldehyde with cellular DNA is clearly nonlinear, with the dose rate at 10 ppm being about 40- to 50-times higher than at 1 ppm. This nonlinear response is explained by the model in terms of a saturation of metabolic defense mechanisms at high formaldehyde concentrations.

The rat is one of the two major species used for carcinogenicity testing, but it is important to determine whether DNA-protein cross-links are formed in species more closely related to man. To investigate this question, investigators at CIIT exposed Rhesus monkeys to [^{14}C]formaldehyde and determined the concentration of [^{14}C]formaldehyde bound to DNA in the nasal mucosa, nasopharynx, trachea, and major airways (Casanova et al. 1989b). The results showed that the concentration of DNA-protein cross-links was substantially lower in the nasal respiratory mucosa of the monkey than in that of the rat, but that the distribution of affected tissues was greater, with histopathologic damage and DNA-protein cross-linking occurring as deep as the carina of the monkey at 6 ppm (Casanova et al. 1989b; Monticello et al. 1989). Work is now in progress to determine whether differing airflow patterns in the respiratory tract can partially explain the differing susceptibilities of these species to the toxic effects of formaldehyde.

Conclusions

This article summarizes results showing that formaldehyde forms DNA-protein cross-links in the rat nasal mucosa at very low airborne concentrations, and that the amount of binding increases rapidly with increasing concentrations of formaldehyde. At high formaldehyde concentrations, cross-links may not be fully removed before cell replication occurs, and mutations in the DNA may result. Both deletions and base-pair substitutions have been produced in a specific gene of human cells by incubation with formaldehyde (Crosby et al. 1988; Liber et al. 1989). Further research to elucidate the role of DNA repair in defense against mutagenic damage, the importance of oncogenes and suppressor genes in formaldehyde carcinogenesis, and the significance of endogenous formaldehyde in low-dose risk assessment should provide additional insight into the mechanism of formaldehyde carcinogenesis and should lead to more realistic assessments of human risk due to formaldehyde exposure.

References

Beland FA, Fullerton NF, Heflich RH (1984) Rapid isolation, hydrolysis and chromatography of formaldehyde-modified DNA. J Chromatogr Biomed Applic 308:121–131

Casanova M, Heck Hd'A (1987) Further studies on the metabolic incorporation and covalent binding of inhaled [^{3}H]- and [^{14}C]formaldehyde in Fischer 344 rats: effects of glutathione depletion. Toxicol Appl Pharmacol 89:105–121

Casanova M, Deyo DF, Heck Hd'A (1989a) Covalent binding of inhaled formaldehyde to DNA in the nasal mucosa of Fischer 344 rats: analysis of formaldehyde and DNA by high-performance liquid chromatography and provisional pharmacokinetic interpretation. Fundam Appl Toxicol 12:397–417

Casanova M, Heck Hd'A, Steinhagen WH, Everitt JI, Morgan KT, Popp JA (1989b) Species and concentration dependence of DNA-protein cross-link (DPX) formation by formaldehyde (HCHO) in Rhesus monkeys and F-344 rats. Toxicologist 9:35 (Abst 139)

Casanova-Schmitz M, Heck Hd'A (1983) Effects of formaldehyde exposure on the extractability of DNA from proteins in the rat nasal mucosa. Toxicol Appl Pharmacol 70:121–132

Casanova-Schmitz M, Starr TB, Heck Hd'A (1984) Differentiation between metabolic incorporation and covalent binding in the labeling of macromolecules in the rat nasal mucosa and bone marrow by inhaled [^{14}C]- and [^{3}H]formaldehyde. Toxicol Appl Pharmacol 76:26–44

Craft TR, Bermudez E, Skopek TR (1987) Formaldehyde mutagenesis and formation of DNA-protein crosslinks in human lymphoblasts in vitro. Mutat Res 176:147–155

Crosby RM, Richardson KK, Craft TR, Benforado KB, Liber HL, Skopek TR (1988) Molecular analysis of formaldehyde-induced mutations in human lymphoblasts and *E. coli*. Environ Mol Mutagen 12:155–166

Fennell TR, Deal FH, Swenberg JA (1987) Characterization of cross-links between DNA and lysine produced by formaldehyde. Toxicologist 7:172 (Abst 686)

Funderburk LH, Aldwin L, Jencks WP (1978) Mechanisms of general acid and base catalysis of the reactions of water and alcohols with formaldehyde. J Am Chem Soc 100:5444–5459

Goldmacher VS, Thilly WG (1983) Formaldehyde is mutagenic for cultured human cells. Mutat Res 116:417–422

Grafstrom RC, Fornace AJ Jr, Autrup H, Lechner JF, Harris CC (1983) Formaldehyde damage to DNA and inhibition of DNA repair in human bronchial cells. Science 220:216–218

Heck Hd'A, Casanova M (1987) Isotope effects and their implications for the covalent binding of inhaled [^{3}H]- and [^{14}C]formaldehyde in the rat nasal mucosa. Toxicol Appl Pharmacol 89:122–134

Kerns WD, Pavkov KL, Donofrio DJ, Gralla EJ, Swenberg JA (1983) Carcinogenicity of formaldehyde in rats and mice after long-term inhalation exposure. Cancer Res 43:4382–4392

Liber HL, Benforado K, Crosby RM, Simpson D, Skopek TR (1989) Formaldehyde-induced and spontaneous alterations in human hprt DNA sequence and mRNA expression. Mutat Res 226:31–37

McGhee JD, von Hippel PH (1975a) Formaldehyde as a probe of DNA structure. I. Reaction with exocyclic amino groups of DNA bases. Biochemistry 14:1281–1296

McGhee JD, von Hippel PH (1975b) Formaldehyde as a probe of DNA structure. II. Reaction with endocyclic imino groups of DNA bases. Biochemistry 14:1297–1303

McGhee JD, von Hippel PH (1977) Formaldehyde as a probe of DNA structure. 4. Mechanism of the initial reaction of formaldehyde with DNA. Biochemistry 16:3276–3293

Monticello TM, Morgan KT, Everitt JI, Popp JA (1989) Effects of formaldehyde gas on the respiratory tract of Rhesus monkeys. Pathology and cell proliferation. Am J Pathol 134:515–527

Natarajan AT, Darroudi F, Bussman CJM, van Kesteren-van Leeuwen AC (1983) Evaluation of the mutagenicity of formaldehyde in mammalian cytogenetic assays in vivo and vitro. Mutat Res 122:355–360

Ragan DL, Boreiko CJ (1981) Initiation of C3H/10T½ cell transformation by formaldehyde. Cancer Lett 13:325–331

Ross WE, Shipley N (1980) Relationship between DNA damage and survival in formaldehyde-treated mouse cells. Mutat Res 79:277–283

Rusch GM, Clary JJ, Rinehart WE, Bolte HF (1983) A 26-week inhalation toxicity study with formaldehyde in the monkey, rat, and hamster. Toxicol Appl Pharmacol 68:329–343

Sloan KB, Siver KG (1984) The aminomethylation of adenine, cytosine and guanine. Tetrahedron 40:3997–4001

Starr TB, Buck RD (1984) The importance of delivered dose in estimating low-dose cancer risk from inhalation exposure to formaldehyde. Fundam Appl Toxicol 4:740–753

Swenberg JA, Gross EA, Randall HW, Barrow CS (1983) The effect of formaldehyde exposure on cytotoxicity and cell proliferation. In: Clary JJ, Gibson JE, Waritz RS (eds) Formaldehyde: toxicology, epidemiology, and mechanisms. Dekker, New York, pp 225–236

Volkov VS, Poverennyi AM, Sverdlov ED (1985) Interaction of formaldehyde with nucleic acids and their structural components in the presence of amines. II. Structure of adenine derivatives formed upon the reaction with formaldehyde and primary amine. Mol Biol (Moscow) 19:574–581 (transl from Molekulyarnaya Biologiya 19:693–701)

Williams IH (1987) Theoretical modeling of specific solvation effects upon carbonyl addition. J Am Chem Soc 109:6299–6307

Wilmer JWGM, Woutersen RA, Appelman LM, Leeman WR, Feron VJ (1987) Subacute (4-week) inhalation toxicity study of formaldehyde in male rats: 8-hour intermittent *versus* 8-hour continuous exposures. J Appl Toxicol 7:15–16

Section 2. Arsenic

14. Environmental and Occupational Exposure to Arsenic

M.E. Vahter[1]

Occurrence and Exposure

Environmental and occupational standards for arsenic are often expressed as the concentration of the element as such, implying exposure to one and the same agent. However, there are many different arsenic compounds in the environment, the physical, chemical, and toxicological properties of which may vary considerably.

Arsenic is a major constituent of many minerals, especially sulfide ores. In water, arsenic is found mainly as arsenite or arsenate, depending on the pH and the presence of other reducing or oxidizing substances. Microorganisms can methylate inorganic arsenic to methylarsonic acid (MMA), dimethylarsinic acid (DMA) and trimethylarsine oxide (TMAO), via the corresponding arsines (McBride et al. 1978). Arsenate may be absorbed by algae and converted into arsenic-containing ribofuranosides, which make their way into the food web of marine animals and are transformed into arsenocholine and arsenobetaine (Edmonds and Francesconi 1987). The current major uses of arsenic are in pesticides, cotton desiccants, wood preservatives, certain alloys, semiconductors, feed additives (growth promotion), and in the production of glass and enamel.

Airborne concentrations of arsenic in urban areas may range from a few nanograms to a few tenths of a microgram per cubic meter (WHO 1981). Near point emissions of arsenic, such as smelters, airborne arsenic concentrations may exceed 1 $\mu g/m^3$. The major part of airborne arsenic is in inorganic form such as As(III) and As(V) oxides and is often associated with particles. The burning of fossil fuels, particularly coal, may be an important source of arsenic in the environment (Piver 1983). Depending on the content of arsenic in tobacco, an average smoker may inhale between a few micrograms and 20 μg of arsenic daily (WHO 1981).

Drinking water normally contains a few micrograms of arsenic per liter or less (WHO 1981). Levels exceeding 1 mg/l, recorded in some areas of the world,

---

[1]Institute of Environmental Medicine, Karolinska Institutet, Box 60208, S-104 01 Stockholm, Sweden.

usually originate from an arsenic-rich bedrock or industrial contamination. With the exception of seafood, human food products generally contain less than 0.25 mg arsenic per kilogram (GESAMP 1986). Elevated arsenic levels may be caused by industrial pollution, use of arsenic-containing pesticides, or use of arsenic-containing feed additives. Seafood commonly contains arsenic in the order of milligrams per kilogram, predominantly in the form of organic arsenic compounds such as arsenobetaine and arsenocholine (GESAMP 1986; Edmonds and Francesconi 1987). Individuals, especially of bottom-feeding species, may contain more than 100 mg arsenic per kilogram. Inorganic arsenic normally constitutes only a few percent of the total arsenic in seafood. The total daily intake of arsenic by humans is greatly influenced by the amount of seafood in the diet, but it is usually less than 0.2 mg per day (WHO 1981; GESAMP 1986). In most countries, the average daily exposure to inorganic arsenic is 10–50 μg.

Occupational exposure to arsenic mainly occurs through the inhalation of particles containing arsenic, e.g., among smelter workers, glass workers and workers engaged in the production and use of arsenic-containing pesticides or production of semiconductors. The occupational health standard for arsenic is 0.01–0.05 mg/m^3 in most countries (International Agency for Research on Cancer 1980).

Metabolism

Absorption

When ingested in dissolved form, inorganic arsenic compounds are readily absorbed (Pomroy et al. 1980). A lower degree of absorption has been reported for arsenic selenide (Mappes 1977), arsenic trisulfide, lead arsenate (Marafante and Vahter 1987), and gallium arsenide (Webb et al. 1984; Yamauchi et al. 1986). Following inhalation of soluble arsenic compounds, e.g., arsenic trioxide, most of the deposited arsenic is rapidly absorbed from either the respiratory or the gastrointestinal tract, while particles of low solubility, e.g., arsenic trisulfide, calcium arsenate, lead arsenate, and gallium arsenide, are retained to a great extent in the lungs (Pershagen et al. 1982; Webb et al. 1984; Rosner and Carter 1987; Marafante and Vahter 1987). Wester et al. (1981) reported the presence of very high concentrations of arsenic in the lungs of smelter workers several years after they had stopped working at the smelter. Dissolution of certain arsenic-containing particles, e.g., lead arsenate, by alveolar macrophages may increase the clearance of arsenic from the lungs (Marafante et al. 1987a).

Arsenobetaine in seafood is taken up very efficiently from the gastrointestinal tract (Tam et al. 1982; Vahter et al. 1983; Marafante et al. 1984). In contrast, arsanilic acid and related arsenic compounds, used as feed additives for poultry and swine, seem to be considerably less available following ingestion (Calesnick et al. 1966; Calvert 1975).

Biotransformation

Most mammals are able to methylate inorganic arsenic to MMA and DMA (for review see Vahter and Marafante 1988). The methylation may be considered a detoxication mechanism for inorganic arsenic, since the methylated metabolites have a lower affinity for tissue constituents and lower toxicity than inorganic arsenic, especially the trivalent form (Buchet et al. 1981a; Vahter and Marafante 1983). The methylation takes place mainly in the liver and is enzymatic (Marafante and Vahter 1984; Buchet and Lauwerys 1985). Pentavalent arsenic is reduced in the blood to trivalent, which is taken up and methylated by the hepatocytes (Marafante et al. 1985; Vahter and Marafante 1985).

In subjects exposed to "normal" environmental levels of inorganic arsenic, the urinary excretion consists of 10%–20% inorganic arsenic, 10 %–15% MMA, and 60%–80% DMA (Crecelius 1977; Tam et al. 1979; Buchet et al. 1981a; Vahter 1986). In human volunteers, ingesting arsenite in daily doses of 125, 250, 500, or 1000 μg in the form of arsenite for 5 days, the urinary excretion of DMA leveled off at the highest dose level (Buchet et al. 1981b). This may be due to a saturation of the methylation capacity or to an inhibition of the methyl transfer reactions (Buchet and Lauwerys 1985). In cases of acute arsenic intoxication, it may take several days before DMA becomes the predominant urinary metabolite of arsenic (Mahieu et al. 1981; Lovell and Farmer 1985). However, the distribution of metabolites of arsenic in the urine of smelter workers exposed to arsenic is similar to that in the general population (Smith et al. 1977; Vahter 1986). Studies on experimental animals indicate that diets deficient in methionine, choline, or protein give rise to decreased methylation and increased tissue retention of arsenic (Vahter and Marafante 1987). A similar situation can be expected for subjects with genetically based low methylation capacity, the so-called poor methylators. In the marmoset monkey, which is the only species shown not to methylate inorganic arsenic, arsenic is retained to a much higher degree than in other species (Vahter et al. 1982).

Following exposure to DMA, a small fraction of the dose is further methylated and excreted in the urine as TMAO (Marafante et al. 1987b). However, the major part is excreted in unchanged form in the urine. Following exposure to MMA, about 10% of the dose is excreted in the urine as DMA (Buchet et al. 1981a). Arsenobetaine is excreted in the urine of mammals without being biotransformed (Vahter et al. 1983). Arsenocholine is, to a great extent, oxidized to arsenobetaine in mammals, probably by a mechanism similar to that of choline (Marafante et al. 1984). The metabolism of arsenosugars, present in certain algae, is not yet known.

Tissue Distribution

In human subjects exposed to normal environmental levels of arsenic, hair and nails show the highest tissue concentrations, while fairly high concentrations are found in skin and lungs (WHO 1981). Experimental studies on mammals exposed to inorganic arsenic show specific retention in skin, hair, squamous epithelium of the upper gastrointestinal tract, intestinal walls, epididymis, thyroid, skeleton, and lens (Lindgren et al. 1982; Vahter et al. 1982). All the tissues mentioned, except the skeleton, contain higher concentrations following administration of arsenite than they do after administration of arsenate (Lindgren et al. 1982). Animal experiments have shown that 2,3-dimercapto-1-propanol, which has traditionally been used in the treatment of arsenic poisoning, but not dimercaptosuccinic acid or 2,3-dimercapto-1-propanesulfonic acid, increases the arsenic content in the brain (Aposhian et al. 1984). There are very few data on the accumulation of arsenic after long-term exposure. Some studies indicate that tolerance to arsenic may be developed (Bencko et al. 1973).

Data on the tissue retention of organic arsenic compounds ingested with seafoods are limited. In mice and rabbits the tissues with longest retention of ^{73}As-arsenobetaine were cartilage, epididymis, testes, semen ducts, thymus, and muscles (Vahter et al. 1983). Siewicki (1981) administered a flounder diet containing 29 mg arsenic per kilogram to rats for 42 days. Arsenic levels in the liver and spleen of treated rats were seven times higher for liver and twice as high for spleen compared to levels in control rats.

Excretion and Biological Half-Time

The rate of excretion of arsenic in urine varies with the form of arsenic absorbed. In human subjects who ingested ^{74}As-arsenate (0.01 μg As per man), 38% of the dose was excreted in the urine within 48 h and 58% within 5 days (Tam et al. 1979). The whole-body clearance of arsenic followed a three-component exponential function with 66% being eliminated with a half-time of 2.1 days, 30% with a half-time of 9.5 days, and 4% with a half-time of 38.4 days. In subjects ingesting 500 μg arsenic in the form of arsenite via drinking water, about 33% of the dose was excreted in the urine within 48 h and 45% within 4 days (Buchet et al. 1981a). Following ingestion of the same amounts of arsenic in the form of MMA or DMA, about 75% of the dose was excreted in the urine within 4 days. A similar rate of excretion has been observed following ingestion of arsenic compounds in seafood (Freeman et al. 1979; Tam et al. 1982).

Assessment of Exposure

Owing to the fast clearance of arsenic from the blood, the concentration of arsenic in blood cannot be used to evaluate the degree of exposure to inorganic arsenic (Vahter 1988). Furthermore, the lack of simple methods for the speciation of arsenic compounds in blood makes it difficult to evaluate the exposure form of arsenic. Ingestion of seafood arsenic may greatly influence the blood arsenic levels. Since inorganic arsenic, but not organic arsenic compounds from seafood, accumulates in hair, the concentration of arsenic in hair may be used as an indicator of exposure to inorganic arsenic. The concentration of arsenic in hair in "nonexposed" subjects is normally less than 1 mg/kg, while in subjects exposed to arsenic occupationally or via medicines or drinking water it may be as high as 50 mg/kg (WHO 1981). The time and length of exposure may be determined by analysis of the variation in concentrations along the length of the hair (Smith 1964). However, the usefulness of arsenic levels in hair as an indicator of the degree of exposure, especially on the individual level, is limited owing to the possible influence from external sources of arsenic.

On a group basis, total urinary arsenic may give a rough estimate of exposure to inorganic arsenic, provided that an extensive exposure to seafood arsenic can be excluded. For example, good correlations between arsenic concentrations in drinking water and total urinary arsenic concentrations have been reported (Harrington et al. 1978; Valentine et al. 1979). The urinary concentration of metabolites of inorganic arsenic is probably the best indicator of exposure to inorganic arsenic. Exposure to MMA or DMA will give an erroneous estimate of the exposure to inorganic arsenic, but seems to be insignificant in most countries. Occupational exposure to MMA or DMA may, of course, be estimated by the same method, provided that the exposure to inorganic arsenic is low. On a group basis, there was a fairly good correlation between the concentrations of metabolites of inorganic arsenic in urine and the concentrations of airborne arsenic among smelter workers exposed mainly to arsenic trioxide (Vahter et al. 1986). The data indicate that the concentration of arsenic metabolites in urine provides a better basis for the evaluation of exposure to airborne arsenic, than does the air concentration of arsenic, even when measured by personal sampler. For example, the concentration of urinary metabolites of inorganic arsenic in subjects inhaling air containing about 10 $\mu g/m^3$ arsenic varied from about 40 to 200 μg per gram of creatinine. In several cases much more arsenic was excreted in the urine than could have possibly been inhaled, which indicated an extensive oral intake via contaminated hands, cigarettes, or snuff. Such exposure does not result in an extra risk for lung cancer. In practice, the origin of the urinary arsenic will not be known for each individual case, so one cannot exclude the possibility that high urinary values are caused by inhalation of arsenic. Therefore, it seems prudent to use the urinary concentration for health-based biological exposure limits on an individual basis.

The relationship found between airborne arsenic and urinary excretion of inorganic arsenic metabolites is probably not valid for arsenic compounds such as arsenate, calcium arsenate, gallium arsenide, and arsenic trisulfide, which have a lower rate of absorption than arsenic trioxide from the lung. Therefore, there is a need for data on the relationships between airborne arsenic and urinary excretion of metabolites following exposure to arsenic compounds, other than arsenic trioxide.

References

Aposhian HV, Carter DE, Hoover TD, Hsu C, Maiorino R, Stine E (1984) DMSA, DMPS and DMPA–as arsenic antidotes. Fundam Appl Toxicol 4:58–70

Bencko V, Dvorak V, Symon K (1973) Organ retention of parenterally administered arsenic (labelled with ^{74}As) in mice preliminary exposed to the element in drinking water; a study in arsenic tolerance. J Hyg Epidemiol Microbiol Immunol 17:165–168

Buchet JP, Lauwerys R (1985) Study of inorganic arsenic methylation by rat liver in vitro: relevance for the interpretation of observations in man. Arch Toxicol 57:125–129

Buchet JP, Lauwerys R, Roels H (1981a) Comparison of the urinary excretion of arsenic metabolites after a single dose of sodium arsenite, monomethylarsonate or dimethylarsinate in man. Int Arch Occup Environ Health 48:71–79

Buchet JP, Lauwerys R, Roels H (1981b) Urinary excretion of inorganic arsenic and its metabolites after repeated ingestion of sodium metaarsenite by volunteers. Int Arch Occup Environ Health 48:111–118

Calesnick B, Wase A, Overby LR (1966) Availability during human consumption of the arsenic in tissues of chicks fed arsanilic-^{74}As acid. Toxicol Appl Pharmacol 9:29–30

Calvert CC (1975) Arsenicals in animal feeds and wastes. In: Woolson EA (ed) Arsenical pesticides. American Chemical Society, Washington (ACS symposium series, no 7)

Crecelius EA (1977) Changes in the chemical speciation of arsenic following ingestion by man. Environ Health Perspect 19:147–150

Edmonds JS, Francesconi KA (1987) Transformations of arsenic in the marine environment. Experientia 43. Birkhäuser, Basel

Freeman HC, Uther JF, Fleming RB, Odense PH, Ackman RG, Landry G, Musial C (1979) Clearance of arsenic ingested by man from arsenic contaminated fish. Bull Environ Contam Toxicol 22:224–229

GESAMP (1986) Review of potentially harmful substances—arsenic, mercury and selenium. IMO/FAO/UNESCO/WMO/WHO/IAEA/UN/UNEP joint group of experts on the scientific aspects of marine pollution—GESAMP. World Health Organization, Geneva

Harrington MJ, Middaugh P, Morse DL, Houseworth J (1978) A survey of a population exposed to high concentrations of arsenic in well water in Fairbanks, Alaska. Am J Epidemiol 108:377–385

International Agency for Research on Cancer (1980) Arsenic and arsenic compounds. IARC Monogr Eval Carcinog Risk Chem Hum 23:39–141

Lindgren A, Vahter M, Dencker L (1982) Autoradiographic studies on the distribution of arsenic in mice and hamster administered ^{74}As-arsenite or -arsenate. Acta Pharmacol Toxicol 51:253–365

Lovell MA, Farmer JG (1985) Arsenic speciation in urine from humans intoxicated by inorganic arsenic compounds. Hum Toxicol 4:203–214

Mahieu P, Buchet JP, Roels HA, Lauwerys R (1981) The metabolism of arsenic in humans acutely intoxicated by As_2O_3. Its significance for the duration of BAL therapy. Clin Toxicol 18(9):1067–1075

Mappes R (1977) Experiments on the excretion of arsenic in urine (in German). Int Arch Occup Environ Health 40:267–272

Marafante E, Vahter M (1984) The effect of methyltransferase inhibition on the metabolism of (^{74}As)arsenite in mice and rabbits. Chem Biol Interact 50:49–57

Marafante E, Vahter M (1987) Solubility, retention and metabolism of intratracheally and orally administered inorganic arsenic compounds in the hamster. Environ Res 42:72–82

Marafante E, Vahter M, Dencker L (1984) Metabolism of arsenocholine in mice, rats and rabbits. Sci Total Environ 34:223–340

Marafante E, Vahter M, Envall J (1985) The role of the methylation in the detoxication of arsenate in the rabbit. Chem Biol Interact 56:225–238

Marafante E, Lundborg M, Vahter M, Camner P (1987a) Dissolution of two arsenic compounds by rabbit alveolar macrophages in vitro. Fundam Appl Toxicol 8:382–388
Marafante E, Vahter M, Norin H, Envall J, Sandström M, Christakopoulos A (1987b) Biotransformation of dimethylarsinic acid in mouse, hamster and man. J Appl Toxicol 7(2):111–117
McBride BC, Merilees H, Cullen WR, Pickett W (1978) Anaerobic and aerobic alkylation of arsenic. In: Brinckman FE, Bellama JM (eds) Organometals and organometalloids. American Chemistry Society, Washington, pp 94–115 (ACS symposium series, no 82)
Pershagen G, Lind B, Björklund N-E (1982) Lung retention and toxicity of some inorganic arsenic compounds. Environ Res 29:425–434
Piver WT (1983) Mobilization of arsenic by natural and industrial processes. Top Environ Health 6:1
Pomroy C, Charbonneau SM, McCullough RS, Tam GKH (1980) Human retention studies with ^{74}As. Toxicol Appl Pharmacol 53(3):550–556
Rosner MH, Carter DE (1987) Metabolism and excretion of gallium arsenide and arsenic oxides by hamsters following intratracheal instillation. Fundam Appl Toxicol 9:730–737
Siewicki TC (1981) Tissue retention of arsenic in rats fed with flounder or cacodylic acid. J Nutr 111:602–609
Smith H (1964) The interpretation of the arsenic content of human hair. J Forensic Sci Soc 4:192–199
Smith TJ, Crecelius EA, Reading JC (1977) Airborne arsenic exposure and excretion of methylated arsenic compounds. Environ Health Perspect 19:89–93
Tam GKH, Charbonneau SM, Bryce F, Pomroy C, Sandi E (1979) Metabolism of inorganic arsenic (^{74}As) in humans following oral ingestion. Toxicol Appl Pharmacol 50:319–322
Tam GKH, Charbonneau SM, Bryce F, Sandi E (1982) Excretion of a single oral dose of fish-arsenic in man. Bull Environ Contam Toxicol 28:669–673
Vahter M (1986) Environmental and occupational exposure to inorganic arsenic. Acta Pharmacol Toxicol 59(Suppl 7):31–34
Vahter M (1988) Arsenic—occurrence in the environment. In: Clarkson TW, Friberg L, Nordberg GF, Sager PR (eds) Biological monitoring of toxic metals. Plenum, New York
Vahter M, Marafante E (1983) Intracellular interaction and metabolic fate of arsenite and arsenate in mice and rabbits. Chem Biol Interact 47:29–44
Vahter M, Marafante E (1985) Reduction and binding of arsenate in marmoset monkeys. Arch Toxicol 57:119–124
Vahter M, Marafante E (1987) Effects of low dietary intake of methionine, choline or proteins on the biotransformation of arsenite in the rabbit. Toxicol Lett 37:41–46
Vahter M, Marafante E (1988) In vivo methylation and detoxication of arsenic. In: Craig PJ, Glockling F (eds) The biological alkylation of heavy elements. Royal Society of Chemistry, London.
Vahter M, Marafante E, Lindgren A, Dencker L (1982) Tissue distribution and subcellular binding of arsenic in marmoset monkeys after injection of ^{74}As-arsenite. Arch Toxicol 51:65–77
Vahter M, Marafante E, Dencker L (1983) Metabolism of arsenobetaine in mice, rats and rabbits. Sci Total Environ 30:197–211
Vahter M, Friberg L, Rahnster B, Nygren Å, Nolinder P (1986) Airborne arsenic and urinary excretion of metabolites of inorganic arsenic among smelter workers. Int Arch Occup Environ Health 57:79–91

Valentine JL, Kang HK, Spivey G (1979) Arsenic levels in human blood, urine and hair in response to exposure via drinking water. Environ Res 20:24–32
Webb DR, Sipes IG, Carter DE (1984) In vitro solubility and in vivo toxicity of gallium arsenide. Toxicol Appl Pharmacol 76:96–104
Wester PO, Brune D, Nordberg G (1981) Arsenic and selenium in lung, liver and kidney tissues from lead smelter workers. Br J Ind Med 38:179
WHO (1981) Arsenic – environmental health criteria, vol 18. World Health Organization, Geneva
Yamauchi H, Takahashi K, Yamamura Y (1986) Metabolism and excretion of orally and intraperitoneally administered gallium arsenide in the hamster. Toxicology 40:237–246

15. Is Inhaled Arsenic Carcinogenic for Sites Other Than the Lung?

H. Gibb[1] and C. Chen[1]

Introduction

In March 1984 the United States Environmental Protection Agency (EPA) released its Health Assessment Document on Arsenic (US EPA 1984a). In that report numerous epidemiologic studies of the carcinogenic risk of inhaled and ingested arsenic were reviewed. The lung, not surprisingly, has consistently proven to be a target organ for the carcinogenic effects from inhaled arsenic. While some of the epidemiologic studies of persons exposed to arsenic via inhalation noted excess risks of cancer for organs or systems other than the lung, these results have been given little attention.

Recently, Chen and Chen (1988) presented an integrated quantitative risk assessment of inorganic arsenic and demonstrated that the relative magnitudes of the potencies of internal cancers (lung, liver, kidney, and bladder) induced by arsenic in the drinking water are consistent with the relative magnitudes of the tissue concentration of arsenic in each of these organs. They further demonstrated that the risk of lung cancer induced by arsenic in the drinking water is consistent with the risk observed in the occupational setting. On these bases they hypothesized that arsenic smelter workers may have an elevated risk of cancer for sites other than the lung.

The purpose of this paper is to discuss the evidence for cancer risk at sites other than the lung resulting from arsenic inhalation and to present a dose-response assessment for such data. In discussing the evidence we will use the data obtained from studies of persons ingesting arsenic as well as from studies of persons inhaling arsenic in arriving at our conclusions. We will also discuss the ramifications that the arsenic data has for the carcinogen risk assessment of other substances.

---

[1]Office of Health and Environmental Assessment, US Environmental Protection Agency, 401 M Street, S.W. (RD-689), Washington, DC 20460, USA.

Evidence of the Carcinogenic Risk of Inhaled Arsenic at Sites Other Than the Lung

Numerous studies have reported an excess risk of lung cancer among smelter worker populations (Lee-Feldstein 1983; Enterline and Marsh 1982; Rencher et al. 1977; Tokudome and Kuratsune 1976; Jarup et al. 1988) and workers engaged in arsenical pesticide manufacture (Mabuchi et al. 1979; Hill and Faning 1948; Ott et al. 1974). While little attention has been paid to the association of inhaled arsenic with an excess risk of cancer of organs other than the lung, we believe the data deserve a closer examination.

Lee-Feldstein (1983) found a significant excess of digestive tract as well as respiratory system cancer mortality among 8045 arsenic-exposed smelter workers at the Anaconda smelter in Anaconda, Montana (SMR = 125, 167 observed deaths, 133.58 expected deaths, $p < 0.05$). She also found that men employed prior to 1925 (when exposure to arsenic was most intense) and who worked 25 years or more experienced twice the expected number of bladder cancer deaths (SMR = 200, 13 observed, 6.5 expected, $p < 0.05$).

Wall (1980) found a significant excess of stomach cancer mortality and mortality from tumors of the brain and nervous system ($p < 0.05$) as well as an excess of respiratory system cancer mortality among a group of Swedish smelter workers. The author also found a dose response for stomach cancer mortality by duration of exposure.

Tokudome and Kuratsune (1976) found an excess of both liver cancer mortality (SMR = 337, 11 observed deaths, 3.26 expected deaths, $p < 0.05$) and colon cancer mortality (SMR = 508, three observed deaths, 0.59 expected deaths, $p < 0.05$) as well as lung cancer mortality among a group of arsenic-exposed copper smelter workers in Saganoseki-machi, Japan.

Enterline and Marsh (1982) found a slight excess of digestive tract cancer mortality among a cohort of 2802 arsenic-exposed workers at the ASARCO smelter in Tacoma, Washington (SMR = 109, 65 observed deaths, 59.7 expected deaths, $p > 0.05$). The SMRs were somewhat higher for cancers of the stomach, colon, and rectum (122, 120, and 122, respectively) although none was statistically significant. The SMR for colon cancer for those first employed before 1930 was statistically significant, however (SMR = 208.5, 12 observed deaths, 5.75 expected deaths, $p < 0.05$). This excess in the period before 1930 is probably related to the fact that in the 1940s workers at the smelter began using respirators which were found to be 99% effective. Enterline and Marsh also reported an excess of kidney cancer mortality (SMR = 133), but this excess was not statistically significant ($p > 0.05$). Here again, however, the SMR for workers first employed before 1930, and thus before respirators were employed, was higher (SMR = 201) although still not statistically significant ($p > 0.05$).

A proportional mortality study of sheep dip manufacturing workers (Hill and Faning 1948), in which the study population was exposed to arsenic primarily via inhalation, reported an excess of skin cancer mortality. Limitations of proportional mortality studies are well known (Rothman 1986); nevertheless, the results are consistent with the numerous studies of persons ingesting arsenic and developing skin cancer (US EPA 1988). Furthermore, the results of the proportional mortality study are supported by the results of the subsequent clinical and environmental investigation of the sheep dip factory workers and the factory, respectively, by Perry et al. (1948). Perry et al. found that the workers exposed to arsenic dust had a higher content of arsenic in their hair, excreted more arsenic in their urine, and had a higher prevalence of hyperpigmentation and hyperkeratinization than controls. Hyperpigmentation and hyperkeratinization are hallmarks of arsenic exposure via ingestion and such arsenic-induced hyperkeratinization may be precancerous (US EPA 1988).

Ott et al. (1974) reported that workers exposed to arsenic in a pesticide-manufacturing plant had a significant excess risk of mortality from neoplasms of the lymphatic and hematopoietic tissues other than leukemia (SMR = 385, five observed deaths, 1.3 expected deaths, $p < 0.01$). Axelson et al. (1978) reported that patients with leukemia and myeloma were significantly more likely to be exposed to arsenic than controls ($p < 0.02$).

A conclusion of an excess cancer risk from arsenic inhalation for sites other than the lung is consistent with the results of studies of persons exposed to arsenic by ingestion. Chen et al. (1985) found that SMRs for cancer of the lung, bladder, kidney, skin, liver, and colon were significantly increased in an area of Taiwan where there was endemic arsenic contamination of the water supply. In a follow-up case-control study in the arsenic endemic area, Chen et al. (1986) found an association of arsenic ingestion with bladder, lung, and liver cancer. In both of these studies and a subsequent SMR study (Chen et al. 1988), the authors found evidence of a cancer dose response to arsenic.

In addition to the studies by Chen et al., there are numerous case reports of persons ingesting arsenic and developing cancer of the internal organs. Fierz (1965), Regelson et al. (1968), Lander et al. (1975), Popper et al. (1978), Falk et al. (1981), and Roat et al. (1982) reported cases of persons who had ingested arsenical medicinals and developed liver cancer. Sommers and McManus (1953) and Nagy et al. (1980) reported cases of bladder cancer that had been chronically exposed to arsenic by ingestion. Nurse (1978) reported on a patient who developed kidney cancer after being treated with an arsenical medicinal.

Studies and case reports supporting an association of arsenic exposure with an excess risk of cancer at sites other than the lung are summarized in Table 15.1.

Dose-Response Analysis

As can be seen from Table 15.1, numerous studies support an association of inhaled arsenic with an excess risk of cancer of systems other than the respiratory system. The evidence appears to be the strongest for an association of arsenic exposure with cancer of the gastrointestinal and urinary systems.

Table 15.1. Epidemiologic studies and case reports supporting an association of arsenic exposure and excess risk of cancer at sites other than the lung by route of exposure and by body organ or system.

Neoplasms of organ and/or system:	Route of exposure	
	Inhalation	Ingestion
Digestive organs and peritoneum (ICD 150–159)	Lee-Feldstein (1983)	
Stomach (ICD 151)	Wall (1980)	
Colon (ICD 153)	Tokudome and Kuratsune (1976) Enterline and Marsh (1982)	Chen et al. (1985, 1988)
Liver (ICD 155)	Tokudome and Kuratsune (1976)	Chen et al. (1985, 1986, 1988) Roat et al. (1982) Falk et al. (1981) Popper et al. (1978) Lander et al. (1975) Regelson et al. (1968) Fierz (1965)
Genitourinary system (ICD 179–189)		
Bladder (ICD 188)	Lee-Feldstein (1983)	Chen et al. (1985, 1986, 1988) Nagy et al. (1980) Sommers and McManus (1953)
Kidney (ICD 189)	Enterline and Marsh (1982)	Chen et al. (1985, 1988) Nurse (1978)
Skin[a] (ICD 173)	Hill and Faning (1948)	Chen et al. (1985, 1988) Cebrian et al. (1983) Tseng (1977) Tseng et al. (1968) Zaldivar and Guiller (1977) Fierz (1965) Bergoglio (1964) Sommers and McManus (1953)
Lymphatic and hematopoietic tissue (ICD 200–208)		
Lymphatic and hematopoietic and tissue other than leukemia (ICD 200–203)	Ott et al. (1974)	
Myeloma and leukemia (ICD 204–209)	Axelson et al. (1978)	
Brain and nervous system (ICD 191–192)	Wall (1980)	

[a] The studies demonstrating an association of arsenic ingestion and skin cancer are too numerous to mention here. Refer to US EPA (1988) for a complete description of studies and reports of skin cancer induced by arsenic.

The largest cohort study of persons exposed to arsenic by inhalation is that by Lee-Feldstein (1983). Dr. Lee-Feldstein has provided us with an updated data tape of her study which we have used for a dose-response assessment of inhaled arsenic and gastrointestinal cancer, the results of which we present here. The method used is similar to that used by the EPA to estimate a unit risk for lung cancer from inhaled hexavalent chromium (US EPA 1984b). The only significant difference is that for this assessment we counted person-years of observation beginning at the end rather than at the beginning of employment. For dose-response assessment this approach better characterizes individual exposure, as noted by Enterline (1976).

From the updated Lee-Feldstein data, we estimated the lifetime gastrointestinal and respiratory cancer risk due to 1 $\mu g/m^3$ of arsenic in air to be 4.3×10^{-4} and 1.7×10^{-3}, respectively. We note that the unit-risk estimate for respiratory cancer risk presented here is almost identical to that estimated by the US EPA (1984a) using the Lee-Feldstein (1983) data and counting person-years since the beginning of employment.

We also note that the unit-risk estimate for gastrointestinal cancer presented here is not insignificant when compared to the unit cancer risk estimates that the EPA has calculated for other chemicals. The inhalation unit-risk estimate for arsenic for gastrointestinal cancer alone is higher than the inhalation unit-risk estimates calculated by the EPA for several chemicals including acrylonitrile, benzene, trichloroethylene, perchloroethylene, vinyl chloride, and ethylene oxide.

Discussion

It is well established that exposure to arsenic is causally associated with excess risks of cancer of the respiratory system and skin (US EPA 1984a, 1988; Health and Safety Executive 1986; International Programme on Chemical Safety 1981). We believe that sufficient information exists to conclude that exposure to arsenic is also associated with excess risks of cancer of the gastrointestinal system, including the stomach, colon, and liver, and the urinary system. Excess risks of cancer of the liver and of the urinary system are quite plausible given that arsenic is distributed widely throughout the body by the blood and excreted primarily in the urine. Excess risk of cancer of the digestive tract from inhaled arsenic may result from the swallowing of arsenic dust and/or transport of arsenic by the blood to the digestive tract from the respiratory system.

In evaluating carcinogenic risk, one must consider cancer response by all routes of exposure, as well as information on distribution within the body. Vahter (1983) reported that arsenic, absorbed from the gastrointestinal tract or the lungs of experimental animals, is transported by the blood and distributed widely throughout the body. Chen and Chen (1988) reported that arsenic appears to have a similar distribution to the lung, liver, and kidney in experimental animals whether the exposure be by inhalation or ingestion.

Arsenic is unique in that ample data exist to demonstrate that it is carcinogenic in humans via both inhalation and ingestion routes of exposure. That this is the case is not particularly surprising because, as previously indicated, arsenic is known to be absorbed by both inhalation and ingestion with similar distribution in the body. The observation that arsenic is carcinogenic by both routes of exposure focuses on an important issue for the risk assessment of other metals, however. Other metallic compounds (e.g., nickel, cadmium, and hexavalent chromium), while known to be carcinogenic by inhalation, have been studied very little or not at all by the ingestion route of exposure. Nickel, cadmium, and hexavalent chromium, like arsenic, are absorbed by both ingestion and inhalation. Recently, Zhang and Li (1987) reported that stomach cancer mortality was increased in an area of China where there was known to have been hexavalent chromium pollution of the water supply for the previous 25 years. Thus it seems prudent, from a public health perspective, that these metallic compounds should be considered as having the potential to be carcinogenic by both routes of exposure while noting that inherent potency, route absorption differences, or the lack of adequate bioassay testing may explain the failure to observe carcinogenic activity.

Although the arsenic unit cancer risk for gastrointestinal cancer, estimated from the Lee-Feldstein study and presented in this paper, is about four times less than that calculated for the lung cancer response, it is of a magnitude that should be of concern. Dose-response estimates for urinary tract and skin cancer from arsenic inhalation should also be attempted if the data are available.

Conclusions

From the discussion presented here, we draw the following conclusions:

1. Inhalation exposure to arsenic increases the risk of gastrointestinal and urinary tract cancer, as well as lung and skin cancer.
2. As an assumption that favors public health protection, carcinogenic risk from a substance should not be considered route-specific until adequate testing can validate a route-specific response.
3. The unit-risk estimate for gastrointestinal cancer from inhaled arsenic presented here is of concern and should be further investigated.

Acknowledgments. We wish to thank Mr. Nasser Noori for his programming assistance and Mrs. Michele Ranere, Ms. Marlene Morris, and Ms. Zena Brooks for their assistance in the preparation of the manuscript.

References

Axelson O, Dahlgren E, Jansson CD, Rehnlund SO (1978) Arsenic exposure and mortality: case-referent study from a Swedish copper smelter. Br J Ind Med 35:8–15.

Bergoglio RM (1964) Mortality from cancer in regions of arsenical waters of the province of Cordoba, Argentine Republic. Pren Med Argent 51:994–998.

Cebrian ME, Albores A, Aquilar M, Blakley E (1983) Chronic arsenic poisoning in the north of Mexico. Human Toxicol 2:121–133

Chen CJ, Chuang YC, Lin TM, Wu HY (1985) Malignant neoplasms among residents of a Blackfoot disease-endemic area in Taiwan: high-arsenic artesian well water and cancers. Cancer Res 45:5895–5899

Chen CJ, Chuang YC, You SL, Lin TM, Wu HY (1986) A retrospective study on malignant neoplasms of bladder, lung, and liver in Blackfoot disease endemic area in Taiwan. Br J Cancer 53:399–405

Chen CJ, Kuo TL, Wu MM (1988) Arsenic and cancers. Lancet 1:414–415

Chen CW, Chen CJ (1988) Integrated quantitative cancer risk assessment of inorganic arsenic. Proceedings of a Symposium on Health Risk Assessment of Environmental, Occupational and Life Style Hazards. December 20–22, 1988, Academia Sinica, Taipei, Taiwan

Enterline PE (1976) Pitfalls in epidemiological research. J Occup Med 18(3):150–156

Enterline PE, Marsh GM (1982) Mortality among workers exposed to arsenic and other substances in a copper smelter. Am J Epidemiol 116:895–910

Falk H, Caldwell GG, Ishak KG, Thomas LB, Popper H (1981) Arsenic related hepatic angiosarcoma. Am J Ind Med 2:43–50

Fierz U (1965) Catamnestic investigations of the side effects of therapy of skin diseases with inorganic arsenic. Dermatologica 131:41–58

Health and Safety Executive (1986) Inorganic arsenic compounds. Her Majesty's Stationery Office, London

Hill AB, Faning EL (1948) Studies on the incidence of cancer in a factory handling inorganic compounds of arsenic. I. Mortality experience in the factory. Br J Ind Med 5:1–6

International Programme on Chemical Safety (IPCS) (1981) Arsenic. Environmental Health Criteria 18. World Health Organization, Geneva

Jarup L, Pershagen G, Wall S (1989) Cumulative arsenic exposure and lung cancer in smelter workers: a dose-response study. Am J Ind Med 15:31–41

Lander JJ, Stanley RJ, Sumner HW, Dee C, Boswell DC, Arch RD (1975) Angiosarcoma of the liver associated with Fowler's solution (potassium arsenite). Gastroenterology 68:1582–1586

Lee-Feldstein A (1983) Arsenic and respiratory cancer in man: follow-up of an occupational study. In: Lederer W, Fensterheim R (eds) Arsenic: industrial, biomedical, and environmental perspectives. Van Nostrand Reinhold, New York, pp 245–254

Mabuchi K, Lilienfeld A, Snell L (1979) Lung cancer among pesticide workers exposed to inorganic arsenicals. Arch Environ Health 312–319

Nagy G, Nemeth A, Bodor F, Ficsor E (1980) Cases of bladder cancer caused by chronic arsenic poisoning. Orv Hetil 121:1009–1011

Nurse DS (1978) Hazards of inorganic arsenic. Med J Aust 1:102

Ott MG, Holder BB, Gordon HI (1974) Respiratory cancer and occupational exposure to arsenicals. Arch Environ Health 29:250–255

Perry K, Bowler RG, Buckell HM, Druett HA, Shilling RSF (1948) Studies in the incidence of cancer in a factory handling inorganic compounds of arsenic. II. Clinical and environmental investigations. Br J Ind Med 5:6–15

Popper H, Thomas LB, Telles NC, Falk H, Selikoff IJ (1978) Development of hepatic angiosarcoma in man induced by vinyl chloride, thorotrast, and arsenic. Am J Pathol 92:349–376

Regelson W, Kim U, Ospina J, Holland JF (1968) Hemangioendothelial sarcoma of liver from chronic arsenic intoxication by Fowler's solution. Cancer 21:514–522

Rencher AC, Carter MW, McKee DW (1977) A retrospective epidemiological study of mortality at a large western copper smelter. J Occup Med 19:754–758

Roat JW, Wald A, Mendelow H, Pataki KI (1982) Hepatic angiosarcoma associated with short-term arsenic ingestion. Am J Med 73:933–936

Rothman KJ (1986) Modern epidemiology. Little Brown, Boston

Sommers SC, McManus RG (1953) Multiple arsenical cancers of the skin and internal organs. Cancer 6:347–359

Tokudome S, Kuratsune M (1976) A cohort study on mortality from cancer and the other causes among workers at a metal refinery. Int J Cancer 17:310–317

Tseng WP (1977) Effects and dose-response relationships of skin cancer and Blackfoot disease with arsenic. Environ Health Perspect 19:109–119

Tseng WP, Chu HM, How SW, Fong JM, Lin CS, Yen S (1968) Prevalence of skin cancer in an endemic area of chronic arsenicism in Taiwan. J Natl Cancer Inst 40(3):453–463

US Environmental Protection Agency (EPA) (1984a) Health assessment document for inorganic arsenic. Final report. Office of Health and Environmental Assessment. EPA-600/8-83-021F. NTIS PB84-190891

US Environmental Protection Agency (EPA) (1984b) Health assessment document for chromium. Final report. Environmental Criteria and Assessment Office. EPA-600/8-83-014F. PB85-115905

US Environmental Protection Agency (EPA) (1988) Special report on ingested inorganic arsenic: skin cancer; nutritional essentiality. Risk Assessment Forum. EPA/625/3-87/013. PB88-174503

Vahter M (1983) Metabolism of arsenic. In: Fowler BA (ed) Biological and environmental effects of arsenic. Elsevier, Amsterdam

Wall S (1980) Survival and mortality pattern among Swedish smelter workers. Int J Epidemiol 9(1):73–87

Zaldivar R, Guiller A (1977) Environmental and clinical investigation on epidemic chronic arsenic poisoning in infants and children. Zentralbl Bakteriol Parasitenkd Infektionskr Hyg Abt 1: Orig Reihe B 165:226–243

Zhang JD, Li XL (1987) Chromium pollution of soil and water in Jinzhow Chung Hua Yu Fang I Hsuch Tsa Chih 21(5):262–264

Section 3. Butadiene

16. Toxicology and Carcinogenicity of 1,3-Butadiene

R.L. Melnick,[1] J.E. Huff,[1] and R.A. Miller[2]

Introduction

1,3-Butadiene (CAS No. 106-99-0) is a colorless, flammable, noncorrosive gas (boiling point: $-4.4°C$; vapour pressure at 20°C: 1900 mmHg) produced largely as a coproduct of ethylene production (Kirshenbaum 1978). Butadiene is a reactive material and can dimerize to 4-vinylcyclohexene. Because butadiene can form explosive peroxides in air, it is shipped with an inhibitor of peroxide formation, such as *t*-butylcatechol.

The major uses of butadiene are in the manufacture of synthetic rubber (styrene-butadiene rubber or polybutadiene rubber) and of thermoplastic resins (acrylonitrile-butadiene-styrene), for the manufacture of rubber tires, footwear, sponges, hoses and piping, luggage, packaging, and a variety of other molded products (Miller 1978). Butadiene has also been identified in automobile exhaust and cigarette smoke, and in the incineration products of fossil fuels. The annual production volume of butadiene is approximately 12 billion pounds worldwide, and 3 billion pounds in the United States (Morrow 1989).

Butadiene has long been considered to have a low and noncumulative toxicity in animals and humans. The median lethal concentration (LC_{50}) in mice for a 2-h exposure was 270000 mg/m^3 ($\sim$ 122000 ppm) and in rats for a 4-h exposure was 285000 mg/m^3 ($\sim$ 129000 ppm) (Shugaev 1969). Carpenter et al. (1944) exposed groups of 24 rats, 12 guinea pigs, four rabbits, and one dog to 0, 600, 2300, or 6700 ppm butadiene for 7.5 h a day, 6 days a week for 8 months. The highest exposure concentration caused slight growth retardation and a mild reversible degenerative change (reported as a slight cloudy swelling) in livers of some animals. There were no observable treatment-related effects in hematologic parameters and blood or urine chemistries, nor were there pathologic changes in the eyes, adrenal gland, heart, kidney, skeletal muscle, pancreas, spleen, testis,

[1]National Institute of Environmental Health Sciences, P.O. Box 12233, Research Triangle Park, NC 27709, USA.
[2]Battelle Pacific Northwest Laboratories, P.O. Box 999, Richland, WA 99352, USA.

or ovary. Exposure of rabbits to 250000 ppm butadiene for 2 min induced light anesthesia, while exposure for 8–10 min induced deep anesthesia; death due to respiratory paralysis occurred after 25–35 min of exposure to this concentration of butadiene. Human volunteers exposed to 2000, 4000, or 8000 ppm butadiene for 6–8 h experienced minor irritation to the eyes and difficulty in focusing (Carpenter et al. 1944). The early studies on butadiene indicated that this compound only caused irritation to mucous membranes, skin, and eyes, or narcosis at high concentrations, and formed the basis for the 8-h time-weighted average work-room exposure concentration standard of 1000 ppm promulgated by the Occupational Safety and Health Administration in the United States. An International Symposium on the Toxicology, Carcinogenesis, and Human Health Aspects of 1,3-Butadiene was held at the National Institute of Environmental Health Sciences in North Carolina in 1988. The proceedings of this symposium will be published in 1989 in *Environmental Health Perspectives*.

Genotoxicity of 1,3-Butadiene

Butadiene was mutagenic towards *Salmonella typhimurium* strains TA1530 and TA1535 (strains that are sensitive to base-pair substitution mutagens) when incubation mixtures contained S-9 fractions prepared from livers of Wistar rats pretreated with phenobarbital or Aroclor 1254 (de Meester et al. 1980). In addition, enhanced reversion rates were observed when plates containing only strain TA1530 were co-incubated in desiccators in an atmosphere of 16% butadiene with separate plates containing only the S-9 activation mixture. This indirect mutagenic effect was attributed to the formation of a volatile, mutagenic intermediate generated by the S-9 mixture which transferred to the *Salmonella*-containing plate. The necessity of a liver microsomal activation system indicates that butadiene is not a direct-acting mutagen to *Salmonella*. The mutagenic activity of butadiene may be due to butadiene monoxide (1,2-epoxybutene-3) and/or diepoxybutane; these oxidative intermediates of butadiene biotransformation (Malvoisin et al. 1979; Malvoisin and Roberfroid 1982) have been shown to be mutagenic for *Salmonella typhimurium* strains that are sensitive to base-pair substitution mutagens (de Meester et al. 1978; Wade et al. 1979).

Butadiene is a potent in vivo genotoxic agent to mouse bone marrow cells (Tice et al. 1987). Exposure of male $B6C3F_1$ mice for 6 h a day over 10 exposure days to 0, 6.25, 62.5, or 625 ppm butadiene induced chromosomal aberrations (CA) and sister chromatid exchange (SCEs) in bone marrow cells, and micronuclei in peripheral blood. The percentage of cells with at least one CA (primarily chromatid breaks or gaps) was increased in the 625-ppm exposure group (14.0%) compared to controls (2.3%). There was a concentration-related increase in SCEs per cell, with values at all exposure levels (6.5, 10.9, and 35.1, respectively) greater than those of controls (5.5). An increase in micronucleated polychromatic erythrocytes per 1000 cells (MN-PCE) was observed in the 62.5- and

625-ppm exposure groups (8.4 and 30.0, respectively) compared to controls (3.8), and the frequency of micronucleated normal chromatic erythrocytes (MN-NCE) was increased in the 625-ppm exposure group (7.2/1000 NCE) compared to controls (2.7/1000 NCE). Butadiene was not genotoxic to bone marrow cells of Sprague-Dawley rats exposed to 100–10000 ppm, 6 h a day for 2 days (Cunningham et al. 1986).

Exposure of C57B1/6 mice to butadiene monoxide by intraperitoneal injection induced dose-related increases in SCEs in bone marrow cells (effective dose range: 10–150 mg/kg) and in the percentage of cells with CA (effective dose range: 25–150 mg/kg) (Sharief et al. 1986). Intraperitoneal injection of Swiss Webster mice with diepoxybutane at doses of 10 μmol/kg (0.86 mg/kg) induced increases in the frequencies of SCEs in bone marrow cells, alveolar marcophages, and regenerating liver cells (Conner et al. 1983).

Studies on the Toxicity/Carcinogenicity of Butadiene in Rats

Crouch et al. (1979) exposed male and female Sprague-Dawley rats to 0, 1000, 2000, 4000, or 8000 ppm butadiene for 6 h a day, 5 days a week for 13 weeks. In this study there were no treatment-related gross or microscopic changes, or effects on growth, survival, hematologic and blood biochemical parameters, urinary measurements, or neuromuscular functions. Based on these results, exposure concentrations of 0, 1000, and 8000 ppm were selected for a 2-year study of butadiene in Sprague-Dawley rats. In the carcinogenicity study (Owen et al. 1987), groups of 100 rats were exposed for 6 h a day, 5 days a week for 111 weeks (males) or 105 weeks (females). Increased mortality in male rats exposed to 8000 ppm butadiene compared to controls was attributed to a more severe nephropathy in this exposure group. Butadiene was carcinogenic in rats, as evidenced by the increased incidences of pancreatic exocrine adenomas (control group: three; 1000-ppm group: one; 8000-ppm group: ten) and Leydig cell tumors of the testis (none, three, and eight, respectively) in males, and increased incidences of uterine stromal sarcomas (one, four, and five, respectively), Zymbal gland carcinomas (none, none, and four, respectively), mammary gland fibroadenomas and carcinomas (50, 79, and 81, respectively) and thyroid follicular cell neoplasms (none, four, and 11, respectively) in females (Owen et al. 1987).

Groups of ten male and ten female Fischer 344 rats were exposed to 0, 200, 625, 2000, or 6250 ppm butadiene for 6 h a day, 5 days a week for 13 weeks (National Toxicology Program, unpublished results). There were no deaths in any group during this exposure period, and mean body weights of animals exposed to butadiene were similar to those of controls. Furthermore, there were no hematologic changes or exposure-related gross or microscopic lesions. A slight increase in the percentage of abnormal sperm was observed in animals exposed to 2000 or 6250 ppm.

Table 16.1. Tumor incidence in $B6C3F_1$ mice exposed to 1,3-butadiene for 60–61 weeks. (From Melnick et al. 1988)

	Sex		Exposure concentration	
		0 ppm	625 ppm	1250 ppm
Primary tumors:	Male (*n*)	10/50	44/50[a]	40/49[a]
	(%)	20	88	82
	Female (*n*)	6/50	40/49[a]	46/49[a]
	(%)	12	82	94
Malignant tumors:	Male (*n*)	3/50	40/50[a]	40/49[a]
	(%)	6	80	82
	Female (*n*)	2/50	35/49[a]	36/49[a]
	(%)	4	71	73

[a] Increased compared to chamber control (0 ppm), $p < 0.01$.

Studies on the Toxicity/Carcinogenicity of Butadiene in Mice

The long-term effects of inhalation exposure to butadiene in $B6C3F_1$ mice were studied using concentrations of 625 and 1250 ppm butadiene (National Toxicology Program 1984; Huff et al. 1985). These exposure concentrations were selected based on increased mortality and decreased body weight gains in mice exposed to 2500 ppm butadiene or higher for 14 weeks (6 h a day, 5 days a week). There were no gross or histopathologic lesions attributed to exposure to butadiene in the 14-week study (National Toxicology Program 1984). In the carcinogenicity study, groups of 50 male and 50 female mice were exposed for 6 h a day, 5 days a week to air containing 0 (chamber control), 625, or 1250 ppm butadiene. This study, intended to last for 103 weeks, was terminated after 60–61 weeks because of reduced survival at both exposure concentrations due to malignant neoplasms involving multiple organs in both sexes. At the end of the study, there were tumors in 82%–94% of the mice exposed to butadiene, compared to 20% and 12% of the control males and females, respectively (Table 16.1). The percentage of animals bearing malignant tumors was 4%–6% in the controls and 71%–82% in the exposed groups. The comparison of total tumor rates as an index of carcinogenic response is possible for this particular study because the background tumor incidences were low when the study was stopped (Haseman et al. 1986).

Malignant lymphomas, hemangiosarcomas of the heart, and alveolar-bronchiolar neoplasms occurred with positive trends in male and female mice, and the incidences of these neoplasms at both exposure concentrations were higher than those of controls (Table 16.2). Early induction and increased incidences of squamous cell neoplasms of the forestomach were also observed in males and females, and increased incidences of acinar cell carcinomas of the mammary gland, granulosa cell neoplasms of the ovary, and hepatocellular neoplasms were observed in females.

Table 16.2. Incidence of primary tumors in B6C3F_1 mice exposed to 1,3-butadiene for 60–61 weeks. (From Huff et al. 1985)

		Exposure concentration					
		Males			Females		
Target	Neoplasm	0 ppm	625 ppm	1250 ppm	0 ppm	625 ppm	1250 ppm
	Malignant lymphoma	0/50[a]	23/50[b]	29/50[b]	1/50[a]	10/49[b]	10/49[b]
Heart	Hemangiosarcoma	0/50[a]	16/49[b]	7/49[b]	0/50[a]	11/48[b]	18/49[b]
Lung	Alveolar-bronchiolar neoplasm	2/50[a]	14/49[b]	15/49[b]	3/49[a]	12/48[b]	23/49[b]
Forestomach	Squamous cell neoplasm	0/49	7/40[b]	1/44	0/49[a]	5/42[b]	10/49[b]
Mammary gland	Acinar cell carcinoma	0/50	0/50	0/50	0/50[a]	2/49	6/49[b]
Ovary	Granulosa cell neoplasm	–	–	–	0/49[a]	6/45[b]	12/48[b]
Liver	Hepatocellular neoplasm	8/50	6/49	2/49	0/50[a]	2/47	5/49[b]

[a] Increasing trend, $p < 0.05$.
[b] Increased compared to chamber control (0 ppm), $p < 0.05$.

Malignant lymphomas, which appeared to originate in the thymus and were observed as early as week 20, were considered to be the major cause of early deaths. Hemangiosarcomas of the heart, which were the second major cause of death, are uncommon neoplasms in B6C3F_1 mice, occurring at a spontaneous rate of about 0.04% in untreated mice and they have rarely been induced in long-term studies.

Nonneoplastic lesions associated with exposure of mice to butadiene (Table 16.3) included epithelial hyperplasia of the forestomach, endothelial hyperplasia of the heart, alveolar epithelial hyperplasia, hepatocellular necrosis, testicular atrophy, ovarian atrophy, and a variety of lesions in tissues of the nasal cavity (chronic inflammation, fibrosis, osseous and cartilagenous metaplasia, and atrophy of the olfactory epithelium).

A second long-term inhalation study of butadiene in B6C3F_1 mice was performed because the first study had been terminated early, and dose-response relationships for various lesions were sometimes unclear (e.g., hemangiosarcomas of the heart in male mice), and because butadiene is an important high production volume industrial chemical with potential human exposure (Melnick et al. 1988). In this study, male and female mice (n = 70–90 per group) were exposed for 6 h a day, 5 days a week for up to 2 years to 0, 6.25, 20, 62.5, 200, or 625 ppm butadiene (Melnick et al. 1989). Interim sacrifices of up to ten animals per group were performed after 40 and 65 weeks of exposure.

Hematologic changes in male mice exposed to 62.5, 200, or 625 ppm butadiene for 40 weeks included decreases in red blood cell counts, hemoglobin concentration and packed red cell volume, and an increase in mean corpuscular

Table 16.3. Incidence of nonneoplastic lesions in B6C3F$_1$ mice exposed to 1,3-butadiene for 60–61 weeks. (From Melnick et al. 1988)

		Exposure concentration					
		Males			Females		
Target	Lesion	0 ppm	625 ppm	1250 ppm	0 ppm	625 ppm	1250 ppm
Forestomach	Epithelial hyperplasia	0/49	5/40[a]	7/44[a]	0/49	5/42[a]	9/49[a]
Heart	Endothelial hyperplasia	0/50	5/49[a]	2/49	0/50	5/48[a]	8/49[a]
Lung	Alveolar epithelial hyperplasia	1/50	5/49	2/49	0/49	8/48[a]	7/49[a]
Liver	Necrosis	1/50	8/49[a]	8/49[a]	6/50	15/47[a]	6/49
Gonad	Atrophy	0/50	19/47[a]	11/48[a]	2/49	40/45[a]	40/48[a]
Nasal cavity	Inflammation, fibrosis, metaplasia	0/50	0/50	37/50[a]	0/50	0/49	2/49

[a] Increased compared to chamber control (0 ppm), $p < 0.05$.

volume (Table 16.4). Similar changes occurred in female mice exposed to 625 ppm butadiene. These changes were not accompanied by increases in reticulocyte counts or in the frequency of polychromatophilic cells in peripheral blood, and may indicate a partial or poorly regenerative response in the bone marrow of butadiene-exposed mice to reduced levels of circulating erythrocytes.

In the groups of mice exposed to butadiene for up to 2 years, survival was reduced for males and females exposed to 20 ppm butadiene or higher (Table 16.5). As in the first study in mice, reduced survival was due to the development of compound related fatal tumors. The percentage of animals bearing malignant tumors increased from about 30% in the controls to nearly 90% in the 625-ppm exposure groups. Lymphocytic lymphomas, of thymic origin, occurred as early as week 23 and were the major cause of death for male or female mice exposed to 625 ppm (Tables 16.6 and 16.7); the incidence of these neoplasms was greater in males (69%) than in females (40%). All of the mice exposed to 625 ppm died by week 65. The incidence of lymphocytic lymphomas was also increased in

Table 16.4. Hematologic changes in male B6C3F$_1$ mice ($n = 10$) exposed to 1,3-butadiene for 40 weeks. (From Melnick et al. 1989)

Exposure concentration (ppm)	RBC ($\times 10^6/\mu l$)	HGB (g/dl)	VPRC (ml/dl)	MCV (fl)
0	10.4 ± 0.3	16.5 ± 0.4	48.1 ± 1.5	46.3 ± 0.8
6.25	10.3 ± 0.3	16.4 ± 0.5	47.8 ± 1.7	46.4 ± 1.0
20	10.4 ± 0.4	16.7 ± 0.7	48.2 ± 2.2	46.3 ± 0.8
62.5	9.9 ± 0.4[a]	15.9 ± 0.6[a]	45.9 ± 2.1[a]	46.7 ± 1.2
200	9.6 ± 0.5[a]	15.6 ± 0.9[a]	45.4 ± 2.7[a]	47.2 ± 1.0
625	7.6 ± 1.2[a]	13.5 ± 1.8[a]	39.9 ± 5.3[a]	53.2 ± 2.9[a]

[a] Different from chamber control (0 ppm), $p < 0.05$.

RBC, red blood cell count; HGB, hemoglobin concentration; VPRC, volume of packed red cells; MCV, mean corpuscular volume.

Table 16.5. Survival and frequency of malignant tumors in $B6C3F_1$ mice exposed to 1,3-butadiene for 2 years.

	Exposure concentration					
	0 ppm	6.25 ppm	20 ppm	62.5 ppm	200 ppm	625 ppm
Males						
Survival[a] (%)	70	78	48[b]	44[b]	8[b]	0[b]
Malignant tumors (%)	33	55	62[c]	68[c]	73[c]	86[c]
Females						
Survival[a] (%)	74	66	48[b]	22[b]	0[b]	0[b]
Malignant tumors (%)	24	63[c]	68[c]	63[c]	80[c]	88[c]

[a]Percentage of animals surviving 2 years of exposure, excluding those deaths due to interim sacrifices at 40 and 65 weeks.
[b]Decreased compared to chamber control (0 ppm), $p < 0.05$.
[c]Increased compared to chamber control (0 ppm), $p < 0.05$, based on adjustment for intercurrent mortality by logistic regression analyses.

Table 16.6. Incidence of lymphoma, hemangiosarcoma of the heart, and alveolar-bronchiolar neoplasms in male mice exposed to 1,3-butadiene for 2 years.

			Exposure concentration					
Target	Neoplasm		0 ppm	6.25 ppm	20 ppm	62.5 ppm	200 ppm	625 ppm
	Lymphocytic lymphoma	(*n*)	2/70	1/60	2/60	4/69	2/70	62/90[a]
		(%)	3	2	3	6	3	69
	All malignant lymphomas	(*n*)	4/70	3/60	8/60	11/69[a]	9/70[a]	69/90[a]
		(%)	6	5	13	16	13	77
Heart	Hemangiosarcoma	(*n*)	0/70	0/60	1/60	5/58[a]	20/70[a]	6/90[a]
		(%)	0	0	2	9	29	7
Lung	Alveolar-bronchiolar	(*n*)	22/70	23/60	20/60	33/69[a]	42/70[a]	12/88[a]
	neoplasm	(%)	31	38	33	48	60	14

[a]Increased compared to chamber control (0 ppm), $p < 0.05$, based on adjustment for intercurrent mortality by logistic regression analyses.

Table 16.7. Incidence of lymphoma, hemangiosarcoma of the heart, and alveolar-bronchiolar neoplasms in female mice exposed to 1,3-butadiene for 2 years.

			Exposure concentration					
Target	Neoplasm		0 ppm	6.25 ppm	20 ppm	62.5 ppm	200 ppm	625 ppm
	Lymphocytic lymphoma	(*n*)	2/70	4/60	6/60	3/70	11/70[a]	36/90[a]
		(%)	3	7	10	4	16	40
	All malignant lymphomas	(*n*)	10/70	14/60	18/60[a]	10/70	19/70[a]	43/90[a]
		(%)	14	23	30	14	27	48
Heart	Hemangiosarcoma	(*n*)	0/70	0/60	0/60	1/59	20/70[a]	26/90[a]
		(%)	0	0	0	2	29	29
Lung	Alveolar-bronchiolar	(*n*)	4/70	15/60[a]	19/60[a]	27/70[a]	32/70[a]	25/88[a]
	neoplasm	(%)	6	25	32	39	46	28

[a]Increased compared to chamber control (0 ppm), $p < 0.05$, based on adjustment for intercurrent mortality by logistic regression analyses.

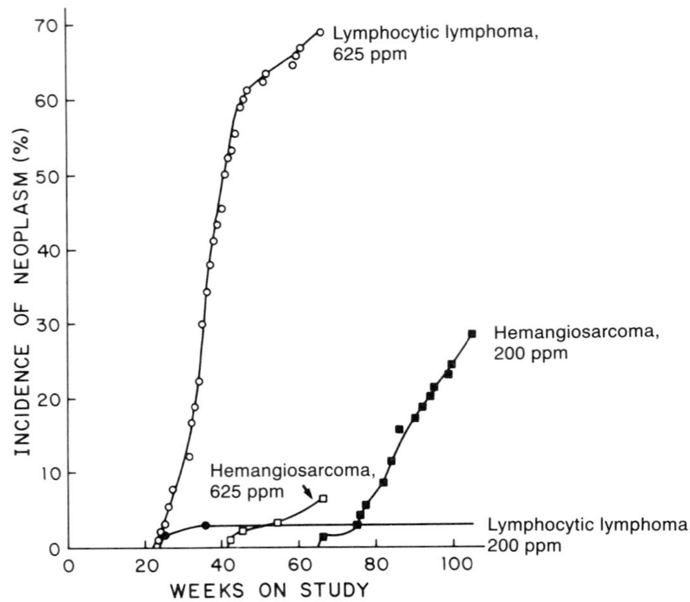

Figure 16.1. Incidence of lymphocytic lymphomas and hemangiosarcomas of the heart versus weeks on study for male $B6C3F_1$ mice exposed to 200 or 625 ppm 1,3-butadiene

females exposed to 200 ppm, but not at lower concentrations. After week 65, malignant lymphomas had less thymic involvement, and the other histologic types more commonly associated with the spontaneous lymphoma of $B6C3F_1$ mice were observed in all groups.

The incidence of hemangiosarcomas of the heart was increased in male mice exposed to 62.5, 200, or 625 ppm, and in female mice exposed to 200 or 625 ppm (Tables 16.6 and 16.7). In addition, two animals (one from the 20-ppm male group and one from the 62.5-ppm female group) were observed with this uncommon endothelial cell tumor. Interestingly, the incidence of hemangiosarcomas of the heart was greater in male mice exposed to 200 ppm than in those exposed to 625 ppm. It is clear from Fig. 16.1, where the incidence of lymphocytic lymphoma or hemangiosarcoma of the heart is plotted against the number of weeks on study, that for male mice exposed to 200 or 625 ppm butadiene, the early and extensive development of lymphocytic lymphoma at 625 ppm resulted in a reduced number of mice at risk for the later developing hemangiosarcomas. For mice exposed to 200-ppm or lower concentrations of butadiene, the dose response for butadiene-induced hemangiosarcomas of the heart is more clearly demonstrated.

The incidence of alveolar-bronchiolar neoplasms in male mice was increased at 62.5 ppm and higher concentrations compared to controls. Significance at 625 ppm was obtained when tumor rates were adjusted for early mortality. The inci-

Table 16.8. Gonadal toxicity in mice exposed to 1,3-butadiene for 2 years.

Lesion		Exposure concentration					
		0 ppm	6.25 ppm	20 ppm	62.5 ppm	200 ppm	625 ppm
Testicular atrophy	(*n*)	1/70	3/50	4/51	2/48	6/69	63/89[a]
	(%)	1	6	8	4	9	71
Ovarian atrophy	(*n*)	8/69	20/59[a]	36/59[a]	48/70[a]	60/70[a]	78/89[a]
	(%)	12	34	61	69	86	88

[a] Increased compared to chamber control (0 ppm), $p < 0.05$.

dence of testicular atrophy was increased only at the 625-ppm exposure level compared to controls (Table 16.8).

In female mice, the incidence of alveolar-bronchiolar neoplasms was increased at all exposure concentrations compared to controls (Table 16.7). Thus, even at 6.25 ppm, butadiene is carcinogenic to B6C3F$_1$ mice. The reduced incidence of lung neoplasms at 625 ppm compared to the incidence at 200 ppm is attributed to the high rate of early deaths due to lymphocytic lymphoma of female mice exposed to 625 ppm (Fig. 16.2). For alveolar-bronchiolar neoplasms, the time-to-tumor detection was slightly shorter with exposure to 625 ppm; however, because all of the female mice exposed to this concentration of butadiene died by week 65, the final incidence of this neoplastic lesion was less than that for female mice exposed to 200 ppm. The rate of alveolar-bronchiolar neoplasms in female mice after adjustment for intercurrent mortality is control: 11%; 6.25 ppm: 40%; 20 ppm: 64%; 62.5 ppm: 80%; 200 ppm: 100%; 625 ppm: 100%. Ovarian

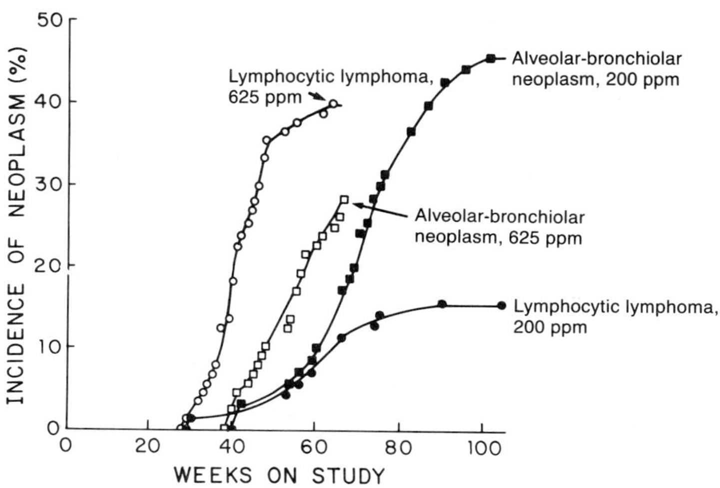

Figure 16.2. Incidence of lymphocytic lymphomas and alveolar-bronchiolar neoplasms versus weeks on study for female B6C3F$_1$ mice exposed to 200 or 625 ppm 1,3-butadiene

atrophy was also increased at all exposure concentrations compared to controls (Table 16.8). Thus, no-effect levels for alveolar-bronchiolar neoplasms and ovarian atrophy were not achieved in this study even at concentrations as low as 6.25 ppm.

Discussion

Inhalation studies of butadiene in rats and mice have shown that this high production volume chemical induces multiple organ carcinogenicity in laboratory animals. Sprague-Dawley rats exposed to 1000 or 8000 ppm butadiene had increased incidences and/or positive trends for pancreatic exocrine adenomas, uterine sarcomas, Zymbal gland carcinomas, mammary gland neoplasms, thyroid follicular cell neoplasms, and Leydig cell neoplasms of the testis (Owen et al. 1987). In $B6C3F_1$ mice, exposure to 625 or 1250 ppm butadiene caused early induction of malignant lymphomas, hemangiosarcomas of the heart, alveolar-bronchiolar neoplasms, squamous cell neoplasms of the forestomach, acinar cell carcinomas of the mammary gland, granulosa cell neoplasms of the ovary, and hepatocellular neoplasms (National Toxicology Program 1984; Huff et al. 1985). In the follow-up study in $B6C3F_1$ mice exposed to 6.25–625 ppm butadiene, the incidences of neoplastic lesions at the organ sites identified in the first study were again increased, and the Harderian gland was identified as an additional site of tumor induction (Melnick et al. 1989).

Lymphocytic lymphoma was the major cause of early death for mice exposed to 625 ppm butadiene in both experiments (Huff et al. 1985; Melnick et al. 1989). Irons (1989) and coworkers (Irons et al. 1987) compared the induction of thymic lymphomas in NIH Swiss mice and $B6C3F_1$ mice exposed to butadiene, and suggested that activation of an endogenous ecotropic retrovirus may play a role in the $B6C3F_1$ strain of mice. However, the finding that exposure of NIH Swiss mice to 1250 ppm butadiene for 1 year also caused a 14% incidence of thymic lymphoma indicates that butadiene can induce this neoplasm independently of an activated retrovirus; the NIH Swiss mouse rarely expresses any type of endogenous retrovirus and has a background rate of zero for thymic lymphoma.

The early and extensive mortalities for mice exposed to 625 ppm butadiene appeared to limit the expression of neoplasms at other sites. At lower exposure concentrations where lymphocytic lymphoma was not a major cause of early death, the dose responses for neoplastic lesions at other organ sites were more clearly demonstrated (Fig. 16.1 and 16.2). For example, in males, the incidence of the uncommon hemangiosarcomas of the heart was increased at the 62.5- and 200-ppm concentrations, and in female mice dose-related increased incidences of alveolar-bronchiolar neoplasms were observed at all exposure concentrations.

Butadiene was toxic to the bone marrow of mice, causing a partially or poorly regenerative anemia at 62.5 ppm and higher concentrations. Furthermore, a potent genotoxic response was observed in bone marrow cells of mice but not in rats. Butadiene was a gonadal toxicant in mice, inducing testicular atrophy in males at 625 ppm and ovarian atrophy in females at exposure concentrations of 6.25 ppm and higher.

The mechanism of butadiene-induced carcinogenicity is not known; however, butadiene monoxide and/or diepoxybutane may be the ultimate carcinogenic intermediates of butadiene biotransformation. These oxidative metabolites are mutagenic in *Salmonella typhimurium* and have been shown to induce local neoplasms when applied three times per week to the skin of Swiss mice (Van Duuren et al. 1963, 1965) or when administered once a week to Swiss mice or Sprague-Dawley rats by subcutaneous injection (Van Duuren et al. 1966). The ultimate mechanism of butadiene carcinogenicity must account for the numerous affected sites in laboratory animals, and the magnitude of the different concentration-dependent responses between rats and mice.

The carcinogenicity, genotoxicity, and gonadal atrophy produced by butadiene in laboratory animals raises concerns of risk for humans exposed to this chemical. Results of epidemiology studies indicate an association between occupational exposure to 1,3-butadiene and the development of lymphatic and hematopoietic cancers (Matanoski et al. 1989). Because of the apparent correspondence between the animal data and the epidemiology findings, there exists a worldwide public health need to reevaluate current workplace exposure standards for butadiene.

References

Carpenter CP, Shaffer CB, Weil CS, Smyth HF jr (1944) Studies on the inhalation of 1:3-butadiene; with a comparison of its narcotic effect with benzol, toluol, and styrene, and a note on the elimination of styrene by the human. J Ind Hyg Toxicol 26:69–78

Conner MK, Juo JE, Gutierrez de Gotera O (1983) Induction and rapid repair of sister-chromatid exchanges in multiple murine tissues in vivo by diepoxybutane. Mutat Res 108:251–263

Crouch CN, Pullinger DH, Gaunt IF (1979) Inhalation toxicity studies with 1,3-butadiene 2. 3 month toxicity study in rats. Am Ind Hyg Assoc J 40:796–802

Cunningham MJ, Choy WN, Arce GT, Rickard LB, Vlachos DA, Kinney LA, Sharff AM (1986) In vivo sister chromatid exchange and micronucleus induction studies with 1,3-butadiene in $B6C3F_1$ mice and Sprague-Dawley rats. Mutagenesis 1:449–452

de Meester C, Poncelet F, Roberfroid M, Mercier M (1978) Mutagenicity of butadiene and butadiene monoxide. Biochem Biophys Res Commun 80:298–305

de Meester C, Poncelet F, Roberfroid M, Mercier M (1980) The mutagenicity of butadiene towards Salmonella typhimurium. Toxicol Lett 6:125–130

Haseman JK, Tharrington EC, Huff JE, McConnell EE (1986) Comparison of site-specific and overall tumor incidence analyses for 81 recent National Toxicology Program carcinogenicity studies. Regul Toxicol Pharmacol 6:155–170

Huff JE, Melnick RL, Solleveld HA, Haseman JK, Powers M, Miller RA (1985) Multiple organ carcinogenicity of 1,3-butadiene in $B6C3F_1$ mice after 60 weeks of inhalation exposure. Science 227:548–549

Irons RD (1989) Studies on the mechanism of butadiene-induced leukemogenesis: the potential role of endogenous murine leukemia virus. Environ Health Perspect (in press)

Irons RD, Stillman WS, Cloyd MW (1987) Selective activation of endogenous ecotropic retrovirus in hematopoietic tissues of B6C3F$_1$ mice during the preleukemic phase of 1,3-butadiene exposure. Virology 161:457–462

Kirshenbaum I (1978) Butadiene. In: Kirk-Othmer encyclopedia of chemical technology, vol 4, 3rd edn. Wiley, New York, pp 313–337

Malvoisin E, Roberfroid M (1982) Hepatic microsomal metabolism of 1,3-butadiene. Xenobiotica 12:137–144

Malvoisin E, Lhoest G, Poncelet F, Roberfroid M, Mercier M (1979) Identification and quantitation of 1,2-epoxybutene-3 as the primary metabolite of 1,3-butadiene. J Chromatogr 178:419–425

Matanoski GM, Santos-Burgoa C, Schwartz L (1989) Mortality of a cohort of workers in the styrene-butadiene polymer (SBR) manufacturing industry 1943–1982. Environ Health Perspect (in press)

Melnick RL, Huff JE, Haseman JK, McConnell EE (1988) Chronic toxicity results and ongoing studies of 1,3-butadiene by the National Toxicology Program. Ann NY Acad Sci 534:648–662

Melnick RL, Huff JE, Roycroft JH, Chou BJ, Miller RA (1989) Inhalation toxicology and carcinogenicity of 1,3-butadiene in B6C3F$_1$ mice following 65 weeks exposure. Environ Health Perspect (in press)

Miller LM (1978) Investigations of selected potential environmental contaminants: butadiene and its oligomers. EPA-560/2-78-008. US Environmental Protection Agency, Washington DC

Morrow NL (1989) Industrial production and use of 1,3-butadiene. Environ Health Perspect (in press)

National Toxicology Program (1984) Toxicology and carcinogenesis studies of 1,3-butadiene (CAS no 106-99-0) in B6C3F$_1$ mice (inhalation studies). NTP TR no 288, Research Triangle Park, NC

Owen PE, Glaister JR, Gaunt IF, Pullinger DH (1987) Inhalation toxicity studies with 1,3-butadiene 3. Two year toxicity/carcinogenicity study in rats. Am Ind Hyg Assoc J 48:407–413

Sharief Y, Brown AM, Backer LC, Campbell JA, Westbrook-Collins B, Stead AG, Allen JW (1986) Sister chromatid exchange and chromosome aberration analyses in mice after in vivo exposure to acrylonitrile, styrene, or butadiene monoxide. Environ Mutagen 8:439–448

Shugaev BB (1969) Concentrations of hydrocarbons in tissues as a measure of toxicity. Arch Environ Health 18:878–882

Tice RR, Boucher R, Luke CA, Shelby MD (1987) Comparative cytogenetic analysis of bone marrow damage induced in male B6C3F$_1$ mice by multiple exposures to gaseous 1,3-butadiene. Environ Mutagen 9:235–250

Van Duuren BL, Nelson N, Orris L, Palmes ED, Schmitt FL (1963) Carcinogenicity of epoxides, lactones, and peroxy compounds. J NCI 31:41–55

Van Duuren BL, Orris L, Nelson N (1965) Carcinogenicity of epoxides, lactones, and peroxy compounds, pt II. J NCI 35:707–717

Van Duuren BL, Langseth L, Orris L, Teebor G, Nelson N, Kuschner M (1966) Carcinogenicity of epoxides, lactones, and peroxy compounds. IV. Tumor response in epithelial and connective tissue in mice and rats. J NCI 37:825–838

Wade MJ, Moyer JW, Hine CH (1979) Mutagenic action of a series of epoxides. Mutat Res 66:367–371

17. Species Differences in Pharmacokinetics, Metabolism, and DNA Binding of Inhaled 1,3-Butadiene

R.J. Laib[1], S. Deutschmann[1], B. Jelitto[1], R.R. Vangala[1], and R. Kreiling[1]

Introduction

1,3-Butadiene, a major component in synthetic rubber and adiponitrile manufacture, has been found to produce cancer at multiple anatomic sites in two species of experimental animals. In an inhalation bioassay with Sprague-Dawley rats (exposure to 1000 and 8000 ppm for 110 weeks), butadiene affected exclusively organs with endocrine functions and was considered a weak oncogen (Hazleton Laboratories Europe 1981). A recent long-term study with $B6C3F_1$ mice (exposure to 625 and 1200 ppm for 60 weeks) revealed a considerably higher carcinogenic activity (Huff et al. 1985). Target organs were lymphatic tissue, heart, lung mammary gland, and possibly liver.

In microsomal incubates butadiene is metabolized by cytochrome P-450 enzymes to 1,2-epoxybutene-3 (I). Further metabolic transformation of epoxybutene by epoxide hydrolase and/or monooxygenase would lead to 3,4-epoxy-1,2-butanediol (III, via 3-butene-1,2-diol) and to diepoxybutane (II, see Fig. 17.1). Epoxybutene is conjugated with glutathione and can be metabolized by glutathione-*S*-transferase (Malvoisin et al. 1979, 1982; Malvoisin and Roberfroid 1982; Bolt et al. 1983). To investigate the species differences observed in butadiene-induced carcinogenesis, comparative studies on inhalation pharmacokinetics, metabolism, and DNA binding of inhaled butadiene were conducted in rats and mice.

Pharmacokinetics and Metabolism of Butadiene and Epoxybutene in Rats and Mice

The pharmacokinetics of distribution and metabolism of butadiene and epoxybutene in mice were investigated by inhalation pharmacokinetics (Kreiling et al. 1986b, 1987). Kinetic parameters were determined based on a pharmacokinetic

[1]Institut für Arbeitsphysiologie, Abteilung für Toxikologie und Arbeitsmedizin, Ardeystr. 67, D-4600 Dortmund 1, FRG.

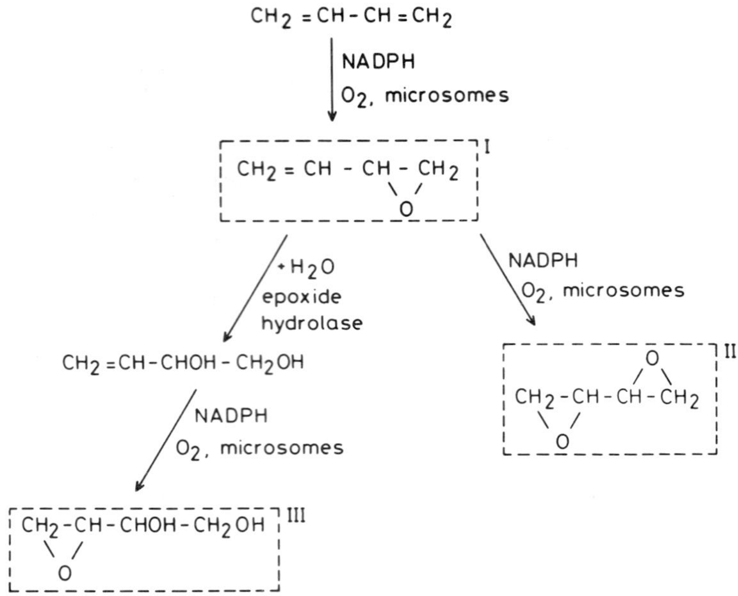

Figure 17.1. Metabolism of butadiene according to Malvoisin and Roberfroid (1982). The potential reactive intermediates are *framed*

model developed by Filser and Bolt (1981). A comparison of these results with the pharmacokinetic parameters of both compounds in rats was based on previous studies (Bolt et al. 1984; Filser and Bolt 1984).

Metabolism of Butadiene

The calculated metabolic elimination rates of butadiene for rats and mice depending on the atmospheric concentration of the compound are shown in Fig. 17.2. Up to ambient concentrations of about 1000 ppm butadiene, metabolic elimination of butadiene is proportional to the exposure concentration in mice and rats. Above 1000 ppm saturation kinetics of butadiene metabolism become apparent in both species. Figure 17.2 shows that the metabolic elimination rate of butadiene in mice is about twice that in rats, both under conditions of low and high exposure concentrations.

Metabolism of Epoxybutene

Figure 17.3 shows the calculated metabolic elimination rates of epoxybutene for rats and mice depending on the atmospheric concentration of the compound. A comparison of both species reveals that at lower exposure concentrations mice show a higher metabolic rate for epoxybutene than rats. Epoxybutene metabo-

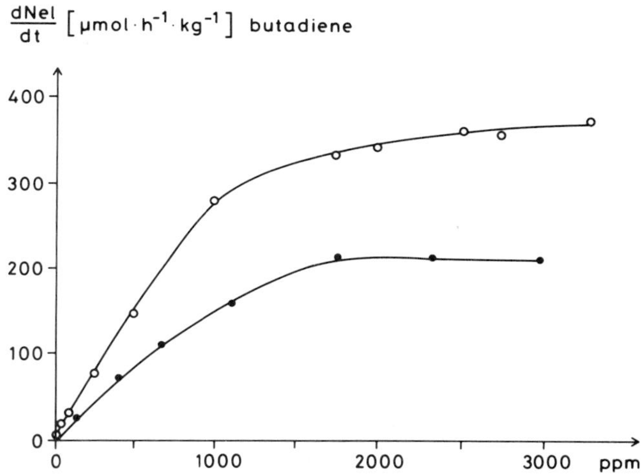

Figure 17.2. Metabolic elimination rate of butadiene in mice (B6C3F_1; *open circles*), and rats (Sprague-Dawley; *solid circles*) depending on the atmospheric concentration of the compound

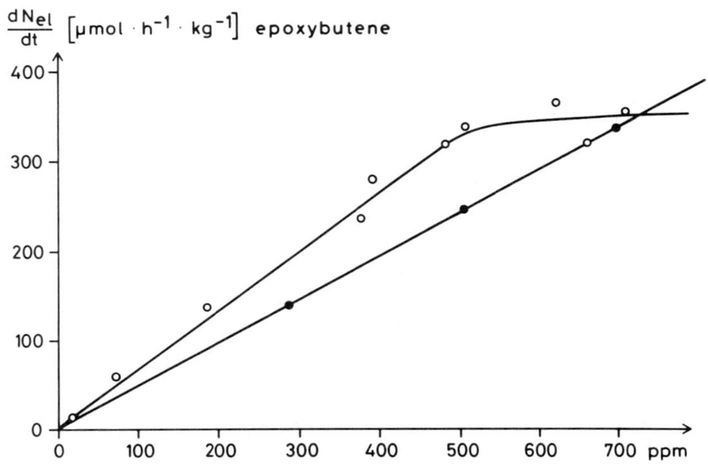

Figure 17.3. Metabolic elimination rate of epoxybutene in mice (B6C3F_1; *open circles*), and rats (Sprague-Dawley; *solid circles*) depending on the atmospheric concentration of the compound. Metabolism of epoxybutene in rats is linearly dependent on the concentration of the compound up to 5000 ppm epoxybutene

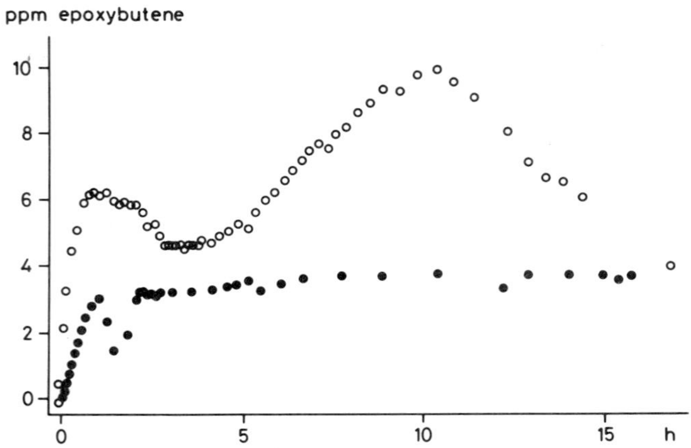

Figure 17.4. Time courses of epoxybutene concentrations exhaled into the atmosphere of a closed exposure system (6.4 liters) by two Sprague-Dawley rats (*solid circles*) or six B6C3F$_1$ mice (*open circles*) during continuous exposure to high butadiene concentrations (> 2000 ppm)

lism in rats is linearly dependent on the atmospheric concentration of the compound up to exposure concentrations of about 5000 ppm (Kreiling et al. 1987). In mice saturation of epoxybutene metabolism can be observed at about 500 ppm. Thus, with increasing exposure concentration the metabolic capacity for epoxybutene becomes rate limiting in mice but not in rats.

Exhalation of Epoxybutene During Exposure to Butadiene

Exhalation of epoxybutene into the atmosphere of a closed exposure system can be measured when mice or rats are exposed continuously to high butadiene concentrations (above 2000 ppm; Filser and Bolt 1984; Kreiling et al. 1987). Figure 17.4 shows that exhalation of epoxybutene by mice leads to an increase of epoxybutene in the atmosphere of the system, progressing with exposure time until a peak concentration of about 10 ppm is reached after 10 h. Epoxybutene exhaled by rats reaches a plateau concentration of about 4 ppm after 2 h of butadiene exposure. The subsequent decline in epoxybutene concentration in the experiment with mice is the result of a decreased butadiene metabolism. From about 12 h onwards mice showed signs of acute toxicity and liver non-protein sulfhydryl (NPSH) content of the animals was practically depleted. In rats using the same protocol the hepatic NPSH content showed no major depletion and no toxicity was observed (Kreiling et al. 1988). The difference in epoxybutene exhalation between both species indicates that in mice the elimination process of epoxybutene proceeds under saturation conditions even when the compound is metabolically generated during exposure to high butadiene concentrations.

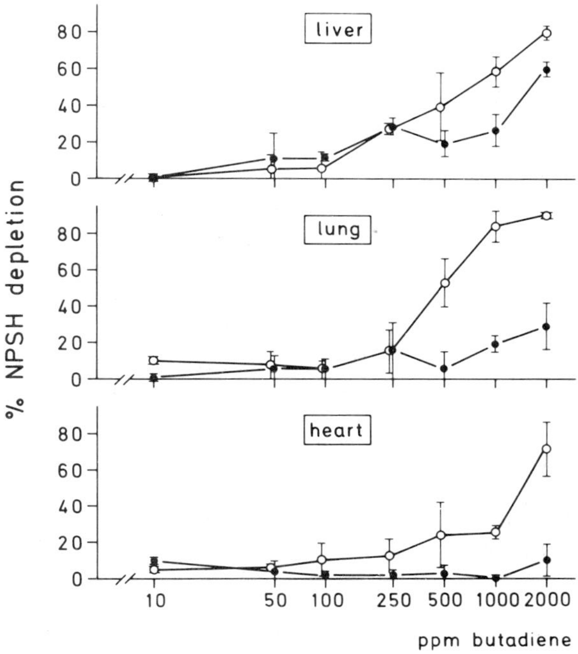

Figure 17.5. Depletion of NPSH content in liver, lung, and heart tissue of Sprague-Dawley rats (*solid circles*) and B6C3F$_1$ mice (*open circles*) after a 7-h exposure to butadiene. Depletion of NPSH content is expressed as a percentage of the corresponding control value (± SD) and plotted as a function of the logarithm of the exposure concentration

Butadiene-Induced Depletion of NPSH Content in Different Tissues of Rats and Mice

The effects of different exposure concentrations of butadiene on the cellular NPSH content of liver, lung, and heart tissue were investigated in mice and rats (Deutschmann and Laib 1989). For a comparison of the data with the animal bioassays, the experimental outline of the study was chosen as close as possible to the long-term inhalation studies. Groups of male animals of both species were exposed for 7 h in an "open" exposure system to different concentrations of butadiene. Tissue NPSH content was determined immediately after exposure according to a modified Ellman procedure. A comparison of both species is shown in Fig. 17.5. In liver and lung of rats only moderate effects of butadiene could be measured. A comparison of the dose-dependent depletion of tissue NPSH with the actual exposure concentrations used in the bioassay with rats should be possible under the assumption that butadiene metabolism and NPSH resynthesis remain unchanged during chronic exposure. On this basis, a recurrent daily depletion of NPSH content in liver of about 25% or 60%, and in lung of about 20% or 30% can be estimated for the low (1000 ppm) and high (6000 ppm) exposure concentration, respectively. Drastic effects of butadiene on NPSH content in tissues of mice could be demonstrated! Under conditions of the animal bioassay with mice (and with the assumptions detailed above), the estimated recurrent daily depletions of NPSH content were about 50% and 70% in liver, about 70% and 90% in lung, and about 25% and 40% in heart for the low (625 ppm) and high (1250 ppm) exposure concentration, respectively. Cardiac tissue does not significantly contribute to xenobiotic metabolism. NPSH depletion in heart tissue of the mouse can thus be regarded as an indictor for systemically available epoxide intermediates. That the efferent blood flows of two butadiene-metabolizing organs pass through the heart may explain its special susceptibility as a target organ of butadiene carcinogenesis in the mouse.

Alkylation of Nuclear Proteins and DNA by Reactive Butadiene Metabolites

When mice ($B6C3F_1$) and rats (Wistar) were exposed to [1,4-^{14}C]1,3-butadiene (specific radioactivity 11.2 mCi/mmol), butadiene-derived radioactivity was covalently bound to liver nucleoprotein fractions and to total liver DNA of both species (Kreiling et al. 1986a). Covalent binding of reactive butadiene metabolites to liver nucleoproteins of mice was about twice as high as in rats. This shows that, in parallel to the approximately two fold higher metabolic rate of butadiene in mice, the formation of reactive protein-binding metabolites is proportionally increased in this species. Comparable amounts of [^{14}C]butadiene-derived radioactivity were associated with total liver DNA of both species.

Formation of 7-*N*-(1-hydroxy-3-buten-2-yl)guanine and of 7-*N*-(2,3,4-trihydroxybutyl)guanine

7-*N*-(2-Hydroxy-3-buten-1-yl)guanine (I) and 7-*N*-(1-hydroxy-3-buten-2-yl) guanine (II), as expected reaction products of guanine residues with epoxybutene, were synthesized according to Citti et al. (1984). 7-*N*-(2,3,4-Trihydroxybutyl)guanine (III), as an expected reaction product of guanine residues with diepoxybutane, was synthesized by reaction of diepoxybutane with guanosine and subsequent acid hydrolysis. The compounds (see Fig. 17.6) were characterized by physicochemical methods. Male Wistar rats and male $B6C3F_1$ mice were exposed to initial concentrations of 500 ppm [1,4-^{14}C]1,3-butadiene (specific radioactivity 11.2 mCi/mmol) in a closed exposure system. Liver DNA was isolated and purified by hydroxylapatite chromatography. DNA with I, II, and III added as nonradioactive markers was subjected to acid depurination and the DNA hydrolysates were analyzed by column chromatography (Jelitto et al. 1989). Analysis of mouse liver DNA revealed the elution of radioactivity within

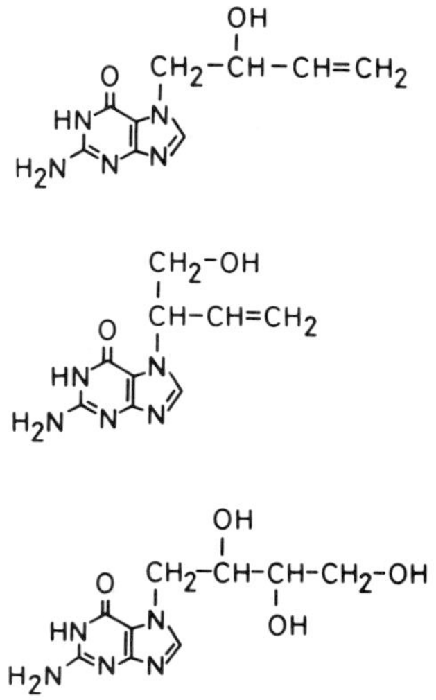

I 7-N-(2-hydroxy-3-buten-1-yl)guanine

II 7-N-(1-hydroxy-3-buten-2-yl)guanine

III 7-N-(2,3,4-trihydroxybutyl)guanine

Figure 17.6. Reaction products of metabolic intermediates of butadiene with guanine residues

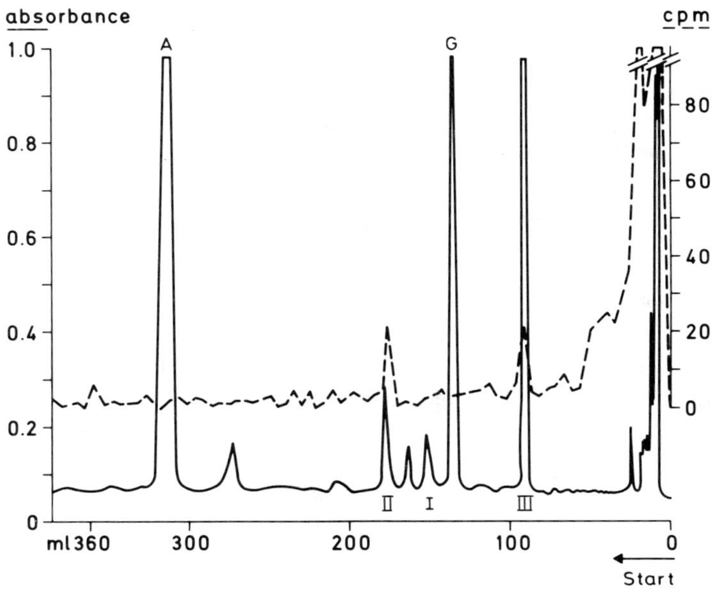

Figure 17.7. Analysis on Aminex-A6 of a hydrolysate of liver DNA from mice (B6C3F$_1$) exposed to [^{14}C]butadiene. The nonradioactive markers I, II, and III had been added prior to hydrolysis. Elution of radioactivity (*top*) and of ultraviolet-absorbing material (*bottom*). *A*, adenine; *G*, guanine

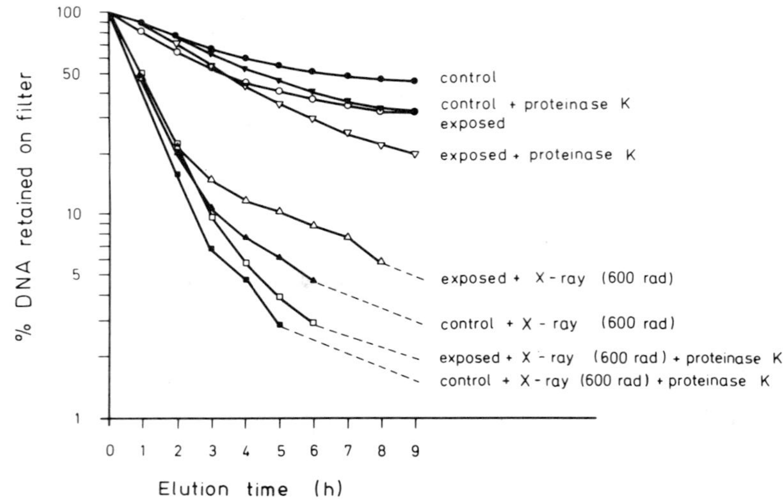

Figure 17.8. Alkaline elution curves of liver DNA from B6C3F$_1$ mice exposed to 500 ppm butadiene for 7 h. Controls were exposed to air only

the first fractions and two further peaks of radioactivity associated with the added marker compounds II and III (see Fig. 17.7). Upon analysis of rat liver DNA, only radioactivity associated with the first fractions of the column was eluted. This shows that 7-*N*-(1-hydroxy-3-buten-2-yl)guanine and 7-*N*-(2,3,4-trihydroxybutyl)guanine reaction products of epoxybutene and diepoxybutane can be identified in liver DNA of mice, but not of rats after exposure of the animals to [^{14}C]butadiene.

Formation of DNA-DNA and DNA-Protein Cross-links

Male Sprague-Dawley rats and male B6C3F$_1$ mice were exposed for 7 h to 250, 500, and 1000 butadiene. Immediately after exposure cell nuclei of liver and lung tissue were isolated and subjected to alkaline elution according to a modification of the method of Sterzel et al. (1984). Alkaline elution curves obtained from the tissues of mice show the occurrence of protein-DNA and DNA-DNA cross-links from about 250 ppm butadiene onwards (Fig. 17.8). No cross-linking activity of butadiene was observed for rats (Jelitto et al. 1989). The cross-linking activity of butadiene in the mouse can be attributed to its reactive and bifunctionally alkylating intermediate diepoxybutane.

Conclusions

Investigations on inhalation pharmacokinetics of butadiene and its primary reactive intermediate epoxybutene in mice and rats have reasonably demonstrated

Table 17.1. Ratios of specific enzyme activities in subcellular preparations of rodent and human liver and lung. (Calculated from Lorenz et al. 1984)

	Monooxygenase		Epoxide-hydrolase		Glutathione-*S*-transferase	
	Liver	Lung	Liver	Lung	Liver	Lung
NMRI mouse	4	1220	0.25	0.7	4	10
Sprague-Dawley rat	2	180	1	0.7	0.8	1
Man	1	1	1	1	1	1

that the species differences observed in butadiene carcinogenicity are related to species differences in metabolism of this compound. In addition to the higher metabolic rate of butadiene in mice, limited detoxification and accumulation of its reactive epoxide intermediate epoxybutene must be major determinants for the higher susceptibility of mice. The detection of alkylation products of epoxybutene and diepoxybutane with guanine residues in liver DNA of mice exposed to butadiene indicates that epoxybutene is further biotransformed to diepoxybutane in this species. This view is supported by the cross-linking activity of butadiene towards DNA and proteins in mice which can be attributed to the bifunctionally alkylating diepoxybutane. The quantitative differences in butadiene metabolism and in biological effectivity of the reactive epoxide intermediates between rats and mice reflect the different enzyme activities involved in butadiene metabolism. Table 17.1 shows the ratios of specific enzyme activities in subcellular preparations of rodent and human liver and lung calculated from Lorenz et al. (1984). The estimated ratios of specific enzyme activities of hepatic enzymes in mice vs. rats are for monoooxygenases 2:1, for epoxide hydrolases 1:4, and for glutathione-*S*-transferases 5:1. For lung tissue these ratios are about 7:1 (monooxygenases), 1:1 (epoxide hydrolases) and 10:1 (glutathione-*S*-transferases). Although these specific enzyme activities were not established with butadiene or its epoxides as substrates, the ratios demonstrate that the higher capacity for epoxidation in mice vs. rats cannot be adequately counterbalanced by epoxide detoxification through epoxide hydrolase in this species. Thus the major part of epoxide detoxification in mice should proceed via glutatione-*S*-transferase-mediated pathways with the result of glutathione depletion and a subsequently increased toxicity and covalent binding of reactive butadiene intermediates. The drastic depletion of tissue NPSH observed in mice but not in rats after acute exposure of the animals to butadiene is supportive of this view. All the available data demonstrate that, from a quantitative point of view, the long-term studies with mice do not adequately reflect the human situation. Additional studies in different species on indicators of genotoxic exposure, e.g., hemoglobin or DNA adducts, to better define dose-response relations and on biomonitoring of humans occupationally exposed to butadiene are needed before the available animal data can be used for risk assessment.

Acknowledgments. This work was supported by the Deutsche Forschungsgemeinschaft grant No. La 515/1-3. The authors wish to thank Ms. W. Zölffel for typing of the manuscript.

References

Bolt HM, Schmiedel G, Filser JG, Rolzhäuser HP, Lieser K, Wistuba D, Schuring V (1983) Biological activation of 1,3-butadiene to vinyl oxirane by rat liver microsomes and expiration of the reactive metabolite by exposed rats. J Cancer Res Clin Oncol 106:112–116

Bolt HM, Filser JG, Störmer F (1984) Inhalation pharmacokinetics based on gas uptake studies. V. Comparative pharmacokinetics of ethylene and 1,3-butadiene in rats. Arch Toxicol 55:213–218

Citti L, Gervasi PG, Turchi G, Belluci G, Bianchini R (1984) The reaction of 3,4-epoxy-1-butene with deoxyguanosine and DNA in vitro: synthesis and characterization of the main adducts. Carcinogenesis 5:47–52

Deutschmann S, Laib RJ (1989) Concentration-dependent depletion of non-protein sulfhydryl (NPSH) content in lung, heart and liver tissue of rats and mice after acute inhalation exposure to butadiene. Toxicol Lett 45:145–183

Filser JG, Bolt HM (1981) Inhalation pharmacokinetics based on gas uptake studies I. Improvement of kinetic models. Arch Toxicol 47:279–292

Filser JG, Bolt HM (1984) Inhalation pharmacokinetics based on gas uptake studies VI. Comparative evaluation of ethylene oxide and butadiene monoxide as exhaled reactive metabolites of ethylene and 1,3-butadiene in rats. Arch Toxicol 55:219–223

Hazleton Laboratories Europe (1981) 1,3-butadiene. Inhalation teratogenicity study in the rat. Final report and addendum no 27800-522/3. Hazleton Labs, Harrowgate, England

Huff JE, Melnick RL, Solleveld HA, Hasemann JK, Power M, Miller RA (1985) Multiple organ carcinogenicity of 1,3-butadiene in $B6C3F_1$ mice after 60 weeks of inhalation exposure. Science 277:548–549

Jelitto B, Vangala RR, Laib RJ (1989) Species-differences in DNA damage by butadiene: role of diepoxybutane. Arch Toxicol (Suppl) 13:246–279

Kreiling R, Laib RJ, Bolt HM (1986a) Alkylation of nuclear proteins and DNA after exposure of rats and mice to $(1,4\text{-}^{14}C)$1,3-butadiene. Toxicol Lett 30:131–136

Kreiling R, Laib RJ, Filser JG, Bolt HM (1986b) Species differences in butadiene metabolism between mice and rats evaluated by inhalation pharmacokinetics. Arch Toxicol 58:35–238

Kreiling R, Laib RJ, Filser JG, Bolt HM (1987) Inhalation pharmacokinetics of 1,2-epoxybutene-3 reveal species differences between rats and mice sensitive to butadiene induced carcinogenesis. Arch Toxicol 61:7–11

Kreiling R, Laib RJ, Bolt HM (1988) Depletion of hepatic non-protein sulfhydryl content during exposure of rats and mice to butadiene. Toxicol Lett 41:209–214

Lorenz J, Glatt JR, Fleischmann R, Ferlintz R, Oesch F (1984) Drug metabolism and its relationship to that in three rodent species; monooxygenase, epoxide hydrolase and glutathione-S-transferase activities in subcellular fractions of lung and liver. Biochem Med 32:43–56

Malvoisin E, Roberfroid M (1982) Hepatic microsomal metabolism of 1,3-butadiene. Xenobiotica 12:137–144

Malvoisin E, Lhoest G, Poncelet F, Roberfroid M, Mercier M (1979) Identification and quantitation of 1,2-epoxybutene-3 as the primary metabolite of 1,3-butadiene. J Chromatogr 178:419–425
Malvoisin E, Mercier M, Roberfroid M (1982) Enzymic hydration of butadiene monoxide and its importance in the metabolism of butadiene. Adv Exp Med Biol 38A:437–444
Sterzel W, Bedford P, Eisenbrand G (1984) Automated determination of DNA using the fluorochrome Hoechst 33258. Anal Biochem 147:462–467

18. Epidemiologic Data Related to Health Effects of 1,3-Butadiene

G.M. Matanoski,[1] C. Santos-Burgoa,[2] S.L. Zeger[1], and L. Schwartz[1]

To understand the potential health effects of a chemical on humans, epidemiologists, industrial engineers, toxicologists, physiologists, as well as other laboratory scientists must interrelate the information gathered from their areas of expertise in order to interpret the overall risks related to chemicals. Before presenting the data from epidemiology studies of 1,3-butadiene, a brief summary of the types of studies which have been done and the weight of evidence which can be determined from each type will be discussed.

The usual epidemiologic study collects information on mortality for all workers in an industry in which the chemical is used, in this case, the rubber polymer production industry. This design addresses the question: given that workers in this industry are similar to the comparison general population, as all males in the United States, with regard to the usual risks for the disease of interest, how much does work in the industry increase the risk of disease in workers? A parallel in toxicology might be the determination of the risk of a common tumor among all mice in one animal quarters, a small percentage of which had had butadiene exposure, as compared to the risk for all mice from all other populations, many of which had also had potential carcinogenic exposures. Several factors determine whether any risk can be seen using such an analysis:

1. The study and comparison population must be similar in regard to their risk of disease. In almost all cases, workers are healthier than the general public, and this reduces their risks of all diseases including cancer.
2. Is the population identified from the time of first exposure in an industry or do we start with a group currently working, some of whom have had long employment and "survived" a period of exposure? Selection for long survival in the industry may lower risks.

[1]Johns Hopkins School of Hygiene and Public Health, 615 North Wolfe Street, Baltimore, Maryland 21205, USA.
[2]Instituto Nacional de Salud Publica, Escuela de Salud Publica de Mexico, Francisco P Miranda No 177-6[0] piso, Mexico, D.F. 01480, Mexico.

3. Has the population been followed long enough to identify the risk? Short follow-up or competing risks of death will lower the risk estimate for the study population.
4. Related to the question above is whether the study population is "fixed" or "dynamic"? A dynamic population continues to add new young workers selected to be healthy and who have short follow-up. This will tend to lower the risk. If this group constitutes a high percentage of the population, the overall risk will be low. A fixed population adds no new workers but simply allows the whole group to age and to be followed for the latency of the disease.
5. Whether a total industrial population will identify a risk from a specific agent will depend on the proportion of the population exposed to that toxic substance as well as the level of risk from such exposure. If there are only a few exposed at a high level, then the overall population is unlikely to show a risk. Among all the above considerations, the proportion of all workers actually exposed and the similarity of workers to the comparison group may be the most important.

Table 18.1 shows the results of three studies which examined the mortality in cohorts who had manufactured rubber polymer (Meinhardt et al. 1982; Matanoski and Schwartz 1987) or who had manufactured butadiene monomer (Downs et al. 1987). All three of these studies had low overall risks of mortality, as might be predicted from the "healthy worker effect." Only the study in the butadiene manufacturing facility showed a significant increased risk of cancer in the total population—a 2.35 increase in the risk of lymphosarcoma. For the total population, Matanoski and Schwartz (1987) showed a significant 28% increase in mortality from arteriosclerotic heart disease in Blacks. Most of these cohorts were dynamic so that recent entrants into employment had only worked and been followed for 1 or 2 years. One of the cohorts suffered from survival selection as well (Matanoski et al. 1982). Thus, many factors contributed to an overall low mortality in these cohorts.

The investigators examined subsets which might have had higher exposure and longer follow-up than the total population and thus may have had higher standardized mortality ratios (SMRs). Meinhardt et al. (1982) examined the risks in a subset of the population who had long follow-up and who had worked in the industry in earlier periods when exposures to potentially hazardous substances may have been less well controlled. Only one of the two plants under study showed a 2.8-fold increased risk of leukemia, which was of borderline significance. The other plant showed no risk. The butadiene levels measured at the time of the study were higher in the plant without a risk.

Downs et al. (1987) selected subgroups who had routine and nonroutine exposure to butadiene and found excess risks of lymphomas and leukemias, respectively, in the subsets, but these excesses were not significant. In discussing a recent update of the original population, Divine (1988) indicated that the SMR for lymphosarcoma was higher than in the original study. Matanoski et al. (1982)

Table 18.1. Epidemiologic studies possibly related to carcinogenicity of butadiene.

Study	Sites	(*n*)	Risk ratios	Notes
McMichael et al. 1976	Lymphatic and hematopoietic	–	5.6 (S)	Identified by > 5 years' work in SBR latex
	Lymphatic leukemia		3.7 (S)	Increases with duration worked
	Stomach		2.1 (S)	No excess of other cancers
	Ischemic heart disease		2.9 (S) (age 40–54)	
Andjelkovich et al. 1977	Lung	3	4.34 (S)	Synthetic latex
Meinhardt et al. 1982	Lymphatic and hematopoietic	9 (plant A)	1.6 2.1 (early)	Risks in early workers in plant A are significant when one-tail test is used
	Leukemia	5 (plant A)	2.8 (early)	
	Lymphatic and hematopoietic	2 (plant B)	0.8	
Matanoski and Schwartz 1987	Lymphatic and hematopoietic	15 (white) 2 (black)	0.87 1.52	None of the cancer ratios was significant
	Other lymphatic	5 (production)	2.02	Work areas other than production had SMRs above 1.00 (NS)
	Stomach	21 (white) 6 (black)	0.94 1.19	Significant SMR in arteriosclerotic heart disease in Blacks only
	Large intestine	39 (white) 3 (black)	1.06 0.92	
	Arteriosclerotic heart disease	86 (black)	1.28 (S)	
Downs et al. 1987	Lymphosarcoma	8	2.35 (S)	Total population
	Lymphomas (not leukemia or Hodgkin's)	4	2.82 (NS)	Routine exposure
	Leukemia	5	2.01 (NS)	Nonroutine exposure

(S), Significant at $p \leq 0.05$; (NS), nonsignificant.

and Matanoski and Schwartz (1987) divided the population into general work areas based on the last job held and found a nonsignificant twofold excess risk of other lymphatic cancers. This category included non-Hodgkin's lymphomas and multiple myeloma. Although none of these studies alone would suggest a risk, the fact that all indicated slight excesses of hematologic neoplasms was interesting in view of the animal data.

Epidemiologic studies can take the form of case-control studies in either the general population or in an industry. In this situation, one estimates the exposure for each case with disease as well as each comparable control without disease. If exposures are examined for subjects in the general population, one would not expect a high proportion to give a history of butadiene exposure. But one can increase the probability that individuals will have exposure by conducting the study within an industrial population which has a high probability of exposure. The McMichael et al. (1976) study selected cases and controls from within a cohort of rubber workers. These cases and controls are assumed to be very similar with regard to factors such as health, personal habits, and socioeconomic factors; and then one determines whether the cases are more likely to have had butadiene exposure than controls. There are problems which may result from examining potential exposures without a preconceived hypothesis. By chance alone, some of the many substances in use in the industry may be associated with the disease. So the epidemiologist must interpret the data with caution. As with any "naturally" occurring disease, not all cases will be due to the exposure. To relate this to toxicology studies, the situation would be identical to mixing exposed and nonexposed mice together, selecting those which have disease and a sample of those without disease, and determining whether the probability of exposure is greater in those with disease compared to those without disease.

Using this design and job histories to identify work exposures, McMichael et al. (1976) found 5.6-fold risk of lymphatic and hematopoietic cancers in association with work in the styrene-butadiene rubber (SBR) latex manufacturing units of the rubber industry. Work in the same area was associated with a 3.7-fold increased risk of lymphatic leukemia. Among the other five cancers investigated, only stomach cancer showed a significant twofold association with work in the latex plants. There was no association between these work areas and colorectal, respiratory, bladder, or prostate cancers. The risk of mortality from ischemic heart disease in 40–54-year-olds who worked in synthetic latex production was three times that expected. All of these age-adjusted ratios of work exposures in cases compared to other workers were significant. Specific exposures were not identified, but only a work area called "synthetic plant" where elastomers of SBR and other rubbers were synthesized. Andjelkovich et al. (1977) compared the mortality for workers whose most representative department was the synthetic latex plant to all other workers and found a significant excess only of respiratory cancers based on three deaths.

The early studies of synthetic rubber protection suggested some excess cancer risks among workers, with cancers of the lymphatic and hematopoietic system most frequently identified. However, the excesses were not always significant; the type of hematologic neoplasm was not always the same; and, with the exception of the butadiene-manufacturing cohorts, no specific exposure was associated with the risks. Since exposure of animals to butadiene had resulted in hematologic neoplasms, the presence of this cancer risk in several of the studies suggested that further investigations were warranted.

The cohort of 12110 workers in eight SBR rubber production plants in the United States and Canada was followed for an additional 3 years until 1982, contributing a total of about a quarter of a million person-years of experience (Matanoski et al. 1989). Only workers with 1 year of employment were included in seven of the eight plants. The Canadian population was restricted to workers with 10 or more years of employment in order to assure complete death ascertainment.

Table 18.2. Standardized mortality ratios in all work areas for selected causes, adjusted for age, calendar time, and race (reference population: United States males).

Cause	White (10915) Observed	SMR	Black (1195) Observed	SMR	Total SMR
All causes	2061	0.79[a]	380	0.95	0.81[a]
Circulatory disease	1063	0.80[a]	214	1.18[b]	0.85[a]
ASHD	772	0.84[a]	125	1.48[b]	0.89[a]
Vascular lesions of CNS	141	0.85	47	1.18	0.92
All cancers	454	0.85[a]	64	0.92	0.85[a]
Digestive	137	0.94	21	0.92	0.93
Esophagus	17	1.34	0	–	1.00
Stomach	25	0.95	9	1.45	1.05
Large intestine	45	0.93	3	0.67	0.91
Liver	8	0.75	3	1.44	0.86
Pancreas	24	0.83	3	0.80	0.83
Lung cancer	146	0.82[a]	20	1.00	0.84[a]
Bladder cancer	11	0.72	1	0.67	0.71
Kidney cancer	15	1.11	0	–	1.03
Lymphopoietic	48	0.92	7	1.46	0.97
Lymphosarcoma	6	0.56	1	1.32	0.61
Hodgkin's	8	1.31	0	–	1.20
Leukemia	18	0.86	4	2.18	0.96
Other lymphatic	15	1.10	2	1.16	1.11

[a] Significantly low, $p \leq 0.05$.
[b] Significantly high, $p \leq 0.05$.
ASHD, arteriosclerotic heart disease; CNS, central nervous system.

For all workers (Table 18.2), the only significantly high SMR is the one for arteriosclerotic heart disease in Blacks which is 48% higher than that expected, based on rates for black males in the United States. This ratio might be artificially inflated because of unknown race in 15% of the population. Many other SMRs are significantly lower than expected. Some SMRs exceed 1.00 including a two-fold excess risk of leukemia in Blacks and some digestive system cancers but these are not significant. These latter sites are of interest only because of the possible relationship to stomach cancers in animals.

The population was divided into four work areas on the basis of the job longest held. In about 20% of the population there was insufficient information to classify the individual. The three major groups shown are production, maintenance, and the other work sites predominantly representing laboratories and administrative services. (Utilities included few workers and is not presented.) Among black workers in the production area, deaths from lymphopoietic cancers occur 5.1 times more frequently and deaths from leukemia 6.6 times more frequently than expected, based on population mortality rates (Table 18.3), and both ratios are significantly high. Both white and black workers in this area have high risks of other lymphomas with a significantly high total SMR of 2.6. None of the SMRs for digestive system cancers are high for this work area.

Table 18.3. Standardized mortality ratios by work area for selected causes, adjusted for age, calendar time, and race (reference population: United States males); Work area: production.

Cause	White (2753) Observed	White (2753) SMR	Black (371) Observed	Black (371) SMR	Total SMR
All causes	502	0.87[a]	92	0.93	0.88[a]
Circulatory disease	254	0.89	49	1.16	0.93
ASHD	182	0.91	29	1.47	0.96
Vascular CNS	37	1.08	9	0.99	1.06
RHD	5	0.93	1	0.69	0.99
All cancers	105	0.88	6	1.23	1.06
Digestive	25	0.79	3	0.56	0.75
Esophagus	3	1.08	0	–	0.79
Stomach	3	0.53	1	0.70	0.57
Large intestine	10	0.95	1	0.96	0.95
Liver	0	–	1	1.98	0.36
Pancreas	5	0.79	0	–	0.69
Lung cancer	41	1.03	6	1.23	1.06
Kidney cancer	5	1.66	0	–	1.53
All lymphopoietic	13	1.10	6	5.07[b]	1.46
Lymphosarcoma	0	–	1	5.32	0.38
Hodgkin's	2	1.31	0	–	1.20
Leukemia	4	0.84	3	6.56[b]	1.34
Other lymphatic	7	2.30	2	4.82	2.60[b]

[a] Significantly low, $p \leq 0.05$.
[b] Significantly high, $p \leq 0.05$.
RHD, rheumatic heart disease.

Workers in maintenance have no increase in deaths from the lymphopoietic cancers (Table 18.4), but white workers have nonsignificant excesses of digestive system cancers including the sites esophagus, stomach, and large intestine. Black workers in both production and maintenance have excess risks of arteriosclerotic heart disease. Workers in the other areas generally have few sites with high SMRs, with the exception of rheumatic heart disease and liver cancer (Table 18.5). None of the work areas indicate impressively high SMRs for lung cancer, even though this cancer reportedly occurs at low doses in animals.

The work sites were grouped according to suspected exposures. However, personal monitoring samples for butadiene taken in 1976–1981 from various plants within the industry indicated that, while most jobs with the high levels occurred in the production area, there was opportunity for serious misclassification. Laboratory workers in the "other work area," for example, had high exposures, and many jobs in production had low exposure.

Not only was there a need for better classification of the exposures of individuals with and without disease, but also for an attempt to separate styrene

Table 18.4. Standardized mortality ratios by work areas for selected causes, adjusted for age, calendar time, and race (reference population: United States males); work area: maintenance.

	White (2828)		Black (443)		
Cause	Observed	SMR	Observed	SMR	Total SMR
All causes	751	0.87[a]	186	1.05	0.90
Circulatory disease	381	0.84[a]	114	1.38[b]	0.92
ASHD	287	0.91	67	1.76[b]	1.01
Vascular CNS	43	0.72[a]	24	1.30	0.86
RHD	5	0.93	1	0.69	0.99
All cancers	174	0.99	24	0.78	0.95
Digestive	55	1.11	8	0.78	1.06
Esophagus	6	1.44	0	–	1.00
Stomach	15	1.66	3	1.04	1.51
Large intestine	18	1.11	1	0.50	1.04
Liver	1	0.28	0	–	0.22
Pancreas	8	0.83	1	0.60	0.80
Lung	59	1.03	8	0.92	1.01
Kidney	5	1.15	0	–	1.04
All lymphopoietic	14	0.85	0	–	0.75
Lymphosarcoma	2	0.59	0	–	0.53
Hodgkin's	3	1.70	0	–	1.51
Leukemia	6	0.88	0	–	0.79
Other lymphatic	2	0.46	0	–	0.39

[a] Significantly low, $p \leq 0.05$.
[b] Significantly high, $p \leq 0.05$.

and butadiene exposure. A nested case-control study has been completed which attempts to identify estimated exposures to styrene and butadiene in cases with lymphopoietic cancers and worker controls within the cohort which did not have cancer. The number of cases was expanded from the 55 included in the mortality study to 59 by re-evaluating the data for inclusion and by adding cases in which lymphopoietic cancer was mentioned anywhere on a death certificate (Santos-Burgoa 1988; Matanoski et al. 1988). The design of these case-control studies can differ. In this design, controls were selected to work for the same period as the cases. Since an individual's dose in human data is a measure of level of exposure multiplied by the time during which that level occurred, to match the population by duration of work may partially control the time variable which is part of the dose measure. In fact, duration of exposure is often the predominant factor in the dose. Matching for total duration of work is not an absolute control on duration of exposure since individuals may hold jobs with different levels for different periods during their total employment. The purpose of controlling for employment was to avoid an over-weighting of the comparison sample with short-term workers and to emphasize the comparison of levels of exposure. This method

Table 18.5. Standardized mortality ratios by work area for selected causes, adjusted for age, calendar time, and race (reference population: United States males); work area: other work areas.

	White (2630)		Black (237)		
Cause	Observed	SMR	Observed	SMR	Total SMR
All causes	545	0.85[a]	79	0.90	0.85[a]
Circulatory disease	301	0.92	40	1.01	0.93
ASHD	208	0.92	21	1.12	0.93
Vascular CNS	47	1.13	12	1.38	1.18
RHD	12	2.09[b]	0	–	1.90
All cancers	105	0.80[a]	19	1.23	0.84
Digestive	35	0.97	8	1.60	1.04
Esophagus	2	0.64	0	–	0.50
Stomach	5	0.75	4	2.97	1.12
Large intestine	10	0.85	1	1.00	0.86
Liver	5	1.88	2	4.39	2.24
Pancreas	9	1.27	1	1.20	1.26
Lung cancer	29	0.67[a]	6	1.35	0.73
Kidney cancer	1	0.30	0	–	0.28
All lymphopoietic	11	0.86	1	0.94	0.87
Lymphosarcoma	1	0.38	0	–	0.36
Hodgkin's	2	1.30	0	–	1.22
Leukemia	6	1.16	1	2.46	1.25
Other lymphomas	2	0.61	0	–	0.54

[a] Significantly low, $p \leq 0.05$.
[b] Significantly high, $p \leq 0.05$.

of measuring dose should be somewhat similar to that used in animal experiments where time of exposure is always fixed. The controls were also selected to be similar by plant, birth date, year of hire, and survival, at least to the point of death of the case as well as duration of employment as noted.

All jobs within the rubber industry were reviewed by a group of experts who ranked each of the jobs from 0 to 10 for both butadiene and styrene. The estimates were made solely on the basis of the personal experience of these senior engineers, most of whom had worked in the industry from its inception. Any factors such as plant or calendar time which they thought might influence the score were included in the ranking. Only 5% of 464 nonsupervisory jobs were classified at a rank of 10 for butadiene; none was ranked between 6 and 9. However, some exposures were ranked as an average of a work area and might fall in the 6–9 ranks. One-third of the jobs had no exposure. An attempt was made to evaluate intermittency of exposure, but, since almost all jobs were judged to have intermittent exposure, this dimension in exposure assessment could not be used.

All jobs and the time during which that job was held were abstracted for all subjects. The cumulated dose of both styrene and butadiene was calculated based on the score and duration held for each job in an individual's work history. The

Table 18.6. Odds ratios for lymphopoietic cancers and exposure to butadiene and styrene, matched analysis (exposure based on level above mean log cumulated rank for each group).

Cancers	Cases	Controls	Butadiene		Styrene	
			OR	95% CI[a]	OR	95% CI
All lymphopoietic[b]	59	193	2.30	1.13–4.71	1.91	0.91–4.0
Leukemia	26	84	9.36	2.05–22.94	3.13	0.84–11.6
Other lymphatic	18	56	1.54	0.50–4.80	1.35	0.25–7.4
Lymphosarcomas	6	23	0.50	0.06–4.18	1.33	0.11–16.6
Hodgkin's disease	8	29	1.10	0.24–5.17	0.40	0.05–3.3

[a] Odds ratios (OR) based on maximal likelihood estimate (MLE) and test-based confidence limits; CI, confidence interval.
[b] One case of poycythemia vera and its one control were omitted from subgroups.

analysis was done using a logarithmic transformation of the exposure score because of skewing.

A total of 59 cases and 193 controls were included in the analysis. There were 26 cases of leukemia, ten of multiple myeloma, and seven of non-Hodgkin's lymphoma included. (Slightly over three controls per case were selected.) The population was first examined to determine the differences in risk by "ever/never exposed" to butadiene and styrene. There is an excess risk of leukemia in the population exposed to butadiene and to styrene. A second method of evaluating the population was to combine the cases and controls in each subset of cancers and to determine the mean log butadiene and styrene in order to dichotomize each cancer group into high- and low-exposure categories. The odds ratios for matched sets (Table 18.6) are based on the maximal likelihood estimate (MLE) of the odds ratio and the test-based confidence limits around the MLE (Miettinen 1970). Using this method, there is a 9.4-fold increased risk of leukemia in association with exposure to butadiene, which is significant, and a threefold increased risk of leukemia in association with styrene exposure, which is not significant. The risk of all lymphopoietic cancers is also significantly elevated in association with butadiene, but this may be a reflection of the leukemia risk since other cancers in this group do not show significantly high odds ratios.

As might be expected, there is significant correlation between the log cumulated dose of butadiene and styrene. Conditional logistic regression models were used in an attempt to separate the risks associated with each of these agents (Breslow and Day 1980). The risk was estimated for each chemical either as a continuous dose variable or as a dichotomous variable using the mean log cumulated dose within each subset of cases and controls as a cut-off point between high and low exposure. The models using a dichotomous measure of dose will be presented.

As shown in Table 18.7, there is a significant 7.6-fold excess risk of leukemia associated with exposure to butadiene, as estimated by the odds ratio, and a lower and nonsignificant association of leukemia with styrene. When the exposures for

Table 18.7. Maximal likelihood estimate of parameters relating butadiene and styrene with leukemia using a dichotomous exposure variable in a conditional logistic regression model (mean log cumulative rank butadiene or styrene).

Model	Parameters	OR	95% CI	Beta	SE	−2MLL	LRS *p*
1	xlb	7.61	1.62–35.64	2.029	0.788	63.03	9.63 .002
2	xls	2.92	0.83–10.27	1.073	0.641	69.52	3.15 .076
3	xlb + xls	7.39	1.32–41.33	2.000	0.878	63.02	9.64 .008
		1.06	0.23–4.96	0.0577	0.787		
4	xlb + xls + xlb.xls	No convergence					

LRS, log ratio statistic; xlb, butadiene; xls, styrene; SE, standard error. Model 1. butadiene (above/below log mean cumulative rank); Model 2. styrene (above/below log mean cumulative rank); Model 3. butadiene and styrene.

both chemicals as dichotomous variables are entered into the model, only butadiene exposure continues to be associated with a 7.4-fold risk of leukemia. The best model which explains the deviance between cases and controls with the least parameters is one with butadiene alone.

In an attempt to determine whether specific jobs might explain some of the risk of leukemia, individuals were classified according to the longest job held, and this variable was evaluated using the conditional logistic regression model and ever/never exposed to the work area represented by that job. Any job group for which the odds ratio indicated an increased risk of leukemia after correction for the presence of butadiene and for which the odds ratio had a *p* value of 20% or less was combined into a new job category called "work areas" (Table 18.8). This group included utilities, operation services, and laboratory jobs. From this analysis, the best model to explain the difference between cases of leukemia and controls would be one which includes butadiene and the combined job categories "work areas" (Table 18.8). The risk of leukemia associated with work in these areas is 6.1-fold, and the odds ratio for butadiene is increased to 13.3 from 7.6 when only butadiene is included in the model. This model provides a significantly better fit to the data than does the model with butadiene alone. Both of these two factors appear to be important risk factors for leukemia in this population.

Another observation which may be worthy of comment relates to the risks of cancers in the other lymphoma category which includes non-Hodgkin's lymphoma and multiple myeloma. When the risk of leukemia is evaluated for butadiene and styrene independently but stratified by the other chemical, the high risk associated with butadiene occurs in the population with a high-level styrene exposure as well, although the data based on small numbers are compatible with no difference in the butadiene effect regardless of the styrene level. However, when the odds ratios for the other lymphomas are evaluated through stratification for each chemical, a significantly increased risk of other lymphomas from butadiene occurs only when styrene is low, and a similar sixfold but nonsignificant risk from styrene occurs only when butadiene is low (Table 18.9). With the small numbers in this group and a category which combines cancers which potentially

Table 18.8. Maximal likelihood estimate of odds ratio associating combined work areas, butadiene, styrene and leukemia using conditional logistic regression model.

Model	Parameter	OR	95% CI	LRS	p
1	xlb[a]	7.61	1.62–35.64	9.63	.002
2	Three areas[b]	3.78	1.20–11.91	5.38	.020
3	Work areas	6.08	1.56–23.72	17.55	<.001
	xlb	13.26	2.24–78.55		
4	Work areas	4.53	1.35–15.22	9.57	.008
	xls	3.89	0.95–15.90		
5	Work areas	6.11	1.55–24.05	17.56	<.001
	xlb	12.35	1.61–94.67		
	xls	1.13	0.20–1.86		

LRS, log ratio statistic.

[a] Butadiene (xlb) and styrene (xls) dichotomized as exposure above or below the mean log cumulated rank.

[b] Three work areas represented by the job longest held by individual include utilities, operation services, and laboratory.

have different etiology, it would be important to increase the numbers in the group and then subgroup them to see if there are differences in response for non-Hodgkin's lymphomas versus multiple myeloma. The suggested protective effect of each chemical in the presence of high levels of the other requires further investigation.

In summary, the epidemiologic data have been examined to see if they support the toxicologic evidence that butadiene is carcinogenic to the bone marrow and other sites in animals, especially the digestive tract and lungs. Some of the early studies of rubber workers suggested risks of lymphopoietic cancers and leukemia, but the results were not consistent. The current update of the previous large synthetic rubber polymer-manufacturing cohort using an external control suggests a significant excess of lymphomas in all production workers and a high risk of leukemia in black production workers. When cases with the hematologic neoplasms are compared to an internal population using a crude estimation of levels of butadiene and styrene associated with jobs in the industry, only the leukemia cases appear to be associated with an exposure specifically to butadiene. There seems to be no association with styrene despite the strong correlation between the chemicals. No similar examination of exposures of other cancers or heart disease cases has been completed to see if they, too, have risks associated with butadiene exposure.

The data raise many issues which still need to be resolved before a conclusion can be reached. The various studies have suggested associations with different hematologic neoplasms. This apparent inconsistency may not be real since there are close interrelationships between many of these tumors. It would be helpful to identify these cancers by their B and T cell characteristics in order to refine the information.

In the original cohort, black workers clearly had a higher risk of leukemia than white workers. This could not be investigated adequately with regard to the

Table 18.9. Odds ratio for other lymphatic cancers associated with butadiene and styrene stratified by other chemical.

Styrene	Butadiene	Cases	Controls	Mantel-Haenszel OR	χ^2
Above mean					
	Above mean	9	30		
	Below mean	2	2	0.30 (0.04–2.4)	1.350
Below mean					
	Above mean	4	4		
	Below mean	3	20	6.67 (1.06–42.07)	4.49
		Adjusted OR		1.68 (0.55–5.15)	5.41[a]

Butadiene	Styrene	Cases	Controls	Mantel-Haenszel OR	χ^2
Above mean					
	Above mean	9	30		
	Below mean	4	4	0.30 (0.6–1.45)	2.35
Below mean					
	Above mean	2	2		
	Below mean	3	20	6.67 (0.67–66.84)	2.97
		Adjusted OR		0.81 (0.28–2.38)	5.59[a]

[a] This refers to Breslow-Day χ^2 test for homogeneity with one degree of freedom.

chemical exposures since the design of the nested case-control study partiallly controlled for race. Since Blacks are thought to have a higher frequency of some retrovirus infections and, since retrovirus infections are known to increase the risk of cancers in animals, it is important to investigate further the issue of risk by race, especially if additional cases could be identified.

The cohort mortality study showed an excess of other lymphomas in relation to work in production, but very little evidence of risk from butadiene or styrene exposure was identified in the case-control study, with the exception of the model in which the two chemicals were treated as categorical variables and an interaction term was added. Even in this model, neither styrene nor butadiene was associated with a significantly high odds ratio, and the model did not explain a significant amount of the deviance. Certainly, the reason for an identified association between butadiene and leukemia and a lack of such an association with other lymphomas is not due to variation in the control samples since there are no differences between the controls selected for each case group. This discrepancy between the identification of a risk of other lymphomas in the production area in the mortality study and no association with the monomers in the case-control study needs to be examined by type of cancer within the group.

From the standpoint of the epidemiologic studies, the major issue for investigation is the dose. These reported levels are based only on crude estimates which have not even been tested for comparability of ranking against monitoring data. Neither butadiene nor styrene shows a smooth increase in risk of leukemia with each increase in log cumulated dose. This is why the model using butadiene dose as a continuous variable has not produced as good a fit as the dichotomous dose model. The models assume a log linear increase in risk with dose, which is not present. At different levels of butadiene scores used to dichotomize the population, the risk changes although at any level there is always an excess in the odds ratio compared to zero exposure and a significant increasing trend. Usually, when these models are used to test dose response from human data using a continuous variable, there is no control for duration of exposure, and this time variable plays an important role in the measure of dose. In the current study design, the matching of cases and controls by total duration of employment may have partially controlled for duration of exposure to the specific substances. For example, individuals with one job during employment will differ from their controls only by estimated exposure for that job. In the total group, 44% of cases and 46% of controls only had one job during employment. So, for almost half of the population, only intensity level can vary. Another concern is that the dose levels which appear to influence the change are small. These concerns require further evaluation of this population using some verification of exposure scores. This is especially true since toxicologic data suggested that hematologic changes were associated with high doses.

Further case-control studies looking at cancers of other anatomic systems which were identified as risk sites in toxicologic data would be valuable. If one examines other sites such as lung and even stomach, as in the case of hematologic neoplasms, precise data from hospital records would be useful to possibly subclassify the cases. To evaluate the risk of lung cancer in association with butadiene, smoking data as well as cell type will be needed. The issue of a potential risk from ischemic heart disease might be worth pursuing further in this population, provided case verification and exposure scores can be obtained.

Acknowledgment. Much of the work reported in this paper was supported by funding from the International Institute of Synthetic Rubber Producers. The authors wish to express their appreciation to the oversight committee of Drs. John Acquavella and Sally Cowles and Mr. Waveland Davis for their thoughtful scientific critiques and advice related to the research work presented. The authors would like to thank all the participating companies and their representatives for their help in the conduct of the research projects which are the basis for this review.

References

Andjelkovich D, Taulbee J, Symons M, Williams T (1977) Mortality of rubber workers with reference to work experience. J Occup Med 19:397–405

Breslow NE, Day NE (1980) Statistical methods in cancer research. I. The analysis of case-control studies. International Agency for Research on Cancer, Lyon (IARC Scientific Publications No. 32)

Divine B (1988) Presentation 1,3-butadiene symposium. National Institute of Environmental Health Sciences, Research Triangle Park, NC

Downs TD, Crane MM, Kim KW (1987) Mortality among workers at a butadiene facility. Am J Ind Med 12:311–329

Matanoski GM, Schwartz L (1987) Mortality of workers in styrene-butadiene polymer production. J Occup Med 29:675–680

Matanoski GM, Schwartz L, Sperrazza J, Tonascia J (1982) Mortality of workers in the styrene-butadiene rubber polymer manufacturing industry. Report to International Institute of Synthetic Rubber Producers, Houston

Matanoski GM, Santos-Burgoa C, Zeger S, Schwartz L (1988) Nested case-control study of lymphopoietic cancers in workers in the styrene-butadiene polymer manufacturing industry. Report to International Institute of Synthetic Rubber Producers, Houston

Matanoski GM, Santos-Burgoa C, Schwartz L (1989) Mortality of a cohort of workers in the styrene-butadiene polymer (SBR) manufacturing industry, 1943–1982, Environ Health Perspect (in press)

McMichael AJ, Spirtas R, Gamble JF, Tousey PM (1976) Mortality among rubber workers: relationship to specific jobs. J Occup Med 18:178–185

Meinhardt TJ, Lemen RA, Crandall MS, Young RJ (1982) Environmental epidemiologic investigation of the styrene-butadiene rubber industry. Mortality patterns with discussion of the hematopoietic and lymphatic malignancies. Scand J Work Environ Health 8:250–259

Miettinen OS (1970) Estimation of relative risk from individually matched series. Biometrics 26:75–86

Santos-Burgoa C (1988) Case-control study of lympho-hemopoietic malignant neoplasms within a cohort of styrene-butadiene polymerization workers. Doctoral thesis, Johns Hopkins School of Hygiene and Public Health

Section 4. Benzene

19. A Review of the Toxicokinetics of Benzene

M.A. Medinsky[1]

Introduction

Benzene is an important, widely used industrial solvent (Yin et al. 1987), and a common pollutant in indoor air (Wallace et al. 1984), cigarette smoke (Wallace et al. 1984; Jermini et al. 1976), and gasoline vapors (Wallace et al. 1984; Kalf 1987). Epidemiology studies have demonstrated an increase in both leukemia and aplastic anemia in exposed people (Aksoy and Erdem 1978; Goldstein 1977). Toxicity has also been demonstrated in laboratory animals. For example, in studies conducted by the National Toxicology Program, rats and mice were exposed orally to benzene for 2 years (National Toxicology Program 1986). Increased incidences of lymphomas and other tumors were noted in benzene-treated mice compared to controls. Rats responded differently to the carcinogenic effects of benzene. One hypothesis for the difference in species response is that there are significant species differences in benzene metabolism. An approach to addressing this hypothesis is to compare the metabolism of benzene in rats and mice and develop toxicokinetic models to describe this metabolism. In studies conducted in rats and mice exposed to a range of oral and inhaled concentrations, distinct species differences in benzene metabolism were noted (Sabourin et al. 1987, 1988a,b, 1989). A toxicokinetic model describing benzene disposition has been developed (Medinsky et al. 1988, 1989a, b). One of the most important observations from both the animal studies and the toxicokinetic model was related to the nonlinear nature of benzene metabolism. This nonlinear metabolism has important implications for risk assessment and is discussed in detail here.

Benzene Toxicokinetic Model

Toxicokinetics includes the absorption, distribution, metabolism, and elimination of a chemical. Research on the toxicokinetics of benzene and its metabolites

---

[1]Inhalation Toxicology Research Institute, Lovelace Biomedical and Environmental Research Institute, P.O. Box 5890, Albuquerque, NM 87185, USA.

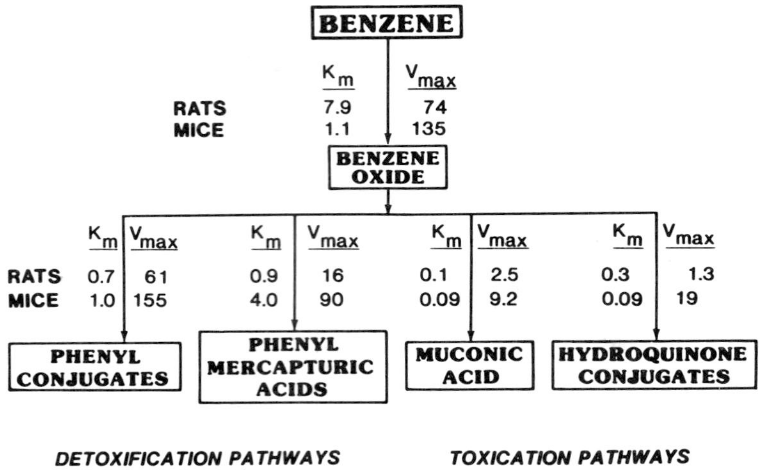

Figure 19.1. Simplified scheme for the metabolism of benzene. Parameters for metabolism of benzene [V_{max} (μmoles/kg h) and K_m (μmoles/L)] for rats and mice were taken from Medinsky et al. (1989b). The phenyl conjugates and the mercapturic acids represent detoxification pathways. Hydroquinone conjugates and muconic acid are markers for the putative toxication pathways in benzene metabolism because they may represent flux of benzene through the pathways leading to the formation of the reactive metabolites benzoquinone and muconaldehyde, respectively

can provide insights into the mechanisms by which benzene may exert its toxic effects and/or the basis for observed differences in species susceptibility to toxicity. Toxicokinetic models may facilitate the development of better estimates of the exposure levels most likely to result in adverse health effects for people. Toxicokinetic models can also facilitate extrapolation from animal studies to human exposures, from high exposure concentrations to low exposure concentrations, and between routes of exposure.

The toxicokinetic model developed for benzene (Medinsky et al. 1988, 1989a,b) was similar to those for other volatile organic compounds. Parameters that were determined to be important in benzene toxicokinetics included those of a physiological nature such as alveolar ventilation, blood flow, cardiac output, tissue volumes, gastrointestinal absorption, and those of a chemical nature such as the blood/air and blood/tissue partition coefficients, and metabolic rate parameters. The metabolic rate parameters (Fig. 19.1) proved to be the most critical parameters for describing total benzene uptake. Both animal studies and results of model simulations indicated that the metabolic parameters could be divided into two types. The first type was a high capacity, low affinity enzyme system that characterized metabolism to the detoxification metabolites: the phenyl conjugates and the mercapturic acid conjugates. The second type was a low capacity, high affinity system. This system characterized metabolism via the two putative

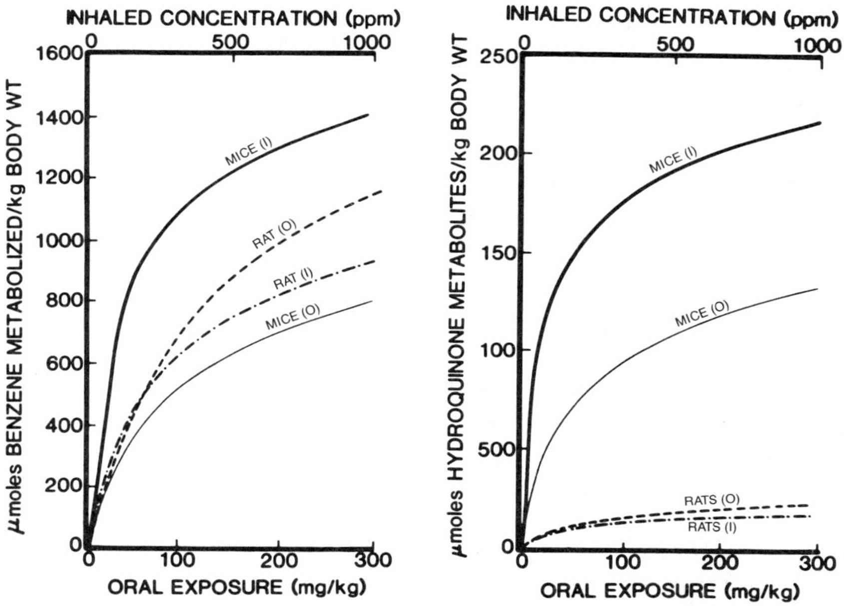

Figure 19.2. Total benzene metabolized (*left*) and total hydroquinone conjugates excreted (*right*) by rats and mice after oral administration (*O*) or a single 6-h inhalation exposure (*I*). *Lines* represent results of model simulations using metabolic parameters in Fig. 20.1

toxication pathways leading to the formation of the reactive metabolites, benzoquinone and muconaldehyde. Ultimately, the stable and water-soluble metabolites, the hydroquinone conjugates and muconic acid, are formed. Thus, hydroquinone conjugates and muconic acid can be viewed as markers for the flux of benzene through these two pathways (Fig. 19.1).

Species Differences in Benzene Toxicokinetics

Using the toxicokinetic model and rate parameters for metabolism outlined in Fig. 19.1, computer simulations of benzene metabolism by rats and mice were conducted for oral exposures and for 6-h inhalation exposures. These simulations have been reported previously (Medinsky et al. 1989a) and adequately describe what was observed in animal studies (Sabourin et al. 1989). As demonstrated in Fig. 19.2(left), after oral exposure rats metabolized benzene to a greater extent than did mice. However, after inhalation exposure, mice metabolized more of the inhaled benzene than did rats (Medinsky et al. 1988). These differences were due to a combination of factors including differences in total metabolic capacity, alveolar ventilation, and gastrointestinal absorption. For example, after inhalation exposure, mice were more efficient in metabolizing benzene because the capacity of the mouse for total metabolism of benzene exceeded that of the rat. However, after an oral exposure, rats metabolized more benzene than mice, primarily because of a combination of the faster gastrointestinal absorption of benzene by mice, which caused saturation of benzene metabolism, and the increased ventilatory rate of the mouse, which permitted increased exhalation of unmetabolized benzene. Because the mouse was identified as a reasonable animal model in the NTP oral toxicity studies, these computer simulations and the experimental data of Sabourin et al. (1988b, 1989) suggested that the total benzene metabolized would not provide insight into species differences in susceptibility to benzene toxicity.

When the toxification metabolites of benzene are examined, a different picture is presented (Fig. 19.2,right). The mice had a higher capacity for formation of hydroquinone metabolites after either oral or inhalation exposure, whereas the rats showed very little capacity for formation of this putative toxic metabolite after either route of exposure. A similar situation was found for production of muconic acid in that mice produce more muconic acid than rats after either oral or inhalation exposure (data not shown). Benzene metabolism in rats was primarily by a detoxification pathway resulting in formation of phenyl conjugates (Sabourin et al. 1988b). These observations are the result of the larger maximum velocity (V_{max}) for metabolism of benzene via the hydroquinone and muconic acid pathways by mice compared to rats (Fig. 19.1). The model-derived values of V_{max} for mice are from 4 to 10 times higher than for rats. Thus, species differences in neoplastic response to benzene appear to correlate with increased production of metabolites associated with the putative toxication pathways in benzene metabolism.

Exposure-Rate Considerations

Examination of the effect of concentration-time products, or time-weighted average (TWA), on the formation of benzene metabolites, blood levels of benzene, and total benzene metabolized demonstrates that parameters selected to measure dose also influence estimates of dose. For example, Fig. 19.3(left) shows the blood levels of benzene predicted by the toxicokinetic model during and after exposures to constant concentration-time products. The highest exposure (600 ppm) resulted in the highest blood levels of benzene. Decreasing the exposure concentration caused a decrease in the peak levels of benzene observed. Thus, this model predicts that peak levels of benzene in blood would be proportional to the inhaled exposure concentration. However, if one looks at the effect of exposure rate on total metabolism of benzene (Fig. 19.3,right), it is evident that decreasing the exposure concentration while keeping the TWA constant resulted in an increasing amount of benzene metabolized. If one extends this observation and examines the effect of exposure rate on the formation of the hydroquinone conjugates, a marker for a putative toxication pathway, it is evident that increasing the time component of the TWA caused an increase in the amount of the

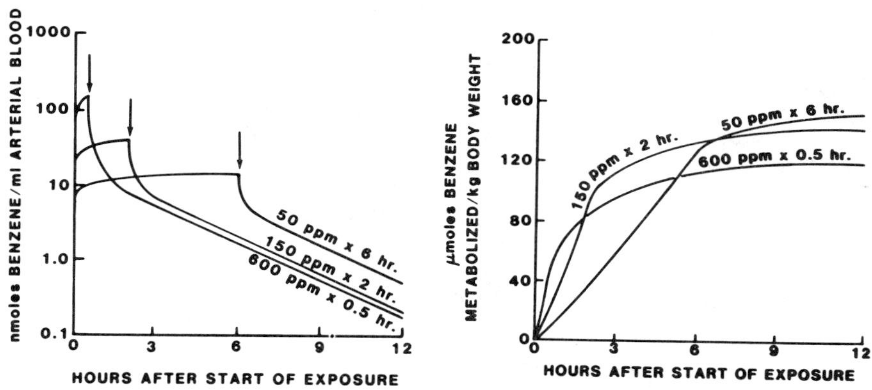

Figure 19.3. Effect of exposure rate on benzene concentration in blood (*left*) and total benzene metabolized (*right*). In all cases, exposures to constant concentration-time products (C × T) were simulated; *arrows* indicate peak blood concentrations of benzene that correspond to the end of exposure

marker for the toxic metabolite formed (Fig. 19.4). Model simulations of a 10 ppm exposure for 30 h produced a 3-fold increase in the formation of this marker compared to a simulated 600 ppm × 0.5 h exposure. The results of these simulations suggest that if a toxic effect is due to a metabolite rather than the parent chemical then measurement of the parent chemical is not always a good indication of dose.

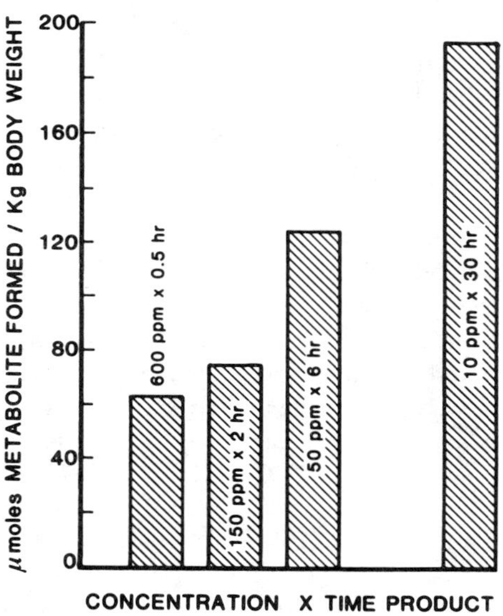

Figure 19.4. Effect of exposure rate on total formation of hydroquinone conjugates. Exposures to constant concentration time products (C × T) were simulated

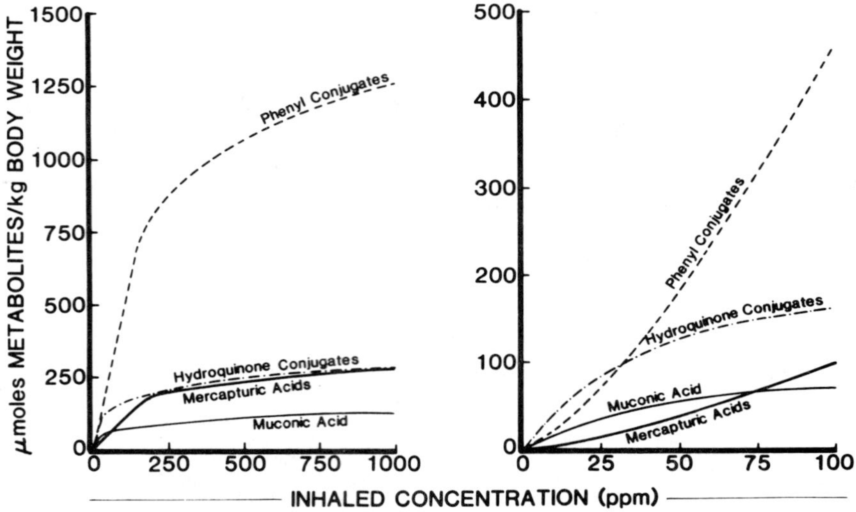

Figure 19.5. Effect of exposure concentration on the formation of benzene metabolites. Simulations of 6-h inhalation exposures of mice to benzene were conducted, and the quantities of individual metaboliltes formed were estimated. Results are shown over a large concentration range (*left*) and at lower concentrations (*right*)

Low Exposure Concentration Extrapolation

The differences in the V_{max} and K_m for the detoxification and toxication pathways outlined in Fig. 19.1 suggest that the metabolic pathways are of two types. The detoxification metabolite pathways can be categorized as high capacity, low affinity, whereas the toxication pathways are high affinity, low capacity. This suggests that metabolism of benzene to the individual metabolites will not be linearly related to inhaled concentration. For example, as shown in Fig. 19.5, model simulations of 6-h inhalation exposures of mice indicate that at high exposure concentrations detoxification metabolites such as the phenyl conjugates are produced in the largest amounts. However, at concentrations below about 25 ppm, there is an increase in the formation of the hydroquinone conjugates compared to the phenyl conjugates. Similarly, at high exposure concentrations, higher levels of the mercapturic acids are produced compared to muconic acid. Yet, at concentrations below about 75 ppm, there is an increased formation of muconic acid compared to mercapturic acids. This nonlinear metabolism has important implications for risk assessment, because it suggests that linear extrapolation of toxicity observed at high exposures may underestimate risk at low exposures.

Leukemia Risk Estimates for Benzene Exposure

Because of the nonlinear nature of benzene metabolism, if one were to estimate leukemia risk for exposure to benzene over a working lifetime, as demonstrated

Table 19.1. Leukemia risk estimates for exposure to benzene over a working lifetime.

Exposure concentration	Metabolite formation (μmoles/kg)		Leukemia risk (no./1000)		
			Linear	Nonlinear	
(ppm)	Hydroquinone conjugates	Muconic acid		Hydroquinone conjugates	Muconic acid
100	30.8	14.8	170[a]	170	170
10	7.18	3.43	17	40	40
1	0.817	0.388	1.7	5.5	4.5

Predicted metabolites formed resulting from simulations of an 8-h exposure to the given concentration of benzene, using metabolite parameters and methods described by Medinsky et al. (1989b). Linear risk estimate was made using the IARC-proposed value of 170 leukemias/1000, and assuming that leukemias at lower levels were proportional to exposure concentration. Nonlinear risk estimate was made using the IARC-proposed value of 170 leukemias/1000, and assuming that leukemias at lower levels were proportional to the amount of metabolite produced.

[a] 170 Leukemias proposed by IARC (1982) as a working life-time risk for benzene exposure to 100 ppm.

in Table 19.1, different estimates of risk could be made depending upon whether the metabolites or the inhaled concentration is used as the estimate of dose. Using the estimated value of 170 leukemias per thousand people proposed by the IARC (1982) as a working lifetime risk for benzene exposures to 100 ppm, a linear risk estimate can be developed using benzene air concentration as an estimate of dose. However, if the quantity of metabolites formed during and after an 8-h exposure to the same benzene air concentrations is used as a dose estimate, then a nonlinear risk estimate results (Table 1). The risk estimate is not linearly related to exposure concentration, but is linearly related to the metabolite formation. Using either hydroquinone conjugates or muconic acid as a dose estimate results in a projected increase in the risk of leukemia at exposures of 1 ppm, compared to risk estimates based on inhaled concentration of benzene. This is due to the nonlinear nature of benzene metabolism.

We have made several assumptions in this model to estimate leukemia risk: (a) Animal models of benzene metabolism can adequately predict benzene metabolism in people. (b) Quantities of hydroquinone conjugates or muconic acid excreted after exposure to different benzene concentrations were proportional to the concentration of the reactive metabolites in target tissue. In other words, whole-body elimination of stable metabolites is a good model for target tissue metabolism of benzene or benzene metabolites. (c) The risk for leukemia would be directly proportional to the concentration of the toxic metabolite in target tissue.

Future Research Directions

The risk estimates presented in Table 1 were based on metabolic rate parameters for mice that were adjusted for humans based solely on species differences in basal metabolism. What is not known is the capacity of human tissue to metabolize benzene. One way to address this question would be to develop a comparative

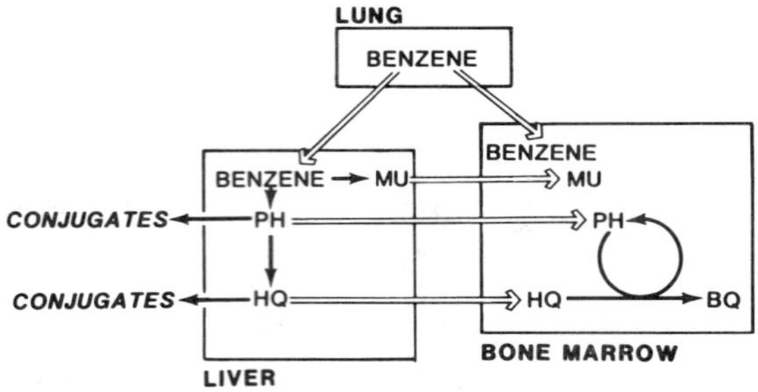

Figure 19.6. Role of tissue-specific metabolism in the formation of putative toxic metabolites of benzene. *HQ*, hydroquinone; *PH*, phenol; *BQ*, benzoquinone; *MU*, muconaldehyde. *White arrows* represent transport of benzene or metabolites in blood. *Black arrows* represent biochemical pathways for metabolism of benzene

approach for estimation of metabolism of benzene. Metabolic parameters for formation of benzene metabolites by rat, mouse, and human tissue could be determined from studies conducted in vitro. The in vitro metabolism of benzene by rat and mouse tissue could be compared to the in vivo metabolism, such as reported in studies of Medinsky et al. (1989b). These data could then be used to better predict the in vivo metabolism of benzene by humans. This approach has been used successfully by Reitz et al. (1989) in estimating human metabolism of methylene chloride in vivo.

What is also not known is the capacity, or the role, of tissue-specific local metabolism in the development of toxicity due to benzene exposure. Rates for formation of toxic metabolites of benzene in the bone marrow by peroxidase-dependent pathways would provide information on the toxicokinetics of hydroquinone and benzoquinone that should be incorporated into models for risk assessment. Information on the extent and rate at which muconaldehyde can leave the liver and enter the bone marrow is important in developing models that use muconic acid as a dose surrogate.

These concepts and the potential role of tissue-specific metabolism in benzene toxicokinetics are shown in Fig. 19.6. Benzene absorbed into the blood can be delivered to the liver and metabolized. Benzene can be metabolized to muconaldehyde, which may then be transported to the bone marrow. Benzene can also be metabolized to phenol and hydroquinone. These metabolites may be transported by the blood to the bone marrow, where they can enter a peroxidase-dependent metabolic pathway, with subsequent formation of benzoquinone (Eastmond et al. 1987). Alternatively, phenol or hydroquinone in the liver can react enzymatically with glucuronic acid or sulfate, subsequently forming conjugated metabolites. These conjugation reactions can be viewed as detoxification pathways, in that the capacity of detoxification enzymes to form glucuronide, sulfate, or glutathione conjugates will ultimately decrease levels of phenol and hydroquinone in bone marrow.

It is likely that both dose and dose-rate changes can affect the balance between detoxification processes, such as glucuronide, sulfate, or glutathione conjugation, and activation processes, such as benzoquinone or muconaldehyde formation. Since the ultimate concentration of the toxic metabolite in the target tissue depends on this balance, both activation and deactivation pathways in benzene metabolism must be taken into account in estimating risks of exposure to benzene.

Acknowledgments. Part of the research reviewed here was conducted under U.S. Department of Energy (DOE) Contract DE-ACO4-76EV01013 through an Interagency Agreement (222-ES-20092) with the National Institute of Environmental Health Sciences (NIEHS). The facilities used for this research were fully accredited by the American Association for Accreditation of Laboratory Animal Care. The author appreciates the useful comments of a number of colleagues at both the LITRI and the NIEHS.

References

Aksoy M, Erdem S (1978) Follow-up study on the development of leukemia in 44 pancytopenic patients with chronic exposure to benzene. Blood 52:285–292

Eastmond DA, Smith MT, Irons RD (1987) An interaction of benzene metabolites reproduces the myelotoxicity observed with benzene exposure. Toxicol Appl Pharmacol 91:85–95

Goldstein BD (1977) Hematotoxicity in humans. J Toxicol Environ Health 2:69–105

IARC (1982) IARC monographs on the evaluation of the carcinogenic risk of chemicals to humans, vol. 29. Some industrial chemicals and dyestuffs. IARC, Lyon, pp 93–148

Jermini C, Weber A, Grandjean E (1976) Quantitative determination of various gas-phase components of the sidestream smoke of cigarettes in room air. Int Arch Occup Environ Health 36:169–181

Kalf GF (1987) Recent advances in the metabolism and toxicity of benzene. CRC Crit Rev Toxicol 18:141–159

Medinsky MA, Sabourin PJ, Henderson RF, Lucier G, Birnbaum LS (1988) Differences in the pathways for metabolism of benzene in rats and mice simulated by a physiological model. Environ Health Perspect 82:43–49

Medinsky MA, Sabourin PJ, Lucier G, Birnbaum LS, Henderson RF (1989a) A physiological model for simulation of benzene metabolism by rats and mice. Toxicol Appl Pharmacol 99:193–206

Medinsky MA, Sabourin PJ, Lucier G, Birnbaum LS, Henderson RF (1989b) A toxicokinetic model for simulation of benzene metabolism. Exp Pathol (submitted)

National Toxicology Program, U.S. Department of Health and Human Services (1986) Toxicology and carcinogenesis studies of benzene in F344/N rats and $B6C3F_1$ mice. National Institutes of Health, National Toxicology Program Technical Report Serial No. 289, Washington, DC

Reitz RH, Mendrala AL, Guengerich FP (1989) In vitro metabolism of methylene chloride in human and animal tissues: use in physiologically based pharmacokinetic models. Toxicol Appl Pharmacol 97:230–246

Sabourin PJ, Chen BT, Lucier G, Birnbaum LS, Fisher E, Henderson RF (1987) Effect of dose on the absorption and excretion of ^{14}C-benzene administered orally or by inhalation in rats and mice. Toxicol Appl Pharmacol 87:325–336

Sabourin PJ, Bechtold WE, Henderson RF (1988a) A high pressure liquid chromatographic method for the separation and quantitation of water-soluble radiolabeled benzene metabolites. Anal Biochem 170:316–327

Sabourin PJ, Bechtold WE, Birnbaum LS, Lucier G, Henderson RF (1988b) Difference in the metabolism of inhaled ^{3}H-benzene by F344/N rats and B6C3F_1 mice. Toxicol Appl Pharmacol 94:128–140

Sabourin PJ, Bechtold WE, Griffith WC, Birnbaum LS, Lucier G, Henderson RF (1989) Effect of exposure concentration, exposure rate, and route of administration on metabolism of benzene by F344 rats and B6C3F_1 mice. Toxicol Appl Pharmacol 99: 421–444

Wallace L, Pellizzari E, Hartwell T, Rosenzweig M, Erickson M, Sparacino C, Zelon H (1984) Personal exposure to volatile organic compounds: direct measurement in breath-zone air, drinking water, food, and exhaled breath. Environ Res 35:293–319

Yin SN, Li Q, Liu Y, Tian F, Du C, Jin C (1987) Occupational exposure to benzene in China. Br J Ind Med 44:192–195

20. Benzene Haematotoxicity and Leukaemia

A. Jacobs[1]

Clinical Effects

Aplastic anaemia due to benzene toxicity was first described at the end of the last century (Santesson 1897), and some 30 years later recognition of the characteristic picture of thrombocytopenia, neutropenia and anaemia, often followed by death from bleeding, led to attempts to reduce exposure and substitute less toxic alternatives for industrial use (McCord 1932; Hamilton 1928). Hamilton (1928) noted that 32% of workers exposed to benzene developed leucopenia and that this could occur even when the concentration was as low as 100 ppm. The first report of leukaemia occurring in a worker exposed to benzene was published by DeLore and Borgomano (1928).

Since that time there have been numerous reports from many countries of cytopenias, aplasia and leukaemia associated with benzene exposure, and these are reviewed by Alderson (1986), the United States Department of Labour (1987) and Aksoy (1988). Despite the many well-documented descriptions of severe bone marrow damage and leukaemia, the effects of low-level exposure to benzene remain uncertain. Episodic exposure may result in transient blood changes with no apparent residual effects, and minimal haematological abnormalities may be difficult to differentiate from unrelated pathological states such as infection, haematinic deficiency or excess alcohol intake. Although removal of a haematologically affected subject from a benzene-containing environment may result in the disappearance of peripheral blood abnormalities the effect on the long-term risk of leukaemia is not known. Patients with refractory anaemia or leukaemia following benzene exposure may have clonal chromosome abnormalities in their bone marrow similar to those seen in other types of chemically induced leukaemia, commonly deletions of chromosomes 5 and/or 7 (Heim and Mitelman 1986). Even in the absence of such specific karyotype abnormalities, there appears to be an increased incidence of aneuploidy and chromosome aberrations

[1]Department of Haematology, University of Wales College of Medicine, Heath Park, Cardiff, CF4 4XN, UK.

such as gaps, breaks and dicentric forms in haematologically normal workers exposed to benzene concentrations of 5–25 ppm when compared to normal controls (Tough and Court-Brown 1965; Forni et al. 1971; Picciani 1979). Although the data are not entirely consistent (Tough et al. 1970), this suggests that genetic damage to haemopoietic cells may occur in the absence of any overt haematological signs and, if an analogy can be drawn with other chemically induced leukaemias and preleukaemic states, particularly those following cytotoxic therapy, a latent period of up to 10 years or more may elapse before the clinical emergence of an aberrant haemopoietic clone.

Epidemiological Evidence

Amongst the many reports of leukaemia in patients previously exposed to benzene are those of Aksoy et al. (1974) who found 26 with acute leukaemia or preleukaemia amongst 28500 shoe workers in Istanbul; Vigliani (1976) who described 66 patients with haematological abnormalities in Milan, 11 of whom developed leukaemia; and Decoufle et al. (1983) who found one case of chronic lymphatic leukaemia, two of acute leukaemia and one of multiple myeloma amongst 259 workers in a United States chemical plant. Aksoy and Erdem (1978) followed 44 pancytopenic patients for periods of between 2 and 17 years and observed six to develop leukaemia, 14 deaths from pancytopenia and one from preleukaemia.

In a number of studies an attempt has been made to compare the incidence of haematological malignancy in exposed workers with a control population. Infante et al. (1977) found a fivefold risk for all leukaemias and a tenfold risk for myeloid and monocytic leukaemia in United States workers exposed during 1940–1949. Vianna and Polan (1979) found a significant excess of deaths from lymphoma in New York State workers exposed to benzene. Rushton and Alderson (1981) studied deaths amongst oil refinery workers in the United Kingdom, and, although there was no excess of deaths from leukaemia compared to national rates, there was an increased risk for those workers with medium or high exposure compared to those with low exposure. Wong (1987a,b) showed an excess mortality of exposed compared to non-exposed United States workers for lymphoma and leukaemia and related this to the cumulative dose. Yin et al. (1987a) found 2676 cases of benzene poisoning amongst 528729 exposed workers in China, aplastic anaemia in shoemakers being 5.8 times as common as in the general population. In a related study the leukaemia mortality was 14 per 100000 in benzene factories compared to 2 per 100000 in control factories (Yin et al. 1987b). The average latency of benzene leukaemia was 11.4 years. Rinsky et al. (1987) followed a cohort of 1165 Ohio men exposed between 1940 and 1965. The standard mortality ratio (SMR) for leukaemia was 337 and for myeloma 409. In both cases mortality was related to cumulative exposure with the SMR for leukaemia increasing from 109 at an annual exposure of 1 ppm to 6639 (CI 1334–

19393) at an annual exposure of 10 ppm. Bond et al. (1986), in their study of 956 Michigan workers, were unable to find any evidence suggesting a causal association between benzene exposure and any particular cause of death.

Austin et al. (1988), in a critical review of the literature, conclude that, while the data support the inference that benzene causes acute myelocytic leukaemia, it is too sparse to substantiate this relationship at 1–10 ppm benzene. They speculate that 1000 men exposed to 10 ppm benzene for a working life of 30 years would suffer about 50 excess deaths due to leukaemia.

A number of case-control studies have been carried out in which prior environmental chemical exposure in patients with haematological disease has been compared with that of control populations. Brandt et al. (1978) showed an excess of patients with acute myeloblastic leukaemia to have had an occupation with exposure to petroleum products. Linet et al. (1989) showed no excess exposure to benzene in patients with aplastic leukaemia, though 41% of the patients in this series were less than 20 years old. In our own study of patients with myelodysplasia, there was a significant association with prior exposure to petrol or diesel compounds (Farrow et al. 1989).

Experimental Data

There is abundant evidence of both benzene haematotoxicity and tumorigenesis in animals at inhaled concentrations down to 100 ppm and less, and this has been recently reviewed (Cronkite 1987; Aksoy 1988). Both leukaemia and lymphoma may be induced experimentally in addition to non-haemopoietic tumours, and marrow aplasia may also be observed. In the haemopoietic system all cell elements can be affected and benzene metabolites may cause both biochemical damage and impairment of proliferation at all levels of the stem cell hierarchy. Inhibition or killing of multipotential stem cells (Green et al. 1981a) or committed erythroid or granulocytic progenitors (Green et al. 1981a; Daudu and Geelhoed 1986) may occur, and the end result is a decrease in numbers of lymphoid, myeloid and erythroid cells in both the bone marrow and the blood (Green et al. 1981b). The decrease in radio-iron incorporation into red cells caused by benzene metabolites (Bolcsak and Nerland 1983) may be the result of impaired proliferation or to a direct effect on haem or globin synthesis (Forte et al. 1976). Similarly, the effect on lymphocytes may be related to impaired protein synthesis (Post et al. 1985).

It is doubtful whether the consequences of benzene exposure in animals can be directly related to humans, either with regard to the precise metabolic derangements produced or with respect to dose-effect relationships. However, the sum of evidence available suggests that genetic damage, haemopoietic damage and malignant transformation may occur in a variety of species including humans. The similarities between species in this respect are perhaps more striking than the differences.

Mechanism of Leukaemogenesis

The mechanism of transforming normal haemopoietic tissue to the leukaemic state is largely unknown, though this process is commonly observed in clinical practice in patients with the myelodysplastic syndrome (MDS). This condition spans the spectrum of haematological abnormalities from minimal cellular changes through hypoplastic states to a preleukaemic proliferation. It may result from chemical exposure either to cytotoxic drugs or to environmental mutagens and is well described as a consequence of benzene exposure (Aksoy 1988). It is characterised by an abnormal clone of haemopoietic stem cells which progresses through a sequence of events in which the earliest stages may be difficult to detect by conventional pathological techniques. The basic abnormality is a genetic change leading to abnormal control of cell proliferation and differentiation. Clonal evolution may be characterised by increasing chromosome abnormalities, increasingly malignant characteristics in the bone marrow, and eventually acute myeloblastic leukaemia (Jacobs (1987).

Three types of "oncogene" may be implicated in MDS, either through mutation, deletion, duplication or abnormal expression. These are genes coding for growth factors and receptors, many of which appear to be directly involved in the regulation of haemopoiesis. A particular concentration of these genes is found on chromosomes 5 and 7 and include GM-CSF, IL3, M-CSF and erythropoietin. The *fms* oncogene codes for the M-CSF receptor and is included in this group. Secondly, the *ras* gene product, a 21-kD protein associated with the inner cell membrane, is concerned with signal transmission from activated receptors and appears to be involved in proliferation control in many different cell types. *Ras* activation by point mutation is found in a wide variety of malignant states, and we have shown this to occur in about 40% of patients with MDS (Padua et al. 1988). *Ras* mutations can be induced experimentally by a number of chemical carcinogens though benzene has not been implicated in this process. Thirdly, those oncogenes such as *myc* and *myb* that code for nuclear-binding proteins playing a direct role in cell cycle control may also be involved in the pathogenesis of MDS. The common occurrence of specific chromosome deletions in MDS raises the possibility of loss of suppressor gene function, but no evidence for this has been forthcoming.

At present we have no precise knowledge of most of the gene defects, their cause, functional significance or the sequence in which they occur. However, it seems likely that the lesions produced by benzene may behave in a similar way to those produced by other chemicals in carrying the transformed haemopoietic cell through this preleukaemic process.

While there are undoubtedly environmental factors responsible for stem cell damage, it is of interest that, amongst 192 cases of MDS analysed in Cardiff, the number of secondary cases seen amongst first-degree relatives was 15 times greater than expected, suggesting that hereditary factors may play a role in predisposition.

Problems

The major problems that remain unsolved in attempting to understand the leukaemogenic role of benzene include:

1. The nature of the genetic lesions caused by benzene metabolites.
2. The relationship between the amount and type of exposure and irreversible stem cell damage.
3. The importance of interaction with other mutagens.
4. Whether hereditary susceptibility is of importance in predisposing to the leukaemogenic effect of benzene.
5. A methodology for the detection of minimal haemopoietic damage.

References

Aksoy M (1988) Benzene carcinogenicity. CRC, Boca Raton

Aksoy M, Erdem S (1978) Follow up study on the mortality and the development of leukaemia in 44 pancytopenic patients with chronic exposure to benzene. Blood 52:285–292

Aksoy M, Erdem S, DinCol G (1974) Leukaemia in shoe workers exposed chronically to benzene. Blood 44:837–841

Alderson M (1986) Occupational cancer. Butterworths, London

Austin H, Delzell E, Cole P (1988) Benzene and leukemia. A review of the literature and a risk assessment. Am J Epidemiol 127:419–439

Bolcsak LE, Nerland DE (1983) Inhibition of erythropoiesis by benzene and benzene metabolites. Toxicol Appl Pharmacol 69:363–368

Bond GG, McLaren EA, Baldwin CL, Cook RR (1986) An update of mortality among chemical workers exposed to benzene. Br J Ind Med 43:685–691

Brandt L, Nilsson PG, Mitelman F (1978) Occupational exposure to petroleum products in men with acute non-lymphocytic leukaemia. Br Med J 1:553

Cronkite EP (1987) Chemical leukemogenesis: benzene as a model. Semin Hematol 24:2–11

Daudu PA, Geelhoed GW (1986) Benzene-induced inhibition of erythroid colony formation in vitro. Am J Med Sci 292:356–362

Decoufle P, Blattner WA, Blair A (1983) Mortality among chemical workers exposed to benzene and other agents. Environ Res 30:16–25

DeLore P, Borgomano C (1928) Leukaemia aigue au cours de l'intoxication benzenique, sur l'origine toxique de certains leukemies aigues et leur relations avec les anemies graves. Journal Medicale Lyons 9:227–233

Farrow A, Jacobs A, West RR (1989) Myelodysplasia, chemical exposure and other environmental studies. Leukaemia 3:33–35

Forni A, Ceppelini E, Pacifico E, Vigliani EL (1971) Chromosome changes and their evaluation in subjects with post exposure to benzene. Arch Environ Health 23:385

Forte FJ, Cohen HS, Rosman J, Freedman FL (1976) Hemin reversal of benzene-induced inhibition of reticulocyte protein synthesis. Blood 47:145–154

Green JD, Snyder CA, LoBue J, Goldstein BD, Albert RE (1981a) Acute and chronic dose/response effects of inhaled benzene on multipotential haemopoietic stem(CFU-S) and granulocyte macrophage progenitor(GM-CFU-C) cells in CD-1 mice. Toxicol Appl Pharmacol 58:492–503

Green JD, Snyder CA, LoBue J, Goldstein BD, Albert RE (1981b) Acute and chronic dose/response effect of benzene inhalation on the peripheral blood, bone marrow, and spleen cells of CD-1 male mice. Toxicol Appl Pharmacol 59:204–214

Hamilton A (1928) The lessening menace of benzol poisoning in American industry. J Ind Hyg 10:227

Heim S, Mitelman F (1986) Chromosome abnormalities in the myelodysplastic syndromes. Clin Haematol 15:1003–1021

Infante PF, Wagoner JK, Rinsky RA, Young RJ (1977) Leukaemia in benzene workers. Lancet ii:76–78

Jacobs A (1987) Human preleukaemia: do we have a model? Br J Cancer 55:1–5

Linet MS, Markowitz JA, Sensenbrenner LL, Warm SG, Weida S, Van Natta ML, Szklo M (1989) A case-control study of aplastic anaemia. Leuk Res 13:3–11

McCord CP (1932) Benzol (benzene) poisoning. A new investigation of the toxicity of benzene and benzene impurities. Industrial Health Consultants Laboratories, Cincinatti

Padua RA, Carter G, Hughes D, Gow J, Farr C, Oscier D, McCormick F, Jacobs A (1988) Ras mutations in myelodysplasia detected by amplification, oligonucleotide hybridisation and transformation. Leukaemia 2:503–510

Picciani D (1979) Cytogenetic study of workers exposed to benzene. Environ Res 19:33

Post GB, Snyder R, Kalf GF (1985) Inhibition of RNA synthesis and interleukin-2 production in lymphocytes in vitro by benzene and its metabolites, hydroquinone and p-benzoquinone. Toxicol Lett 29:161–167

Rinsky RA, Smith AB, Hornung R, Filloon TG, Young RJ, Okun AH, Landrigan PJ (1987) Benzene and leukaemia: an epidemiological risk assessment. N Engl J Med 316: 1044–1050

Rushton L, Alderson MR (1981) A case-control study to investigate the association between exposure to benzene and deaths from leukaemia in oil refinery workers. Br J Cancer 43:77–84

Santesson GG (1897) "Über chronische Vergiftungen mit Steinkohlenbenzin. Vier Todesfälle., Archives of Hygiene 31:336

Tough IM, Court-Brown WM (1965) Chromosome aberrations and exposure to ambient benzene. Lancet i:684

Tough IM, Smith PG, Court-Brown WM, Harnden DG (1970) Chromosome studies on workers exposed to atmospheric benzene. The possible effect of age. Eur J Cancer 6:49–55

US Department of Labor (1987) Occupational exposure to benzene; final rule. Occupational Safety and Health Administration 29 CFR Part 1910 Federal Register, Friday September 11

Vianna NJ, Polan A (1979) Lymphomas and occupational benzene exposure. Lancet i:1394–1395

Vigliani EC (1976) Leukaemia associated with benzene exposure. Ann NY Acad Sci 271:143–151

Wong O (1987a) An industry wide mortality study of chemical workers occupationally exposed to benzene. I. General results. Br J Ind Med 44:365–381

Wong O (1987b) An industry wide mortality study of chemical workers occupationally exposed to benzene. II. Dose response analyses. Br J Ind Med 44:382–395

Yin S-N, Li G-L, Tain F-D, Fu Z-I, Jin C, Chen Y-J, Luo S-J, Ye P-Z (1987a) Leukaemia in benzene workers: a retrospective cohort study. Br J Ind Med 44:124–128
Yin S-N, Li Q, Liu Y, Jin C (1987b) Occupational exposure to benzene in China. Br J Ind Med 44:192–195

Section 5. Mineral Fibers

21. Man-Made Mineral Fiber Exposure: Assessment and a Note on Nonoccupational Exposure to Asbestos in Air

M. Corn[1]

Introduction

"Man-made mineral fibers " (MMMF) is a generic term that refers to fibrous inorganic substances produced by fiberizing molten mixtures composed primarily of rock, clay, slag, glass, or carbon. They are broadly classified as glass fibers (comprising glass wool and glass filament), mineral wool (comprising rock wool and slag wool), ceramic fibers, and graphite fibers. The International Agency for Research on Cancer (IARC) has classified glass wool, rock wool, slag wool, and ceramic fibers as class 2B carcinogens, that is, they are possibly carcinogenic to humans: glass filaments were not classified as carcinogenic to humans, and the carcinogenicity of graphite fibers was not addressed (IARC 1988, p. 152). In the general remarks to the classification document, the following paragraph was included to summarize our current understanding of relevant physical/chemical parameters for assessment of inhalation risk from inhaled MMMF. This paragraph is quoted here because it guided the formulation of this manuscript, which addresses the current knowledge of exposure assessment of MMMF and translation of this knowledge to risk assessments performed in the manner outlined by the National Academy of Sciences (1983):

> Present scientific knowledge indicates that the major determinants of the carcinogenic potential of fibres are biological durability, dimensions (length and diameter) and, as for any other carcinogen, dose to the target organ. In this monograph, specific evidence is evaluated concerning the carcinogenicity of glass fibre, rockwool, slagwool and ceramic fibres as groups. It is conceivable, however, that the fibre characteristics, durability and physical dimensions span the categories of all mineral fibres evaluated here, and that these are the characteristics that are most important in relation to the possible carcinogenicity of a material (IARC 1988, p. 35).

---

[1]Department of Environmental Health Sciences, Division of Environmental Health Engineering, School of Hygiene and Public Health, Johns Hopkins University, Baltimore, Maryland 21205, USA.

The quantities of MMMF produced worldwide are difficult to ascertain. The latest published worldwide figures for glass (textile and insulation) and mineral wool are 4585 million kg in 1973 (WHO 1983). United States production of glass and mineral wool was estimated to be approximately 2100 million kg in 1983. United States ceramic fiber production in 1983 was approximately 41 million kg (USEPA 1986). Ceramic fiber production is increasing at a rapid rate in the United States, and 1988 production may well be twice that of 1983. At this time there are no reliable figures for graphite production.

Clearly, MMMF are versatile and valuable materials. The technical production processes for MMMF will not be described here; they are extensively described elsewhere for mineral wool (Ohberg 1987), glass wool (NIOSH 1977), and ceramic fiber (Mohr and Rowe 1978). The technology of producing graphite and other nonoxide fibers (silicon carbide, boron nitride, silicon nitride) does not differ, in general, from that utilized for glass and mineral wool. Because of the potential for ceramic and graphite fibers in environments with high heat and corrosion (engine components, in particular) and in composites with metals for special applications, we can anticipate future large increases in ceramic and graphite fiber production.

The purpose of this paper is briefly to summarize our knowledge of production worker, secondary user, and nonoccupational exposure to MMMF. Also, because the biological effects of asbestos is the factor which initially stimulated concern for the health effects of MMMF, the present approach to utilization of United States exposure assessments of asbestos in nonoccupational environments will be summarized.

Exposure Assessment: Occupational

Morphology of Fibers and Carcinogenicity

The response of experimental animals to fibrous materials is the basis for our current scientific views on the relevant fiber dimensions for the initiation or promotion of tumorigenic responses in man following inhalation. The studies of Stanton et al. (1977) have been particularly influential; results for MMMF and asbestos are consistent using the same experimental techniques (Stanton et al. 1981). The subject of inhaled fiber dimensions related to the various manifestations of asbestos disease has been reviewed by Lippman (1988). Table 21.1 displays Lippman's conclusions regarding fiber sizes and disease.

Inhalable long thin fibers have the greatest carcinogenic potential. A fiber is defined as a particle with an aspect ratio (length/diameter) of at least 3:1. There are suggestions from the animal experimentation literature that this should be altered to 5:1 for more relevance to the biological activity of fibers. Inhalable fibers are those with diameters less than 3.5 μm (Timbrell 1965).

Instrumental techniques for microscopically viewing fibers cannot be divorced from assessment of the hazard potential of fibers. The phase contrast microscope (PCM) has a limit of resolution of about 0.3 μm, at best, and perhaps 0.5 μm practically speaking, with commonly available microscopes. Thus, PCM count-

Table 21.1. Asbestos exposure indices. (From Lippman 1988. Reprinted with permission.)

Disease	Relevant exposure index
Mesothelioma	Number of fibers with: L > 5 μm; D < 0.1 μm
Lung cancer	Number of fibers with: L > 10 μm; D > 0.15 μm
Asbestosis	Surface area of fibers with: L > 2 μm; 0.15 μm ≤ D ≤ 2 μm

ing and sizing of fibers using United States Occupational Safety and Health Administration procedures includes fibers with diameters between 0.5 μm and 3.5 μm, with lengths greater than 5 μm (US Department of Labor 1986). The PCM cannot differentiate fiber type, so that all fibers, regardless of composition, are counted. The electron microscope (EM), on the other hand, can resolve down to angstrom sizes (1 Å = 10^{-8} cm) and can identify the chemical nature of the fiber. Both of these microscopic techniques were developed for asbestos-in-air assessment and have been co-opted for MMMF assessment.

Another characteristic of MMMF relevant to assessment of exposure and to the methodologies of air sample evaluation is their breakage characteristics in contrast to asbestos, which is composed of fibrous units or fibrils, down to its constituent structural units, and therefore breaks into smaller and smaller fibers with application of degradative energy. MMMF break concoidally, or across their axis (Assunciao and Corn 1975). This means that airborne MMMF are associated with particle size distributions that include fewer fibers not visible in the PCM, as contrasted to asbestos-in-air particle size distributions, which are associated with large numbers of fibers visible only in the EM. The designations of United States glass fiber sizes and their associated diameters are shown in Table 21.2 (Corn 1979). Unfortunately, for fibrous glass and presumably for other MMMF, it has not been possible to correlate the airborne fiber sizes to the nominal fiber

Table 21.2. United States glass fiber size designations and associated diameters. (From Corn 1979. Reprinted with permission.)

	Nominal diameter			Nominal diameter	
Fiber size	Min (μm)	Max (μm)	Fiber size	Min (μm)	Max (μm)
AAAAA	0.05	0.20	J	11.43	12.70
AAAA	0.20	0.50	K	12.70	13.97
AAA	0.50	0.75	L	13.97	15.24
AA	0.75	1.50	M	15.24	16.51
A	1.50	2.52	N	16.51	17.78
B	2.52	3.81	P	17.78	19.05
C	3.81	5.08	Q	19.05	20.32
D	5.08	6.35	R	20.32	21.58
E	6.35	7.62	S	21.58	22.86
F	7.62	8.89	T	22.86	24.13
G	8.89	10.12	U	24.13	25.40
H	10.12	11.43			

Table 21.3. Distribution of measured average employee exposure to fibers, expressed as cumulative percentage of samples less than stated concentrations, in 16 facilities in the United States. (From Corn 1979. Reprinted with permission.)

	Average concentration (fibers/cm³)										
Plant	<0.005	<0.01	<0.05	<0.1	<0.5	<1	<1.5	<2	<5	<10	<20
1	6.1	11.2	72.4	83.7	98.0	98.0	99.0	100			
2	0	0	32.7	63.6	96.4	100					
3	40.8	50.7	87.3	95.8	97.2	100					
4	6.7	27.8	78.9	93.3	100						
5	3.3	13.1	91.8	98.4	100						
6	47.7	76.6	97.3	98.2	100						
7	0	0	3.2	14.3	81.0	92.1	100				
8	17.1	40.0	89.5	99.0	100						
9	16.7	42.2	95.5	98.9	100						
10	82.5	97.9	100								
11	3.9	13.7	66.7	84.3	100						
12	33.2	65.4	98.6	99.1	100						
13	1.3	6.3	60.8	75.9	97.5	100					
14	6.0	27.7	89.2	97.6	98.8	100					
15	0	0	9.3	28.0	69.3	81.3	89.3	93.3	97.3	98.6	100
16	7.8	20.0	87.8	94.4	98.9	100					

sizes present in the bulk materials, thus necessitating the utilization of air sampling to assess each particular exposure situation.

It is also important to understand that the airborne particulate matter associated with MMMF is a mixture of fibrous and nonfibrous particles, with the latter in great abundance. In United States glass and mineral wool plants, it has not been possible to correlate total suspended particulate matter, expressed as milligrams per cubic meter, to fiber concentration, expressed as fibers/cubic centimeter (Esmen et al. 1976). Schneider and Breum (1987) reported total mass/respirable fiber concentrations sufficiently well correlated to suggest the use of a screening limit value for total gravimetric particulate mass in hygienic assessments; this is an approximate, practical tool, at best.

Airborne Concentrations of MMMF in Manufacturing

MMMF personal exposures of production workers were determined in two major epidemiological investigations of the industry. Results in the United States study are comparable to those associated with a European investigation (Schneider and Breum 1987). Table 21.3 indicates the distribution of employee exposures to fibers less than 3 μm diameter based on 8-h personal samples in the United States study of 16 production facilities. The exposure to fibers less than 1 μm in diameter, determined by transmission EM (TEM), is shown in Table 21.4 for the same facilities. The facilities in this study were either mineral wool or glass wool

Table 21.4. Concentration (fibers/cm^3) of fibers <1 μm in diameter in 16 facilities in the United States. [From Esmen et al. 1976. Reprinted with permission by the *American Industrial Hygiene Association Journal* Vol. 40(2):108–117 (1979).]

	Forming		Production		Manufacturing		Maintenance		Quality control		Shipping		Overall	
Plant[a]	Mean	SD	Mean	SD	Mean	SD	Mean	SD	Mean	SD	Mean	SD	Mean	SD
1	0.002	0.000	0.73	2.3	0.01	0.007	0.01	0.03	0.45	0.76	0.002	0.002	0.38	0.17
2	0.04	0.03	0.04	0.02	0.03	0.03	0.04	0.04	0.07	0.08	0.01	0.01	0.04	0.03
3	–	–	0.03	0.03	–	–	0.20	0.71	–	–	0.005	0.003	0.08	0.42
4	0.01	0.005	0.19	0.29	0.12	0.42	0.02	0.02	0.02	0.01	0.009	0.005	0.10	0.28
5	0.01	0.01	0.01	0.007	0.004	0.002	0.01	0.005	0.02	–	0.01	0.002	0.01	0.01
6	0.002	0.001	0.004	0.004	0.004	0.003	0.008	0.03	0.007	0.01	0.004	0.004	0.005	0.02
8	–	–	0.05	0.04	0.02	0.01	0.02	0.02	–	–	0.02	0.01	0.03	0.03
10	0.003	0.002	0.003	0.002	0.004	0.002	0.01	0.01	0.002	0.002	0.002	0.002	0.004	0.007
11	0.12	0.09	0.02	0.02	0.02	0.03	0.01	0.01	0.01	0.01	0.01	0.01	0.03	0.06
12	0.006	0.005	0.004	0.005	0.003	0.003	0.002	0.001	0.005	0.004	0.003	0.001	0.003	0.004
13	0.04	–	0.03	0.03	0.03	0.04	0.02	0.02	–	–	0.01	0.01	0.02	0.04
15	2.0	2.6	6.49	9.37	5.25	14.6	1.3	2.2	12.0	5.83	0.58	0.38	4.4	9.9
16	0.03	0.02	0.04	0.03	0.07	0.04	0.22	0.84	0.03	–	0.01	0.007	0.01	0.04

[a] TEM data are not reported for facilities 7, 9, and 14 because the analytical method used was less reliable than that at other plants.

Table 21.5. Ranges of airborne fiber concentrations in typical exposure situations. (From Cherrie et al. 1986. Reprinted with permission.)

Fiber concentration (fibers/cm³)	Location/use
Ultralow (<0.0001)[a]	Outdoor: rural area
	Buildings: thermal insulation
Extremely low (0.0001–0.001)[a]	Outdoor: large cities
	Buildings: ceiling boards
	Ventilation systems
Very low (0.01–0.1)	
Glass continuous filament	Production and use
Coarse glass fiber	Production and use
Ceiling boards	Buildings: some damage, some ventilation ducts
Low (0.01–0.1)	
Glass wool	Production and most secondary production
Rock wool	Production and most secondary production
Rock/slag wool	Production and most secondary production
Ceiling boards	Buildings: severe damage
Medium (0.1–1.0)	
Fine glass fiber	Production
Rock wool	Some secondary production and user industry
Ceramic fiber	Primary production and user industry
Glass wool	User industry
High (>1.0)	
Very fine glass fiber	Production and use
Glass/rock wool, loose	User industry: blowing into attic
Glass/rock wool, without dust suppressant	Production and use
Ceramic fiber	Secondary production and some user industry

[a] Estimated from TEM measurement.

production plants. In general, concentrations of total and respirable particulate matter, expressed as milligrams per cubic meter, were somewhat higher in mineral wool plants. Exposures also varied by production area, on average, but variation in personal exposures in a given area was large, causing overlap of exposure range between many areas.

With the variation shown in Tables 21.3 and 21.4 in mind, it is possible to very roughly generalize the typical exposure concentrations shown in Table 21.5 with measurements derived from a variety of sources (Cherrie et al. 1986). Investigations to obtain improved estimates of personal exposures to fibers in air for all categories of exposure, as contrasted to airborne concentrations, are in progress in the United States.

Few measurements of exposure are available for graphite fibers in air, and exposures in composite fiber applications are not available. Also, few categories of user exposures have been characterized, particularly in view of the hundreds of known user applications.

Durability

The IARC report (1988, p. 82) cautions that:

While it is important to attempt to predict in-vivo solubility when estimating the possible biological effects of MMMF, it is difficult to reproduce in vitro the varying conditions of pH and concentrations of complexing agents which fibres encounter in the intra- and extracellular environments of the lung. Furthermore, no overall generalization regarding the absolute or relative solubilities of the main families of such fibres can be made on the basis of the results of the studies reported. For example, while most samples of glasswool have proved to be relatively soluble and ceramic fibres relatively insoluble, there have been exceptions – at least one sample of glass fibre was extremely durable and one sample of ceramic fibres relatively soluble.

Nonoccupational Exposure to Asbestos-in-Air and Relevance to Nonoccupational MMMF Exposure

The risk to health of inhaled asbestos to those nonoccupationally exposed has been determined to be a significant one to United States school children by the United States Congress, which has passed legislation to address the risk (US Congr. AHER, 1986). On the other hand, the province of Ontario in Canada and the United Kingdom judge the risk to be insignificant and they have not promulgated regulations for interventions in schools or buildings. The United States approach to the problem is presented in United States Environmental Protection Agency documents (USEPA 1979, 1983, 1985). I have been highly critical of the United States approach (Corn 1986). Among the diverse data relevant to a decision to regulate or not regulate are:

- The asbestiform materials of concern.
- The fiber sizes of significance.
- The concentration of significant fibers in air that is breathed.
- The personal exposure of individuals at risk.
- The methods to be used to assess the risk.

Each of these "inference guidelines," as the United States National Academy of Sciences (1983) refers to them, involves judgment of inadequate data, judgements which may differ among well-meaning scientists and professionals, leading to different estimates of lifetime risks due to mesothelioma and lung cancer. Until agreement can be reached on inference guidelines, different risk estimates will be calculated by different individuals. In the United States, the government has rejected air sampling as an assessment tool, a policy which has led to the public perception of inhalation hazard as existing when the asbestos material is present in bulk materials, i.e., spray fireproofing, acoustic ceilings, lagging, etc. This approach has not utilized risk assessment techniques. It has resulted in the widespread practice of removing asbestos whenever it is present, regardless of type or condition. The ramifications for real estate transfers and building owner potential liability have been great. Business decisions, not scientific health-based decisions are the driving forces for dealing with asbestos in buildings. Table 21.6 indicates indoor and outdoor asbestos concentrations in 387 air samples from the

Table 21.6. Summary statistics for average airborne asbestos fiber concentrations (fibers/cm³). (Battelle Columbus 1987. Reprinted with permission.)

			ACM	
	Outdoor	Category 1	Category 2	Category 3
Median	0.00000	0.00010	0.00040	0.00058
Mean	0.00039	0.00099	0.00059	0.00073
Sample size	48 (sites)	6 (buildings)	6 (buildings)	37 (buildings)
Standard deviation	0.00096	0.00198	0.00052	0.00072

ACM, asbestos-containing material.

Notes: 1. The data points used in the calculation of each statistic are the average concentration within a building (for indoor samples) or the concentration outside each building (for outdoor samples).

2. Category 1—no ACM.
Category 2—ACM primarily in good condition.
Category 3—at least one area of significantly damaged ACM.
3. The data comprises 387 air samples and 48 outdoor air samples.
4. No asbestos fibers detected in 83% of the 387 samples.

United States; air sample analysis was by TEM. Building concentrations are comparable to outdoor air concentrations.

The resolution of public concerns for asbestos in buildings has enormous implications for MMMF utilization. What will be substituted for asbestos-containing materials? Will substitution of many MMMF equally concern segments of the population now concerned with the health risks of asbestos materials? What are the projected risks to health of nonoccupational exposure to MMMF and can scientific consensus be reached with regard to these risks? How will the public be educated and informed of the issues? These questions—and others—are already under discussion in the United States. Thus, MMMF are a perfect example which needs to be examined in the context of utilizing diverse data to formulate public policy with regard to health risk.

References

Assunciao J, Corn M (1975) The effects of milling on diameters and lengths of fibrous glass and chrysotile asbestos fibers. Am Ind Hyg Assoc 38:811

Battelle Columbus Division, Price Associates. Alliance Technologie Corp., Energy Technology Consultants, Midwest Research Institute (1987). Assessing asbestos exposure in public buildings. Prepared for the Exposure Evaluation Division, Office of Toxic Substances, Office of Pesticides and Toxic Substances, US Environmental Protection Agency (Draft Report, November 20)

Cherrie J, Dodgson J, Groat S, McClaren W (1986) Environmental surveys in the european man-made mineral fiber production industry. Scand J Work Environ Health 12[Suppl 1]:18–25

Corn M (1979) An overview of inorganic man-made fibers in man's environment. In: Lemen R, Dement JM (eds) Dusts and disease. Pathotox, Park Forest South IL, pp 23–36

Corn M (1986) Asbestos and disease: an industrial hygienist's perspective. Am Ind Hyg Assoc J 47(9):515–523

Esmen NA, Corn M, Hammad YY, Whittier D, Kostsko N (1976) Summary of measurements of employees' exposure to airborne dust and fiber in sixteen facilities producing man-made mineral fibers. Am Ind Hyg Assoc J 40:108–117

IARC (1988) Man-made mineral fibres and radon. In: IARC Monograph, vol 43. International Agency for Research on Cancer, Lyon, p 152

Lippman M (1988) Asbestos exposure indices. Environ Res 46:81–106

Mohr JG, Rowe WP (1978) Fiber glass. Van Nostrand, New York

National Academy of Sciences (1983) Risk assessment in the federal government: managing the process. National Academy of Sciences, Washington DC

NIOSH (1977) Criteria for a recommended standard . . . occupational exposure to fibrous glass. DHEW, NIOSH publication no 77-152; NTIS publication no PB-274195, National Institute for Occupational Safety and Health, Cincinnati

Ohberg I (1987) Technological development of the mineral wool industry in Europe. Ann Occup Hyg 31(4B):529–545

Schneider T, Breum NO (1987) Screening man-made vitreous fibre exposures by gravimetric dust measurements. Ann Occup Hyg 31:547–555

Stanton MF, Layard ML, Tegeris A, Miller E, May M, Kent E (1977) Carcinogenicity of fibrous glass: pleural response in the rat in relation to fiber dimension. J NCI 58:587–603

Stanton MF, Layard M, Tegeris A, Miller E, May M, Morgan E, Smith A (1981) Relation of particle dimension to carcinogenicity in amphibole asbestos and other fibrous minerals. J NCI 67:965–975

Timbrell V (1965) The inhalation of fibrous dusts. In: Biological effects of asbestos. Ann NY Acad Sci 132:255

US Congr (US Congress Asbestos Hazard Emergency Response Act) AHERA (1966) Publication L, no 99-519, 100 stat 2970. Section two of AHERA added a new Title II, 88201-214 to the Toxic Substances Control Act ("TSCA"), codified at 15 U.S.C. 992641-54 [Suppl IV, 1986]

US Department of Labor (1986) OSHA occupational exposure to asbestos 29CFR1926.58

USEPA (US Environmental Protection Agency) (1979) Asbestos containing materials in school buildings: a guidance document, pt 1. Office of Toxic Substances, Washington DC

USEPA (US Environmental Protection Agency) (1983) Guidance for controlling friable asbestos-containing materials in buildings. Office of Pesticides and Toxic Substances, Washington DC (EPA rep no 560/5-83-002)

USEPA (US Environmental Protection Agency) (1985) Guidance for controlling friable asbestos-containing materials in buildings. Office of Pesticides and Toxic Substances, Washington DC (EPA rep no 560/5-85-002)

USEPA (US Environmental Protection Agency) (1986) Durable fiber industry profile and market outlook. Office of Pesticides and Toxic Substances, Washington DC

WHO (World Health Organization) (1983) Biological effects of man-made mineral fibres. Report on a WHO/IARC meeting. EURO reports and studies 81, Copenhagen

22. Carcinogenicity of Fibers in Experimental Animals – Data and Evaluation

F. Pott[1]

Introduction

The carcinogenicity of asbestos has been recognized for 5 decades. For this reason, measures have been introduced in many countries; at the same time, the need for fibrous materials has grown in many industries. At least 20 years ago, questions were raised as to whether those fibers which do not belong to the asbestos group may cause cancer. In the years which followed, many carcinogenicity studies with natural and man-made mineral fibers (MMMF) were carried out in experimental animals. Because these studies have been recently reviewed (IARC 1987; IARC 1988), only a summary will be given in this paper. Finally, a proposal for the classification of inorganic and organic fibers according to their carcinogenicity will be presented.

Inhalation Studies

The results of inhalation experiments with fibers from glass, rock, slag, and ceramic fibers are described in eight publications (Davis et al. 1984; Goldstein et al. 1983; Le Bouffant et al. 1984; McConnel et al. 1984; Muhle et al. 1987; Pigott and Ishmael 1982; Smith et al. 1987; Wagner et al. 1984). Only one of 26 groups exposed to one of these types of MMMF showed a borderline effect (Davis et al. 1984). However, also some inhalation studies with animals exposed to crocidolite asbestos (see Table 22.1) showed no, or only weak, carcinogenicity of this material, which has been proven to be a very strong carcinogen in humans. High tumor rates were reported in only one of these experiments (Wagner et al. 1974); however, this result could not be confirmed in two further studies by the same research group (Wagner et al. 1985, 1987). In contrast, chrysotile induced high lung tumor incidences in rats (Davis et al. 1978; Davis and Jones 1988; Wagner et al. 1984). This effect can be explained by multiplying the number of chrysotile fibers in the body after splitting the fiber bundles (Bellmann et al. 1987).

---

[1]Medical Institute for Environmental Hygiene at the Heinrich Heine University of Düsseldorf, Auf'm Hennekamp 50, D-4000 Düsseldorf, FRG.

Table 22.1. Inhalation experiments with crocidolite.

Concentration/m^3		Time of exposure				Animals with tumor			
[mg]	Fibers $\times 10^6$	Hours per week	Months	Species	Animals examined (n)	(n)	(Mes)	(%)	Reference
12.5		7	0.03	Rat	43	7	1	16	Wagner et al. 1974
12.6		7 × 5	3	Rat	36	15	1	42	Wagner et al. 1974
10.7		7 × 5	6	Rat	18	4	0	22	Wagner et al. 1974
10.6		7 × 5	12	Rat	26	20	2	77	Wagner et al. 1974
10.3		7 × 5	24	Rat	18	13	0	72	Wagner et al. 1974
Control				Rat	126	7	0	6	Wagner et al. 1974
50	1105	4 × 4	24	Rat	66	5	0	8	Reeves et al. 1974
13.5		7 × 5	35	Baboons	10	0	0	0	Goldstein et al. 1983
5	430	7 × 5	12	Rat	43	3	1	7	Davis et al. 1978
10	860	7 × 5	12	Rat	40	1	0	3	Davis et al. 1978
10	1630	7 × 5	12	Rat	28	1	0	4	Wagner et al. 1985
10		6 × 5	12	Rat	28	1	0	4	Wagner et al. 1987
7		6 × 5	24	Rat	57	3	1	5	Smith et al. 1987
7		6 × 5	24	Hamster	58	0	0	0	Smith et al. 1987
2.2[a]	162	5 × 4	12	Rat	50	1	0	2	Muhle et al. 1987

[a] Sample contained longer fibers than UICC crocidolite; other samples: UICC crocidolite, except Reeves et al. (1974).
Mes, mesothelioma.

Intratracheal Studies

Only four experiments with MMMF in which the intratracheal exposure route has been used have been published. Two groups of hamsters and one group of rats developed tumors after treatment with very thin glass fibers (Pott et al. 1984, 1987); four groups of rats and hamsters did not (Feron et al. 1985; Smith et al. 1987).

Serosa Tests

The majority of the carcinogenicity studies with MMMF were performed by administration of dust suspensions into the pleural space or the peritoneal cavity (IARC 1988). The working groups of Stanton (Stanton et al. 1981) and of Wagner used the pleura, Davis and our group preferred the peritoneum. The intraperitoneal (i.p.) test has been proven to be more simple and more sensitive. Table 22.2 shows recent results which have not yet been reviewed in the IARC Monographs (1988).

Figure 22.1 summarizes the dose-response relationships between the numbers of injected fibers of a certain size range and the incidences of mesotheliomas and sarcomas in the abdominal cavity observed in many experiments carried out over a 15-year period. Remeasurements of all the fiber types applied in these experiments were done by Bellmann and Muhle of the Fraunhofer Institute of Toxicology and Aerosol Research, Hannover (see also Pott et al. 1989a). Figure 1 shows, in principle, a dependency of the tumor incidence on the number of fibers injected. However, there exists a relatively wide variation of the data. This may be caused by small numbers of animals per group, uneven particle size distributions in the dust samples, and difficulties with the fiber measurements. On the other hand, consideration has to be given as to whether a negative or a weak positive result after i.p. injection of 10^9 or more fibers can exonerate a fiber type from the suspicion of being carcinogenic. It is acknowledged that the i.p. test may be not suitable for the examination of plastic fibers because these fibers agglomerate much more than mineral fibers.

Because consideration has to be given to the influences of particle deposition and elimination on the dose-response relationship in the lung, the results from the i.p. test cannot be extrapolated directly to the lung. Qualitatively, however, there seems to be a good analogy between the carcinogenicity of substances in the lung and at the serosa (see Table 22.3). Nevertheless, there are some general objections to the extrapolation of the serosa tests, which can be summarized in two theses:

Thesis 1. Inhalation exposure is the method of choice if the potential risk to humans warrants the effort (Wehner et al. 1986).

Thesis 2. Nonphysiological routes of exposure lead to irrelevant (false-positive) results.

Table 22.2. Recent results of a carcinogenicity test after i.p. injection (Pott et al. 1989b).

Dust	Dose i.p. (mg)	Fibers[a] injected × 10^6 (*n*)	Median fiber size: Length (μm)	Median fiber size: Diameter (μm)	Rats with tumor[b] (*n*)	Total examined (*n*)	(%)	Grade of adhesions
Actinolite	0.01	4	1.1	0.1	8	35	23	0
Actinolite	0.05	20	1.1	0.1	15	36	42	0
Actinolite	0.25	102	1.1	0.1	20	36	56	0
Chrysotile UICC/Can	0.05	40	0.67	0.05	12	36	33	0
Chrysotile UICC/Can	0.25	202	0.67	0.05	23	34	67	0
Chrysotile UICC/Can	1.0	808	0.67	0.05	30	36	83	0–1
Glass fibers 104/475	5	680	2.6	0.15	34	53	64	3
Basalt fibers	75	59	17	1.1	30	53	57	3
Ceramic fibers Ff	45	150	13	0.89	33	47	70	3
Ceramic fibers M	75	21	16	1.4	12	54	22	3
Wollastonite India	100	430	8.1	1.1	0	54	0	1–2
γ-Ferric oxide hydrate	250		0.5	0.03	8	49	16	2
α-Ferric oxide hydrate	250		0.1	0.01	2	51	4	2
Kevlar fibers	20	1260	4.9	0.48	3	53	6	2
Polypropylene	50	409	10	1.1	2	51	4	2–3
Polyvinylchloride	500		(Nonfibrous)		5	51	10	2
Titanium dioxide, anatase	100		(Nonfibrous)		2	53	4	0
Wood dust, beech	250		(Nonfibrous)		0	52	0	2
Sodium chloride solution	10				2	102	2	0

[a] Definition: length > 5 μm, diameter < 3 μm, aspect ratio > 5:1.
[b] Mesothelioma or sarcoma in the abdominal cavity; tumors of the uterus are excluded.

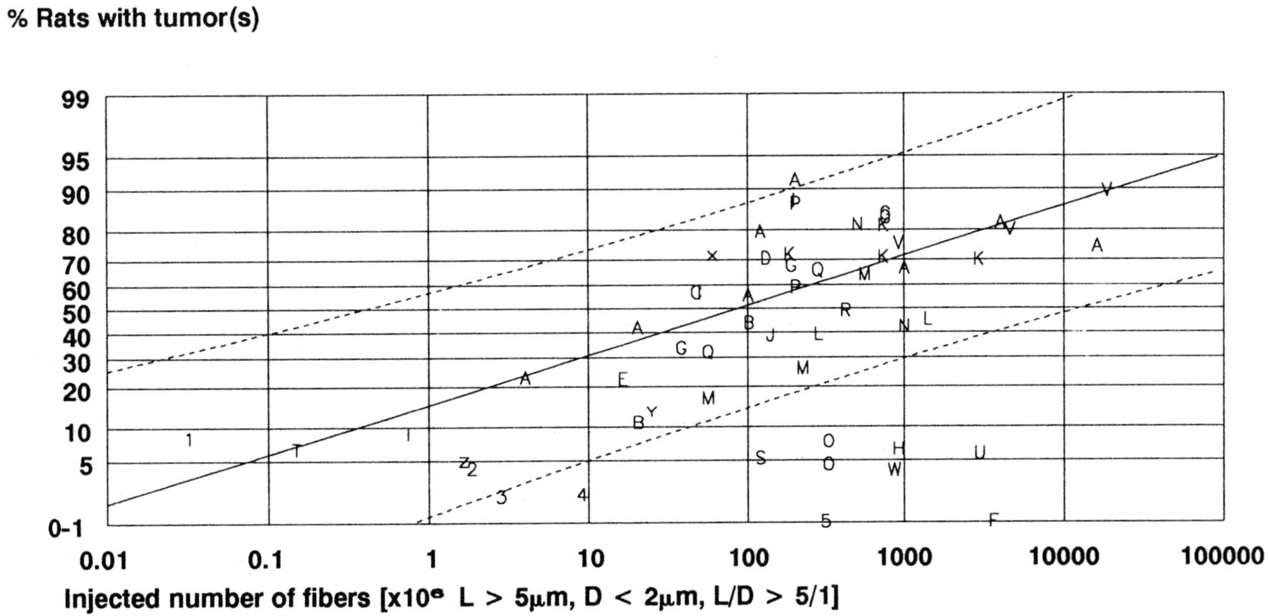

Figure 22.1. Dose-response relationship between the number of fibers injected intraperitoneally and the tumor rate in the abdominal cavity (tumors of the uterus are excluded). Materials: *A*, actinolite; *B*, anthophyllite; *C*, basalt FRG; *D*, ceramic f. Fi; *E*, ceramic f. Ma; *F*, chrys. A HCl; *G*, chrys. B; *H*, chrys. cal.; *I*, crocidolite SA; *J*, crocidolite UICC; *K*, erionite Turkey; *L*, glass 100 Pen; *M*, glass 104/475; *N*, glass 104/E1; *O*, glass 104/E1 HCl; *P*, glass 104/E1 NaOH; *Q*, glass 104/E2; *R*, glass 104/E3a; *S*, glass 106; *T*, glass filam. 3; *U*, kevlar; *V*, nemalite; *W*, polypropylene; *X*, rock S; *Y*, rock S fine; *Z*, slag Ko fine; *1*, slag Rh coarse; *2*, slag Rh fine; *3*, slag Zi coarse; *4*, slag Zi fine; *5*, wollastonite

Table 22.3. Carcinogenicity of hardly soluble dusts in the lung and at the serosa of humans and rats.

	Lung		Pleura/peritoneum	
	Human	Rat	Human	Rat
Asbestos	+	+	+	+
Erionite	+		+	+
Glass fibers		+		+
Other mineral fibers	(+)			+
Nickel	+	+		+
Cadmium	(+)	+		+
Quartz	(+)	+		+
Several other dusts				−

These theses may be correct for some substances, however, do they indeed hold true in the case of inhalable fibers? The findings after inhalation of crocidolite in experimental animals show unequivocally that false-negative results have to be expected when the carcinogenic potency of a fibrous dust is lower than that of crocidolite. Therefore, in contrast to the theses mentioned above, two antitheses can be set up:

Antithesis 1. In the case of mineral fibers, realistic exposure leads to irrelevant (false-negative) results.

Antithesis 2. Nonphysiological exposure leads to relevant results which are suitable for the classification of fibers according to their carcinogenicity.

Classification of Natural and Man-Made Mineral Fibers by IARC

IARC, as well as the United States Environmental Protection Agency (US EPA), also evaluate the carcinogenicity of fibers on the basis of results with injection tests. Table 22.4 shows the evaluations of IARC for some nonasbestos mineral fibers. However, there are some objections to this system, because the mineralogic names or trade names for fibrous materials are only names for a fiber family and do not describe fiber types which have identical properties. Therefore, a different proposal for the classification of fibers has been developed which can be explained here only very briefly.

Proposal for the Classification of Inorganic and Organic Fibers According to Their Carcinogenicity

The fiber properties, rather than fiber names, are the basis of the rationale of this proposal. It can be concluded from the experimental results that all elongated particles are carcinogenic if these fibers are "sufficiently" long and thin and if their durability in the body is "sufficiently" long. Figure 22.2 illustrates that the

Table 22.4. Evaluations of the carcinogenicity of some nonasbestos mineral fibers by IARC (1987, 1988).

Dust	Degree of evidence		Overall evaluation	
	Humans	Animals	Carcinogenic to humans	Group
Erionite	Sufficient	Sufficient	Carcinogenic	1
Wollastonite	Inadequate	Limited	Not classifiable	3
Attapulgite	Inadequate	Limited	Not classifiable	3
Sepiolite	No data	Inadequate	Not classifiable	3
Glass fibers	Inadequate	Sufficient	Possibly	2B
Glass filaments	Inadequate	Inadequate	Not classifiable	3
Rock fibers	Limited	Limited	Possibly	2B
Slag fibers		Inadequate	Possibly	2B
Ceramic fibers	No data	Sufficient	Possibly	2B

carcinogenic potency of fibers is dependent on certain parameters: length, diameter, aspect ratio, durability in the body, and perhaps additionally on surface properties.

As a first step in the classification, all inorganic and organic fibers with an aspect ratio of more than 5 to 1, which are longer than 5 μm and thinner than 2 μm are proposed to be put in the group labeled "possibly carcinogenic to humans." The fiber types should stay in this kind of waiting list (see Table 22.5) until a scientific committee has evaluated their behavior in the body and has

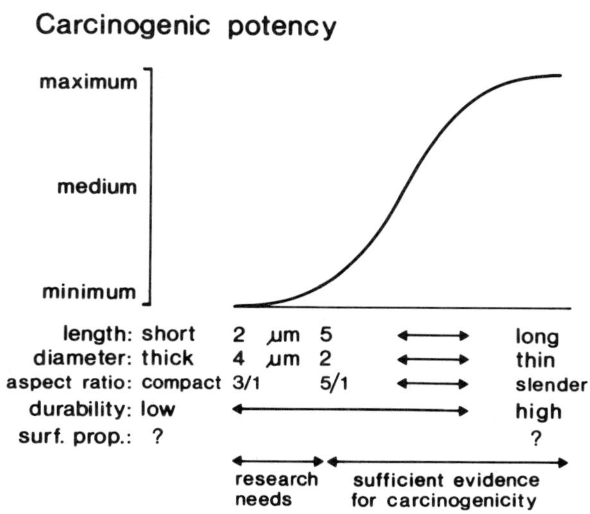

Figure 22.2. Simplified illustration of the carcinogenic potency of fibers dependent on some parameters

Table 22.5. Proposal for the classification of inorganic and organic fibers according to their carcinogenicity.

	Carcinogenic in humans		
	Probably	Possibly	Probably not carcinogenic in humans
Corresponding classification			
of IARC	2 A	2 B	4
US EPA	B 2	C	
MAK FRG	A 2	B	
European Commission	2	3	
General criteria for classification		a. *"Waiting-list"*	
1. According to result(s) of valid carcinogenicity test(s)	1. Fibers L > 5 μm D < 2 μm L/D > 5/1 proven to be carcinogenic in the lung or at the serosa (i.p. test or intrapleural test)	All inorganic and organic fibers or fiber bundles capable of splitting up into fibers L > 5 μm D < 2 μm L/D > 5/1 until evaluation of their behavior in the body and definite decision on whether their durability is sufficient for tumor induction or not b. *"Softies"* Durable fibers either "short" (L 2–5 μm) or "thick" (D 2–4 μm) or "compact" (L/D > 3/1–5/1) (these groups need further investigation)	1. Inorganic fibers L > 5 μm D < 2 μm L/D > 5/1 proven to be not, or only weakly, carcinogenic in the i.p. test after injection of at least 0.5–1 × 10^9 fibers (size range mentioned above). If results on carcinogenicity using other test models are available, they have to be negative. (A reliable test procedure for exoneration of organic fibers from the suspicion of being carcinogenic is not yet available)

2. According to analogy with tested fibers	2. Fibers L > 5 µm D < 2 µm L/D > 5/1 which have not been examined sufficiently in experiments but are considered to act analogously to tested fibers evaluated to be probably carcinogenic in humans	2. Fibers L > 5 µm D < 2 µm L/D > 5/1 which have not been examined sufficiently in experiments but are considered to act analogously to fibers tested and evaluated to be probably noncarcinogenic in humans
Individually evaluated fiber types		
1. According to result(s) of valid carcinogenicity test(s)	1. Attapulgite, Caceres Attapulgite, Torrejon Ceramic fibers, Fiberfrax Ceramic fibers, JM Glass fibers, JM 104/475 Glass fibers, JM 104/E Nemalite Rock fibers, FRG Rock fibers, S Sepiolite, Finland Silicon carbide fibers scw-1-s	1. HCl-leached glass fibers JM 104/E HCl-leached UICC chrysotile Sepiolite, Spain, specimen tested
2. According to analogy with tested fibers	2. All fiber types having a "durable" chemical composition: ceramic, quartz, glass JM "475" and "E," silicon carbide, others	2. Gypsum fibers

L, length; D, diameter.

decided definitely on whether their durability is sufficient for tumor induction or not. Moreover, durable fibers which are relatively thick, short, or compact cannot be excluded from the suspicion of being carcinogenic, although their carcinogenic potency may be low (see Fig. 22.2 and Table 22.5, "Softies"). These groups of fibers need further investigation.

In a second step, fiber types already established to be carcinogenic in experimental animals should be classified as "probably carcinogenic in humans" and those proven to be noncarcinogenic in a sensitive test model would be put in the group classified as "probably not carcinogenic in humans" (see Table 22.5, "Individually evaluated fiber types," point 1). In a third step, fibers which have nof been examined sufficiently by experiments, but are considered to act analogously to tested and classified samples may be classified according to the analogy used with the tested fibers. Examples for such fiber types are proposed in Table 23.5 under "Individually evaluated fiber types," point 2.

It has to be underlined that, regarding risk assessment, the influences of particle deposition and the lung clearance on the dose-response relationships have to be considered carefully.

References

Bellmann B, Muhle H, Pott F, König H, Klöppel H, Spurny K (1987) Persistence of man-made mineral fibres (MMMF) and asbestos in rat lungs. Ann Occup Hyg 31:693–709

Davis JMG, Jones AD (1988) Comparisons of the pathogenicity of long and short fibres of chrysotile asbestos in rats. Br J Exp Path 69:717–737

Davis JMG, Addison J, Bolton RE, Donaldson K, Jones AD, Wright A (1984) The pathogenic effects of fibrous ceramic aluminium silicate glass administered to rats by inhalation or peritoneal injection. In: Biological effects of man-made mineral fibres: proceedings of a WHO/IARC conference in association with JEMRB and TIMA. Copenhagen, April 20–22, 1982, vol. 2. World Health Organization, Regional Office for Europe, Copenhagen, pp 303–322

Davis JMG, Addison J, Bolton RE, Donaldson K, Jones AD (1978) Inhalation and injection studies in rats using dust samples from chrysotile asbestos prepared by a wet dispersion process. Br J Exp Path 67:113–129

Feron VJ, Scherrenberg PM, Immel HR, Spit BJ (1985) Pulmonary response of hamsters to fibrous glass: chronic effects of repeated intratracheal instillation with or without benzo[a]pyrene. Carcinogenesis 6:1495–1499

Goldstein B, Rendall REG, Webster I (1983) A comparison of the effects of exposure of baboons to crocidolite and fibrous-glass dusts. Environ Res 32:344–359

IARC (1987) Silica and some silicates. IARC Monogr Eval Carcinog Risk Chem Hum: 42

IARC (1988) Man-made mineral fibres and radon. IARC Monogr Eval Carcinog Risk Chem Hum: 43

LeBouffant L, Henin JP, Martin JC, Normand C, Tichoux G, Trolard F (1984) Distribution of inhaled MMMF in the rat-lung – long-term effects. In: Biological effects of man-made mineral fibres: proceedings of a WHO/IARC conference in association with JEMRB and TIMA. Copenhagen, April 20–22, 1982, vol. 2. World Health Organization, Regional Office for Europe, Copenhagen, pp 143–168

McConnell EE, Wagner JC, Skidmore JW, Moore JA (1984) A comparative study of the fibrogenic and carcinogenic effects of UICC Canadian chrysotile B asbestos and glass microfibre (JM 100). In: Biological effects of man-made mineral fibres: proceedings of a WHO/IARC conference in association with JEMRB and TIMA. Copenhagen, April 20–22, 1982, vol 2. World Health Organization, Regional Office for Europe, Copenhagen, pp 234–252

Muhle H, Pott F, Bellmann B, Takenaka S, Ziem U (1987) Inhalation and injection experiments in rats to test the carcinogenicity of MMMF. Ann Occup Hyg 31:755–764

Pigott GH, Ishmael J (1982) A strategy for the design and evaluation of a 'safe' inorganic fibre. Ann Occup Hyg 26:371–380

Pott F, Ziem U, Mohr U (1984) Lung carcinomas and mesotheliomas following intratracheal instillation of glass fibres and asbestos. In: Bergbauberufsgenossenschaft Bochum (ed) VI. International Pneumoconiosis Conference 1983. Vol. 2. Wirtschaftverl. NW, Verlag für neue Wissenschaft, Bremerhaven

Pott F, Ziem U, Reiffer FJ, Huth F, Ernst H, Mohr U (1987) Carcinogenicity studies on fibres, metal compounds, and some other dusts in rats. Exp Pathol (Jena) 32:129–152

Pott F, Bellmann B, Muhle H, Rödelsperger K, Rippe RM, Roller M, Rosenbruch M (1989a) Intraperitoneal injection studies for the evaluation of the carcinogenicity of fibrous phyllosilicates. In: Bignon J (ed) First International Conference on Health Related Effects of Phyllosilicates, Paris, March 16–18, 1989. Springer, Berlin Heidelberg New York (in press)

Pott F, Roller M, Ziem U, Reiffer FJ, Bellmann B, Rosenbruch M, Huth F (1989b) Carcinogenicity studies on natural and man-made fibres with the intraperitoneal test in rats. In: Bignon J, Peto J, Saracci R (eds) Non-occupational exposure to mineral fibres. International Agency for Research on Cancer, Lyon, pp 173–179 (IARC Scientific Publications, no 90)

Reeves AL, Puro HE, Smith RG (1974) Inhalation carcinogenesis from various forms of asbestos. Environ Res 8:178–202

Smith DM, Ortiz LW, Archuleta RF, Johnson NF (1987) Long-term health effects in hamsters and rats exposed chronically to man-made vitreous fibers. Ann Occup Hyg 31:731–754

Stanton MF, Layard M, Tegeris A, Miller E, May M, Morgan E, Smith A (1981) Relation of particle dimension to carcinogenicity in amphibole asbestoses and other fibrous minerals. J Natl Cancer Inst 67:965–975

Wagner JC, Berry G, Skidmore W, Timbrell V (1974) The effects of the inhalation of asbestos in rats. Br J Cancer 29:252

Wagner JC, Berry GB, Hill RJ, Munday DE, Skidmore JW (1984) Animal experiments with MMM(V)F—effects of inhalation and intrapleural inoculation in rats. In: Biological effects of man-made mineral fibres: proceedings of a WHO/IARC conference in association with JEMRB and TIMA. Copenhagen, April 20–22, 1982, vol 2. World Health Organization, Regional Office for Europe, Copenhagen

Wagner JC, Skidmore JW, Hill RJ, Griffiths DM (1985) Erionite exposure and mesotheliomas in rats. Br J Cancer 51:727–730

Wagner JC, Griffiths DM, Munday DE (1987) Experimental studies with palygorskite dusts. Br J Ind Med 44:749–753

Wehner AP, Dagle GE, Clark ML, Buschbom RL (1986) Lung changes in rats following inhalation exposure to volcanic ash for two years. Environ Res 40:499–517

23. The Health Effects of Man-Made Mineral Fibers

R. Saracci[1]

Introduction

Twenty-five years ago, the first comprehensive conference on biological effects of asbestos (Whipple 1965) reviewed and summarized the wealth of information already accumulated by that time on the adverse health effects of asbestos in the occupational and in the general environment (the first clearly described case of asbestosis going back to Cooke, 1924). The proceedings of the discussion record a remark (Schefers 1965) calling attention, by analogy to asbestos, ". . . on the number of new synthetic fiber products which need to be studied as potential health hazards. . . ." In the subsequent quarter of a century, the response to this need has materialized in a number of epidemiological studies investigating populations exposed not only to a variety of such synthetic fibers (man-made mineral fibers, ceramic fibers), but also to natural mineral particles other than asbestos: fibrous, such as erionite, attapulgite, sepiolite; or nonfibrous, such as silica, which has been—after extensive studies on silicosis—the object of a renewed wave of interest recently aimed at clarifying a possible carcinogenic role (IARC 1987).

Among these different classes of agents, the interest in man-made mineral fibers stems mainly from two sources; the large use of these materials for acoustic and thermal insulation as well as for reinforcing plastics (entailing a total worldwide production of more than 4 million tons yearly and with an exposed population of workers of 40000 in the production industry alone, IARC 1988) and their synthetic nature. This, in principle, gives room for adapting the characteristics of the material in such a way as to reduce or eliminate unwanted biological effects.

Epidemiological Evidence

The epidemiological evidence on the health effects of man-made mineral fibers has been reviewed on several occasions as new results of studies became available

[1]International Agency for Research on Cancer, 150 Cours Albert-Thomas, F-69372 Lyon Cedex 08, France.

(Saracci 1985, 1986; Saracci and Simonato 1982). Here I will recall some key points, especially as they relate to the activity of the International Agency for Research on Cancer in this field. The results of the epidemiological studies can be considered in respect to three long-term pathological endpoints: chronic pulmonary disease, respiratory (lung) cancer, and mesothelioma.

The issue of *chronic respiratory disease* has been addressed from different perspectives and at different levels of detail. Mortality studies have included, in addition to one very small investigation on autopsied workers (Gross et al. 1976), data from seven historical cohort investigations (Bayliss et al. 1976; Enterline and Henderson 1975; Enterline and Marsh 1984; Enterline et al. 1987; Morgan et al. 1984; Robinson et al. 1982; Saracci et al. 1984; Shannon et al. 1984, 1987; Simonato et al. 1987) providing death figures for a disease category usually labeled as "nonmalignant respiratory disease." This encompasses a heterogenous mix of pathological and clinical conditions and therefore lacks both sensitivity in detecting and specificity in attributing an excess risk. Morbidity studies consist of 16 cross-sectional investigations (Carpenter and Spoylar 1945; Engholm and Schmalensee 1982; Grimm 1983; Hill et al. 1973, 1984; Maggioni et al. 1980; Malmberg et al. 1984; Moulin et al. 1987; Nasr et al. 1971; Siebert 1942; Sixt 1983; Skuric and Stahuljak-Beritic 1984; Utidjian and de Treville 1970; Valentin et al. 1983; Weill et al. 1984; Wright 1968), which have employed questionnaires eliciting symptoms, chest radiography and lung function tests, or a combination of these approaches. All except two (Engholm and Schmalensee 1982; Sixt et al. 1983) have been conducted on workers still employed ("survivors") in the production industry, with the majority of subjects exposed for not more than 20 years. The best-designed of these studies (Weill et al. 1984), involving 1028 male workers in production plants in the United States, found an association between small opacities in chest X-rays (categories 0/1, 1/0, 1/1 of the International Labor Office Classification) and duration of employment in the industry in smokers.

Evidence on *respiratory (essentially, lung) cancer* is derived from six studies investigating mortality (Bayliss et al. 1976; Enterline and Henderson 1975; Enterline and Marsh 1984; Enterline et al. 1987; Morgan et al. 1984; Robinson et al. 1982; Shannon et al. 1984, 1987), one study investigating mortality and cancer incidence (Saracci et al. 1984; Simonato et al. 1987), and one only investigating cancer incidence (Moulin et al. 1986) in workers of the man-made mineral fiber production industry, and from one study in users (building industry) (Engholm et al., 1987). The two largest and most informative studies were carried out, with successive follow-ups (Enterline and Marsh 1984; Enterline et al. 1987; Saracci et al. 1984; Simonato et al. 1987), in the production industry of western Europe and of the United States on a multiplant basis. Both investigations included industrial hygiene surveys, documenting that, under present-day conditions, the concentration of airborne respirable fibers is low, mostly in the range of 0.01–0.1 f/ml. Only a small minority of environmental measurements reached values of 1 f/ml and above. Reconstruction of past exposure, an essential step to investigate the relationships, if any, with lung cancer risk or with other diseases with delayed onset

Table 23.1. Lung cancer mortality by time since first exposure to man-made mineral fibers and by process, using national and local reference populations. Both sexes. (From Simonato et al., 1987)

Type of production process		Years since first exposure to MMMF									
		0–9		10–19		20–29		30+		Total	
		SMR	Exp	SMR	Exp	SMR	Exp	SMR	Exp	SMR	Exp
Total cohort	*National*	107	36.3	127	55.3	119	42.8	173[a]	16.8	125[a]	151.2
	Local	96	40.8	112	62.5	104	48.8	152[a]	19.1	110	171.2
Rockwool/ slagwool	*National*	103	16.5	122	24.6	125	17.6	178	6.7	124	65.4
	Local	104	16.4	122	24.7	124	17.8	185	6.5	124	65.3
Glasswool	*National*	82	15.8	139	24.5	125	23.3	173[a]	9.8	127[a]	73.3
	Local	68	19.3	113	30.0	100	28.9	138	12.4	103	90.5
Continuous filament	*National*	224[a]	4.0	96	6.3	0	2.0	0	0.3	120	12.5
	Local	176	5.1	76	7.9	0	2.1	0	0.3	97	15.4

[a] SMR statistically significant ($p < 0.05$).
MMMF, man-made mineral fibers; Exp, expected numbers of deaths.

required indirect approaches, as virtually no measurements of airborne fibers were available from the past. In the European study, which was coordinated by the International Agency for Research on Cancer, an ad hoc historical environmental investigation was carried out, questioning key informants in a structured way on past plant conditions. The most recent follow-up and analysis of this study (Simonato et al. 1987) includes 21967 workers in 13 plants (seven rockwool/slagwool, four glasswool, two continuous filament) in seven European countries (Denmark, Federal Republic of Germany, Finland, Italy, Norway, Sweden, and the United Kingdom). For no cause of death, except lung cancer, was a consistent excess (within plants or across plants) found in respect to the expectation based on the mortality experience of the general population of the relevant country or region. In particular, mortality for nonmalignant respiratory disease was not found to be in excess. For lung cancer, however, the standardized mortality ratio (SMR) was elevated overall (189 observed deaths, 151.2 expected, SMR 125, $p < 0.05$) and showed an increase with time since first exposure, but not with duration of employment. As shown in Table 23.1, the overall SMR for lung cancer is elevated for each of the three production processes (rockwool/slagwool, glasswool, continuous filament) when compared with national rates. For both the rockwool/slagwool and glasswool subcohorts, a trend (statistically not significant at $p = 0.05$) of increasing SMRs with time since first exposure is present. No such pattern is apparent for continuous filament, for which, however, the number of expected deaths 20 years or more after first exposure is very small. Except for continuous filament, the highest SMRs are found for the periods of 30 or more years since first exposure. When adjustment factors based on the local rather

Table 23.2. Lung cancer mortality by time since first exposure to man-made mineral fibers and by technological phase at first exposure as identified in the historical environmental investigation. Both sexes. (From Simonato et al., 1987)

Production phase		Years since first exposure to MMMF									
		0–9		10–19		20–29		30+		Total	
		SMR	Exp	SMR	Exp	SMR	Exp	SMR	Exp	SMR	Exp
Early											
	National	0	0.2	0	0.6	270	1.5	244	2.5	214[a]	4.7
	Local	0	0.1	0	0.5	317	1.3	295[a]	2.0	257[a]	3.9
Intermediate											
	National	0	1.0	106	2.8	161	4.3	211	1.9	139	10.1
	Local	0	1.0	107	2.8	164	4.3	217	1.8	141	9.9
Late											
	National	111	15.3	127	21.2	93	11.8	84	2.4	113	50.7
	Local	112	15.2	126	21.4	90	12.2	77	2.6	111	51.5

[a] SMR statistically significant ($p < 0.05$).
MMMF, man-made mineral fibers; Exp, expected number of deaths.

than the national lung cancer mortality rates are applied, the SMRs for glasswool and continuous filament no longer appear raised: instead, the SMRs for rockwool and slagwool remain nearly unchanged. When the results of the historical environmental investigation are taken into account, the excess lung cancer in the rockwool/slagwool process is mostly found to be concentrated in the small subcohort of workers (331 subjects) first employed in the "early" technological phase of the rockwool/slagwool production process, defined as the phase when no dust-suppressing agents (oil, binders) were used, and/or the manual batch manufacturing process was used. This phase was ranked as the one with the highest estimated level of airborne fibers.

For this phase, there were ten observed lung cancer deaths against an expectation of 3.9 with an SMR of 257 ($p < 0.05$), all the excess being concentrated in the period of 20 or more years since first exposure. The excess appears to decline (Table 23.2) with the introduction of dust-suppressing agents, which started to a variable extent in the "intermediate" technological phase and became systematic in the "late" technological phase (defined as one where modern production techniques were used to manufacture man-made mineral fibers with oil added). Among the possible explanations for this two- to threefold observed excess of lung cancer, the influence of tobacco smoking should, of course, have been considered (no data were available on the smoking habits of the workers); however, it is most unlikely that a more than twofold excess risk could be accounted for only by an imbalance of smoking habits of the working population in respect to the general population against which they were compared.

Instead, some information was available which enabled classifying the workers by their employment in the plants during periods in which some minor use was

Table 23.3. Mortality from lung cancer[a] by time since first exposure and type of plant. European and United States cohorts. (From Doll, 1987)

Time since first exposure (years)	Rock- or slagwool SMR	Rock- or slagwool Deaths	Fibrous glasswool SMR	Fibrous glasswool Deaths	Glass filaments SMR	Glass filaments Deaths
Less than 10	102	19	81	26	138	15
10–19	130	43	113	88	61	15
20–29	125	42	108	140	111	36
30 or more	148	37[b]	119	125	79	13
All periods	128	141[c]	110	379	93	79

[a] Respiratory cancer for plants in the United States.
[b] $p < 0.05$.
[c] $p < 0.01$.

made of asbestos, or of formaldehyde or bitumens: these analyses did not indicate any effects which could be attributed to these other potential carcinogenic agents. An increased lung cancer mortality 20 or more years since first exposure was associated specifically with the use of slag; but, on the other hand, it did not show any relationship with the extent to which slag was used during the early technological phase. Hence, the conclusion can be drawn that the excess should be attributed to the exposure to the agents, including the rockwool fibers, as they were present in the early technological phase in this type of production process. As to the glasswool production process, no similar excess could be demonstrated: this process was ranked as likely to involve lower levels of airborne respirable fibers than the rockwool/slagwool production process.

These results are consistent with those from the most recent follow-up (Enterline et al. 1987) of the study in the United States, in which 16661 workers were followed up at 17 plants (six rockwool/slagwool, six glasswool, three continuous filament, and two mixed glasswool and continuous filament). An excess of lung cancer was found 20 years or more after first exposure, greater for rockwool/slagwool workers (45 observed, 34.4 expected, SMR 130.8) than for glasswool workers (52 observed, 46.9 expected, SMR 110.8). In addition, in a case-control study within the cohort, taking smoking into account, a weak (statistically significant) relationship was found with time-weighted cumulative exposure in the rockwool/slagwool workers, but not in the glasswool and continuous filament workers.

As to the mesothelioma, there were four cases recorded in the combination of the two studies, indicating no apparent increase in risk.

As indicated in Table 23.3, when combined (Doll, 1987), the results from the United States and Europe reinforce the indication of an excess risk in the rockwool and slagwool section of the industry 30 years or more since first exposure, i.e., among workers whose first exposure took place several decades back in time and when older processes were in operation.

Comments and Conclusions

The results of the epidemiological studies allow a number of comments and conclusions pertinent to the issue of risk assessment for man-made mineral fibers:

1. *Risk identification* represents the first step in the process of risk assessment, evaluating the degree of evidence that an agent is (or is not) capable, under whatever appropriate conditions, of a toxic effect (fibrosis, cancer, etc.) in humans. At this level, no clear conclusion can be drawn for the time being as to *chronic respiratory diseases*: only one adequate cross-sectional study indicates a probable causal link between exposure to fibers and chest X-ray anomalies (small opacities) only in smokers.

As to *mesothelioma*, there is no apparent excess risk. For *lung cancer*, the epidemiological evidence indicates that respirable fibers, as present in the environmental conditions of the early slagwool/rockwool production processes, are likely to have played an etiological role in the production of lung cancer.

This evidence has translated, when combined with the evidence in experimental animals, into the evaluation from the IARC Monographs Working Group, which met in June 1987, classifying rockwool/slagwool as "possibly carcinogenic" for humans (IARC 1988). As shown in Table 23.4, because of sufficient evidence in animals, glasswool and ceramic fibers were, however, also classified as "possibly carcinogenic" for humans, notwithstanding the inadequate epidemiological evidence for the first, and the absence of any epidemiological evidence for the time being for the second (continuous filament simply could not be classified because of the inadequacy of both the epidemiological and animal data). That different types and degrees of scientific evidence may converge into similar classes of strengths of evidence of carcinogenicity in humans is intrinsic to any classification aiming at a synthetic evaluation which, of necessity, must (a) reduce the *multiple dimensions* of the scientific evidence to just one axis of classification; and (b) to have along this axis a restricted number of categories, lest the classification becomes unduly sensitive to variation of assessment when it is applied by different groups of assessors. While this compaction of the evidence is instrumental in making a scientific classification usable for the public health purposes of risk identification, it is wise for the reader not to concentrate exclusively on the final, synthetic evaluation, but to pay attention also to the component elements on which it is based.

2. In respect to the quantitative aspect, namely the second step in the process of risk assessment (*risk estimation*), the situation is, for the moment, quite unclear. No direct estimations of past airborne levels of respirable fibers are available. Within the context of the European study, airborne levels have been simulated and measured in a rockwool factory during the manufacture of a product without the use of oil or binder in an experimental way, giving results ranging from 9.2 to 25.6 f/ml. This "dry" production line was completely mechanized and, for the purpose of the experiment, no worker exposure occurred in the area sampled. If one uses these latter results and the average duration of exposure in the rockwool/slagwool early phase (2.7 years) the estimate of

Table 23.4. Evaluation of the evidence. (From IARC, 1988)

Fiber	Animals	Humans	Overall
Glasswool	Sufficient	Inadequate	2B–possibly carcinogenic
Rockwool	Limited	Limited for	2B–possibly carcinogenic
Slagwool	Inadequate	rock/slagwool	2B–possibly carcinogenic
Continuous filament	Inadequate	Inadequate	3–not classifiable
Ceramic	Sufficient	No data	2B–possibly carcinogenic

the highest cumulative exposure is in the range of 25–70 fibers/year per ml (Simonato et al. 1987). The relationship between such a cumulative fiber dose and the SMR observed in the early phase groups would turn out not to be out of line with the dose-response estimates from other studies on workers exposed to crysotile asbestos (Dement et al. 1987; McDonald et al. 1983; McDonald et al. 1980; Peto et al. 1985; Rubino et al. 1979). However, the exposure estimates for man-made mineral fibers are highly tentative and the dose estimates in the asbestos studies themselves suffer, as has recently been pointed out (Doll and Peto 1985), from severe errors related to the conversion of old particle counts into fiber levels. There is indeed a possibility that past levels of exposure to man-made mineral fibers in the rockwool/slagwool production process may have been lower than those indicated in the simple and single experiment of the "dry" production line and were in fact nearer the present and rather low levels, leading to cumulative exposures of the order of 10 fibers/year per ml or below. This low exposure would then appear discrepant in respect to the size (more than twofold) of the observed excess of lung cancer, unless rockwool/slagwool fibers are regarded as more carcinogenic than crysotile asbestos. While this appears implausible if fiber potency is assessed on a mass basis, it is not incompatible with the available evidence from studies in humans and animals if potency is assessed per unit fiber (Peto 1989). Whatever the correct approach, it appears that, because of the absence of reliable quantitative exposure indicators, no risk estimation can yet be done with confidence for man-made mineral fibers.

References

Bayliss DL, Dement JM, Wagoner JK, Blejer HP (1976) Mortality patterns among fibrous glass production workers. Ann NY Acad Sci 271:324–335

Carpenter JL, Spolyar LW (1945) Negative chest findings in a mineral wool industry. J Ind Med Assoc 38:389–390

Cooke WE (1924) Fibrosis of the lungs due to inhalation of asbestos dust. Br Med J 2:147

Dement JM, Harris RL, Symons JJ, Shy C (1987) Estimates of dose-response for respiratory cancer among chrysotile asbestos textile workers. Ann Occup Hyg 26:869–887

Doll R (1987) Symposium on MMMF, Copenhagen, October 1986: overview and conclusions. Ann Occup Hyg 31 (4B):805–819

Doll R, Peto J (1985) Effects on health of exposure to asbestos. HMSO, London

Engholm G, Schmalensee G (1982) Bronchitis and exposure to man-made mineral fibers in non-smoking construction workers. Eur J Respir Dis [Suppl] 118:73–78

Engholm G, England B, Fletcher BC, Hollin N (1987) Respiratory cancer incidence in Swedish construction workers exposed to man-made mineral fibers and asbestos. Ann Occup Hyg, 31:663–676

Enterline PE, Henderson V (1975) The health of retired fibrous glass workers. Arch Environ Health, 30:113–116

Enterline PE, Marsh GM (1984) The health of workers in the MMMF industry. In: Guthe T (ed) Biological effects of man-made mineral fibres. Proceedings of a WHO/IARC conference in association with JEMRB and TIMA, Copenhagen, 20–22 April 1982, vol 1. World Health Organization, Regional Office for Europe, Copenhagen, pp 311–339

Enterline PE, Marsh GM, Henderson V, Callahan C (1987) Mortality update of a cohort of US man-made mineral fibre workers. Ann Occup Hyg, 31 (4B):625–656

Grimm HG (1983) Berufliche Exposition mit kunstlichen mineralischen Fasern und ihre Auswirkungen auf die Gesundheit. Verlag des Vereins Deutscher Ingenieure, Düsseldorf, pp 153–159 (VDI-Berichte 475)

Gross P, Harley RA, David JMG (1976) The lungs of fibre glass workers: comparison with the lungs of a control population. In: National Institute for Occupational Safety and Health (ed) Occupational exposure to fibrous glass. NIOSH, Washington, DC, pp 249–263 (US Department of Health, Education and Welfare publication no NIOSH 76-151)

Hill JW, Whitehead WS, Cameron JD, Hedgecock GA (1973) Glass fibres: absence of pulmonary hazard in production workers. Br J Ind Med 30:174–179

Hill JW, Rossiter CE, Foden DW (1984) A pilot respiratory morbidity study of workers in a MMMF plant in the United Kingdom. In: Guthe T (ed) Biological effects of man-made mineral fibres. Proceedings of a WHO/IARC conference in association with JEMRB and TIMA, Copenhagen, 20–22 April 1982, vol 1. World Health Organization, Regional Office for Europe, Copenhagen, pp 413–426

IARC (1987) Silica and some silicates. IARC Monogr Eval Carcinog Risk Chem Hum 42:1–289

IARC (1988) Man-made mineral fibres and radon. IARC Monogr Eval Carcinog Risk Chem Hum 43:1–300

Maggioni A, Meregalli G, Sala C, Riva M (1980) Respiratory and skin diseases in glass fibre workers. Med Lav 71:216–227

McDonald AD, Fry JS, Woolley AJ, McDonald J (1983) Dust exposure and mortality in an American chrysotile textile plant. Br J Ind Med 40:361–367

McDonald JC, Liddell FDK, Gibbs GW, Eyssen GE, McDonald AD (1980) Dust exposure and mortality in chrysotile mining, 1910–1975. Br J Ind Med, 37:11–14

Malmberg P, Hedenstrom H, Kolmodin-Hedman B, Krantz S (1984) Pulmonary function in workers in a mineral rock fibre plant. In: Guthe E (ed) Biological effects of man-made mineral fibres. Proceedings of a WHO/IARC conference in association with JEMRB and TIMA, Copenhagen, 20–22 April 1982, vol 1. World Health Organization, Regional Office for Europe, Copenhagen, pp 427–435

Morgan RW, Kaplan SD, Bratsberg JA (1984) Mortality in fibrous glass production workers. In: Guthe T (ed) Biological effects of man-made mineral fibres. Proceedings of a WHO/IARC conference in association with JEMRB and TIMA, Copenhagen, 20–22 April 1982, vol 1. World Health Organization, Regional Office for Europe, Copenhagen, pp 340–346

Moulin JJ, Mur JM, Wild P, Perreaux JP, Pham QT (1986) Oral cavity and laryngeal cancers among man-made mineral fiber production workers. Scand J Work Environ, Health, 12:27–31

Moulin JJ, Pham QT, Mur JM, Meyer-Bisch C, Caillard JF, Massin N, Wild P, Teculescu D, Delepine P, Hunzinger E, Perreaux JP, Muller J, Betz M, Baudin V, Fontana JM, Henquel JC, Toamain JP (1987) Enquête épidémiologique dans deux usines productrices de fibres minérales artificielles. Arch Mal Prof, 48(1):7–16

Nasr ANM, Ditchek T, Scholtens PA (1971) The prevalence of radiographic abnormalities in chests of fiber glass workers. J Occup Med 13:371–376

Peto J (1989) Fibre carcinogenesis and environmental hazards. In: Bignon J, Peto J, Saracci R (eds) Nonoccupational exposure to mineral fibres. International Agency for Research on Cancer, Lyon pp 457–470 (IARC scientific publication no 90)

Peto J, Doll R, Hermon C, Binns W, Clayton R, Goffe T (1985) Relationship of mortality to measures of environmental asbestos pollution in an asbestos textile factory. Ann Occup Hyg, 29:305–355

Robinson CF, Dement JM, Ness GO, Waxweiler RJ (1982) Mortality patterns of rock and slag mineral wool production workers: an epidemiological and environmental study. Br J Ind Med 39:45–53

Rubino GF, Piolatto G, Newhouse ML, Scansetti G, Aresini GA, Murray R (1979) Mortality of chrysotile asbestos workers at the Balangero mine, Northern Italy. Br J Ind Med, 36:187–194

Saracci R (1985) Man-made mineral fibers and health: answered and unanswered questions. Scand J Work Environ, Health, 11:215–222

Saracci R (1986) Ten years of epidemiologic investigations on man-made mineral fibers and health. Scand J Work Environ, Health, 12: [Suppl 1], 5–11

Saracci R, Simonato L (1982) Man-made vitreous fibers and workers' health: an overview of the epidemiologic evidence. Scand J Work Environ Health, 8:234–242

Saracci R, Simonato L, Acheson ED, Andersen A, Bertazzi PA, Claude J, Charnay N, Estéve J, Frentzel-Beyme RR, Gardner MJ, Jensen OM, Maasing R, Olsen JH, Teppo LHI, Westerholm P, Zocchetti C (1984) Mortality and incidence of cancer of workers in the man made vitreous fibres producing industry: an international investigation at 13 European plants. Br J Ind Med, 41:425–436

Schefers GWH (1965) Discussion. In: Whipple HE (ed) Biological effects of asbestos. Ann NY Acad Sciences, 132:601–602

Shannon HS, Hayes MV, Julian JA, Muir D (1984) Mortality experience of glass fibre workers. In: Guthe T (ed) Biological effects of man-made mineral fibres. Proceedings of a WHO/IARC conference in association with JEMRB and TIMA, Copenhagen, 20–22 April 1982, vol 1. World Health Organization, Regional Office for Europe, Copenhagen, pp 347–349

Shannon HS, Jamieson E, Julian JA, Muir DCF, Walsh C (1987) Mortality experience of Ontario glass fibre workers—extended follow-up. Ann Occup Hyg, 31 (4B):657–662

Siebert WJ (1942) Fiberglass health hazard investigation. Ind Med 11:6–9

Simonato L, Fletcher AC, Cherrie JW, Andersen A, Bertazzi P, Charnay N, Claude J, Dodgson J, Estève J, Frentzel-Beyme R, Gardner MJ, Jensen O, Olsen J, Teppo L, Winkelmann R, Westerholm P, Winter PD, Zocchetti C, Saracci R (1987) The International Agency for Research on Cancer historical cohort study of MMMF production workers in seven European countries: extension of the follow-up. Ann Occup Hyg, 31 (4B):603–624

Sixt R, Bake B, Abrahamasson G (1983) Lung function of sheet metal workers exposed to fiber glass. Scand J Work Environ, Health, 9:9–14

Skuric Z, Stahuljak-Beritic D (1984) Occupational exposure and ventilatory function changes in rockwool workers. In: Guthe T (ed) Biological effects of man-made mineral fibres. Proceedings of a WHO/IARC conference in association with JEMRB and TIMA, Copenhagen, 20–22 April 1982, vol 1. World Health Organization, Regional Office for Europe, Copenhagen, pp 436–437

de Treville RTP, Utidjian HMD (1970) Fibrous glass manufacturing and health: report of an epidemiological study, parts I and II. In: Transactions bulletin. Industrial Health Foundation, Pittsburgh

Valentin H, Bost HP, Essing HJ (1983) Sind Glasfaserstaube gesundheitsgefährdend? Die berufliche Exposition mit kunstlichen mineralischen Fasern und ihre Auswirkungen auf die Gesundheit. Verlag des Vereins Deutscher Ingenieure, Düsseldorf, pp 153–159 (VDI-Berichte 475)

Weill H, Hughes J, Hammad YY, Glindmeyer HW, Sharon G, Jones RN (1984) Respiratory health of workers exposed to MMMF. In: Guthe T (ed) Biological effects of man-made mineral fibres. Proceedings of a WHO/IARC conference in association with JEMRB and TIMA, Copenhagen, 20–22 April 1982, vol 1. World Health Organization, Regional Office for Europe, Copenhagen, pp 387–412

Whipple HE (1965) Biological effects of asbestos. Ann NY Acad Sciences, 132:766

Wright GW (1968) Airborne fibrous glass particles: chest roentgenograms of persons with prolonged exposure. Arch Environ Health, 21:175–179

Section 6. Environmental Tobacco Smoke

24. Environmental Tobacco Smoke: Adverse Effects on Respiratory Infection, Respiratory Symptoms, and Lung Function

J.M. Samet[1]

Introduction

Environmental tobacco smoke (ETS) refers to the mixture of sidestream smoke, released from the cigarette's burning end, and mainstream smoke, exhaled by the smoker. Although exposure of nonsmokers to ETS has only recently been investigated as a risk factor for disease, sufficient evidence has accumulated to identify adverse effects causally related to ETS. In 1986, reports of the United States Surgeon General (US DHHS 1986) and of the United States National Research Council (1986) linked ETS to increased respiratory infections and respiratory symptoms, reduced lung growth in children, and lung cancer in adult nonsmokers. The evidence for adults on respiratory morbidity, cancers at sites other than the lung, and ischemic heart disease was not considered conclusive.

The reports emphasized epidemiological data in their findings, but toxicological studies were also considered relevant, indicating potential mechanisms of disease causation by ETS. The widespread exposure of nonsmokers to ETS was demonstrated by questionnaire data, biological markers, and monitoring of indoor air for tobacco smoke components. Quantitative risk assessment was not a primary goal of the reports.

This paper addresses effects other than malignancy of ETS on the respiratory tract. The focus is on lower respiratory infections in infants, respiratory symptoms and illnesses in children, and reduced lung growth in children—outcomes for which the data are most extensive.

Exposure and Dose in Epidemiological Studies of ETS

ETS is a complex mixture of gases and particles; its adverse health effects are probably caused by the uptake of gases and the deposition of particles at multiple

[1]Department of Medicine and the New Mexico Tumor Registry, Cancer Center, University of New Mexico, Albuquerque, NM 87131, USA.

Table 24.1. Health outcomes of ETS exposure, typical exposure measures, and relevant components and sites of injury.

Health effect	Exposure measure	Agent	Site
Irritation	Number of smokers or number of cigarettes	Gas phase irritants and particles	Eyes, nose, throat
Lower respiratory tract illness in children	Maternal smoking or number of smoking parents	Gas phase	Tracheobronchial epithelium
		Particles	Alveolar macrophages
Reduced lung growth	Maternal smoking or extent of maternal or paternal smoking	Uncertain	Airways and alveoli
Respiratory symptoms	Maternal smoking or extent of maternal or paternal smoking	Gas phase and particles	Airways
Ischemic heart disease	Spouse smoking	Nicotine	Airways, alveoli
		Carbon monoxide	Alveoli
Lung cancer	Spouse smoking	Particles	Lower airways

sites in the lungs. Toxicological evidence related to individual components of ETS makes it possible to speculate upon the agents which produce the adverse effects of ETS and the sites in the lung which are damaged. Table 24.1 provides a listing of ETS components probably associated with adverse effects, the exposure measures typically employed in epidemiological investigations of the effects, and the sites of injury.

With regard to lower respiratory infection in infants, impairment of mucociliary clearance and of alveolar macrophage function are plausible mechanisms for the increased risk of infection observed in the children of smokers. Acrolein, a gas phase component, slows ciliary beating; in active cigarette smokers, macrophage function and lung immune responses are altered (US DHHS 1984). Respiratory symptoms probably result from inflammation in the respiratory epithelium of large and small airways. The effect of ETS on lung growth may reflect increased or more severe respiratory infections in the children of smokers, airways injury, or alveolar damage as the alveoli are formed. Specific components of ETS which are likely to contribute to reduced lung growth cannot readily be identified; injury to airways and alveoli by cigarette smoke components, or injury to these structures by respiratory pathogens, might produce this adverse effect.

These toxicological considerations suggest the difficulty of measuring the dose of ETS in the context of an epidemiological study. Multiple components of the complex mixture of ETS are relevant; however, concentrations and characteristics change as the ETS mixture ages (US DHHS 1986), and doses vary with the proximity of the smoker and the ventilation pattern. Moreover, the circumstances of data collection in field studies generally do not permit the estimation of doses of biologically relevant agents to target tissues. In the epidemiological studies of ETS, measures of exposure to ETS have served as surrogates for dose. Exposure is most frequently estimated by questionnaire. Typically, the exposure measures for studies of children have been parental smoking categorized by the number of smokers or the number of cigarettes smoked; for studies of adults, spouse smoking or the extent of spouse smoking has been commonly used.

The validity of these simple questionnaire measures is supported by the results of studies in which homes are monitored for ETS components. For example, Spengler et al. (1981) monitored 80 homes over several years for respirable particles and found that a smoker of one pack of cigarettes daily contributes about 20 $\mu g/m^3$ to 24-h indoor concentrations of particles. Other studies have also documented the effects of cigarette smoking on indoor air quality in homes. Questionnaire descriptions of the extent of smoking provide information about average concentrations of ETS components over longer averaging times, but may not give sufficient temporal resolution to capture variation over shorter averaging times. Because cigarettes are not smoked uniformly throughout the day, higher peak concentrations occur when cigarettes are actually being smoked, and activity patterns that place the nonsmoker in proximity to the smoker during active smoking may strongly influence personal exposure (Spengler et al. 1985).

The validity of the questionnaire approach for establishing exposure is also supported by studies of biological markers in nonsmokers. Levels of biological markers of tobacco smoke, such as cotinine, tend to increase with the number of smokers in the household (Samet et al. 1987). For children, smoking by parents is the predominant determinant of cotinine levels, and levels in adults are also associated with household exposure. While biological markers provide short-term measures of exposure (Samet et al. 1987), the presently available markers may not be valid measures of average exposure unless multiple samples are taken (Coultas et al. 1987).

The monitoring and biological marker data also indicate, however, that questionnaire reports of exposure may introduce misclassification, and that these reports cannot readily be translated to concentrations of ETS components in air. Misclassification of exposure refers to the incorrect assignment of level of exposure to individual subjects. If misclassification affects diseased and nondiseased subjects equally (random misclassification), then estimates of the effects of ETS are biased towards finding no effect. If misclassification affects diseased and nondiseased subjects differentially (differential misclassification), then bias may increase or decrease the effects of ETS.

Because exposure to ETS takes place in many locations, exposure measures based solely on questionnaire descriptions of exposure at home generally misclassify total exposure to ETS. For example, in a survey of ETS exposure of adult nonsmokers in California, only 23.8% reported exposure in the home; about 40% of nonsmokers married to smokers denied exposure at home (Friedman et al. 1983). In a community-based survey in New Mexico, about 35% of children and adults not living with smokers had a detectable level of salivary cotinine (Coultas et al. 1987). Thus, exposure variables that only include the home environment

Table 24.2. Relative risks for lower respiratory tract illness in early childhood by parental smoking.

Reference	Study design	Relative risk by parental smoking				
Harlap and Davies 1974	Hospitalization for bronchitis or pneumonia during first 12 months[a]	0	Ex	1–10	11–20	≥ 21
		1.0	0.8	1.1	1.7	3.3
Colley et al. 1974	Parent report of bronchitis or pneumonia during first 12 months[b]	0	1–14	15–24	≥ 25	
		1.0	1.4	1.6	2.3	
Fergusson and Horwood 1985	Medical consultation for bronchitis during first 24 months[a]	0	1–10	≥ 11		
		1.0	1.4	1.8		
Chen et al. 1986	Hospitalization for respiratory illness during first 18 months[c]	0	1–9	≥ 10		
		1.0	1.1	1.9		
Taylor and Wadsworth 1987	Hospitalization for lower respiratory tract illness during first 5 years[a]	0	1–14	≥ 15		
		1.0	1.2	1.9		

[a] Cigarettes smoked daily by the mother.
[b] Total tobacco smoked by the parents.
[c] Mothers were all nonsmokers; amount smoked by father and others.

create strata of more-exposed and of less-exposed subjects, but not of exposed and unexposed subjects.

Moreover, we lack data to link the questionnaire measures with levels of exposure to specific components of ETS in indoor air. The mean increments in ETS components associated with smoking at home have been described (Spengler et al. 1981; US DHHS 1986), but the relationship of intensity of smoking with indoor air quality is probably highly variable within a home and among homes. We monitored ten homes in New Mexico for respirable particles and nicotine on 10 sampling days: every other day over 10 days, and then 1 day every other week over 10 weeks (Coultas et al. 1988). A linear regression model that included variables for the number of smokers in the home, the number of hours of smoking, and the season explained only 9% and 6% of the variability of the respirable particle and nicotine concentrations, respectively.

The epidemiological evidence on ETS must be interpreted in the context of the limitations posed by questionnaire reports of exposure. Random misclassification with flattening of exposure-response relationships undoubtedly affects the results of all investigations. The extent of differential misclassification probably varies among the types of investigations; case-control studies of the effects of ETS are particularly vulnerable to this type of bias (Lee 1988). It must be recognized, then, that questionnaire-based measures of exposure cannot describe concentrations of ETS components, integrated personal exposures, or doses of biologically relevant agents at target sites.

Health Effects of ETS

Lower Respiratory Tract Illness in Early Childhood

Epidemiological investigations have linked involuntary smoking in children to increased occurrence of lower respiratory tract illness during infancy and childhood (Table 24.2); presumably these illnesses are infectious in etiology and do not represent a direct toxic effect of ETS. In this regard, maternal smoking has been shown to increase the incidence of culture-documented respiratory syncytial (RS) virus infection (Sims et al. 1978; Pullan and Hey 1982). Although the health outcome measures varied somewhat (Table 24.2), the studies showed that the relative risks associated with involuntary smoking were consistent among the studies, and that the relative risk tended to increase with exposure. Heavy smoking by the parents approximately doubled the risk of lower respiratory tract illness. While most of the studies showed that maternal rather than paternal smoking underlies the increased risk of parental smoking, a recent study from China shows that paternal smoking alone can increase the incidence of lower respiratory tract illness in infants (Chen et al. 1986).

Respiratory Symptoms

Data from numerous surveys demonstrate a greater frequency of the most common respiratory symptoms—cough, phlegm, and wheeze—in the children of smokers (US DHHS 1986) (Table 24.3). In these studies the subjects have generally been school children, and the effects of parental smoking have been examined. Thus, the flatter exposure-response relationships for respiratory symptoms, in comparison with the studies of lower respiratory tract illness (Table 24.2), may reflect lower exposure to ETS by older children, who normally spend less time with their parents. While involuntary exposure to tobacco smoke has not been established as a cause of asthma, recent evidence indicates that involuntary smoking worsens the status of those with asthma (US DHHS 1986; Samet et al. 1987). Recent studies have also shown that children exposed to cigarette smoke in the home are at increased risk for middle ear disease (US DHHS 1986).

Only a few cross-sectional investigations provide information on the association between respiratory symptoms in adult nonsmokers and exposure to ETS. These studies have primarily considered exposure to a smoking spouse and have not evaluated sources of exposure outside the home. Consistent evidence of an effect of passive smoking on respiratory symptoms in adults has not been found (US DHHS 1986; Samet et al. 1987).

Lung Function

During childhood, measures of ventilatory function and of lung size increase as the lung grows; lung growth is most rapid during adolescence, and a plateau is

Table 24.3. Prevalence of respiratory symptoms in selected investigations of children by number of smoking parents[a]

Reference	Subjects	Respiratory symptoms	Prevalence (per 100) by number of smoking parents		
			0	1	2
Colley 1974	2426 children aged 6–14 in England	Chronic cough	15.6	17.2	22.2
Bland et al. 1978	3105 non-smoking children aged 12–13 in England	Cough during the day or at night	16.4	19.0	23.5
Weiss et al. 1980	650 children aged 5–9 in Massachusetts	Chronic cough and phlegm	1.7	2.7	3.4
		Persistent wheeze	1.8	6.8	11.8
Schenker et al. 1983	4071 children aged 5–14 in Pennsylvania	Chronic cough	6.3	7.0	8.3
		Chronic phlegm	4.1	4.8	4.0
		Persistent wheeze	7.2	7.7	5.4
Ware et al. 1984	8528 children aged 5–9 in six cities	Chronic cough	7.7	8.4	10.6
		Persistent wheeze	9.9	11.0	13.1
Charlton 1984	1733 nonsmoking children aged 8–10 in England	Frequent cough (boys)	35	42	48
		Frequent cough (girls)	32	40	52

[a] Abstracted from Table 4, pp. 360–362, in US DHHS 1984 and from Charlton 1984.

reached during early adulthood. Numerous cross-sectional studies have shown that the lung function of school-age children with smoking parents is lower than that of children with nonsmoking parents (US DHHS 1986). On average, the decrements in function, as assessed by spirometry, have been several percent of predicted value. Evidence from longitudinal studies suggests that these cross-sectional findings reflect a reduced rate of lung growth associated with ETS exposure.

Based on cross-sectional data from children in East Boston, Massachusetts, Tager et al. (1979) reported in 1979 that the level of forced expiratory flow from 25% to 75% of the forced vital capacity (FEF_{25-75}), a spirometric flow rate sensitive to subtle effects on airways and parenchymal function, declined with the number of smoking parents in the household. On follow-up of these children over a 7-year period, both maternal and active smoking by the child were found to independently reduce the growth rate of the forced expiratory volume in 1 s (FEV_1) (Tager et al. 1983). Other studies have also shown that parental smoking reduces the rate of lung growth (Berkey et al. 1986; Burchfiel 1984). The estimated reduction of lung function resulting across the period of lung growth ranges from several percent up to about 10% (Samet et al. 1987).

With regard to ETS and lung function in adults, exposure has been associated with reduced lung function in some cross-sectional studies (US DHHS 1986; Samet 1988). In the studies of adults, exposure to ETS has generally been assessed by categorical descriptors of the extent of smoking at home and at work; and exposure-response relationships have not been described.

Synthesis

Risk assessment has been used increasingly since the 1970s to characterize the health effects of environmental pollutants. The four steps – hazard identification, dose-response assessment, exposure assessment, and risk characterization – provide a useful framework for evaluating the evidence on ETS. For children, adverse effects have been causally associated with exposure to ETS. The epidemiological evidence and the supporting toxicological data link ETS to increased lower respiratory tract illness, increased respiratory symptoms, and reduced lung growth. Exposure-response relationships have been described for these effects; errors in exposure estimates would tend to lessen rather than to exaggerate these relationships. We lack data from large national samples on the extent of children's exposure to ETS, but information on the prevalence of smoking among adults suggests that ETS exposure poses a substantial disease burden for children. The evidence on the adverse effects of ETS on children is sufficiently compelling to provide a rationale for public health action; mothers should be advised about the adverse effects of their smoking on their children.

By contrast, the evidence for adverse respiratory effects of ETS exposure on adults, other than lung cancer, is presently less conclusive. Misclassification of exposure may be more severe than for children, and the resulting bias towards the null may obscure effects of ETS on respiratory symptoms and lung function. Research on adults is hindered by the difficulty of estimating exposures received in diverse environments over a lengthy period. Continued controversy related to the effects of ETS exposure in the workplace and in public locations provides a rationale for further research. However, more valid methods for assessing the exposures of adults to ETS are needed for such research.

References

Berkey CS, Ware JH, Dockery DW et al. (1986) Indoor air pollution and pulmonary function growth in preadolescent children. Am J Epidemiol 123:250–260

Bland M, Bewley BR, Pollard V et al. (1978) Effect of children's and parents' smoking on respiratory symptoms. Arch Dis Child 53:100–105

Burchfiel CM III (1984) Passive smoking, respiratory symptoms, lung function and initiation of smoking in Tecumseh, Michigan. PhD dissertation, University of Michigan, Ann Arbor

Charlton A (1984) Children's coughs related to parental smoking. Br Med J 288:1647–1649

Chen Y, Wanxian LI, Shunzhang YU (1986) Influence of passive smoking on admissions for respiratory illness in early childhood. Br Med J 293:303–306
Colley JR (1974) Respiratory symptoms in children and parental cigarette smoking and phlegm production. Br Med J 2:201–204
Colley JRT, Holland WW, Corkhill RT (1974) Influence of passive smoking and parental phlegm on pneumonia and bronchitis in early childhood. Lancet 2:1031–1034
Coultas DB, Howard CA, Peake GT et al. (1987) Salivary cotinine levels and involuntary tobacco smoke exposure in children and adults in New Mexico. Am Rev Respir Dis 136:305–309
Coultas DB, Samet JM, McCarthy JF, Spengler JD (1988) Variability of measures of exposure to environmental tobacco smoke in the home. APCA specialty conference on combustion processes and the quality of the indoor environment, Niagara Falls, New York, September 1988
Fergusson DM, Horwood LJ (1985) Parental smoking and respiratory illness during early childhood: a six-year longitudinal study. Pediatr Pulmonol 1:99–106
Friedman GD, Petitti DB, Bawol RD (1983) Prevalence and correlates of passive smoking. Am J Public Health 73:401–405
Harlap S, Davies AM (1974) Infant admissions to hospital and maternal smoking. Lancet 1:529–532
Lee PN (1988) Misclassification of smoking habits and passive smoking. Springer, Berlin Heidelberg New York
National Research Council, Committee on Passive Smoking (1986) Environmental tobacco smoke: measuring exposures and assessing health effects. National Academy Press, Washington, DC
Pullan CR, Hey CN (1982) Wheezing, asthma, and pulmonary dysfunction 10 years after infection with respiratory syncytial virus in infancy. Br Med J 284:1665–1669
Samet JM (1988) Involuntary exposure to tobacco smoke. Ann Sports Med 4:1–15
Samet JM, Marbury MC, Spengler JD (1987) Health effects and sources of indoor air pollution. Part I. Am Rev Respir Dis 136:1486–1508
Schenker MB, Samet JM, Speizer FE (1983) Risk factors for childhood respiratory disease: the effect of host factors and home environmental exposures. Am Rev Respir Dis 128:1038–1043
Sims DG, Downham MAPS, Gardner PS, Webb JKG, Weightman D (1978) Study of 8-year-old children with a history of respiratory syncytial virus bronchiolitis in infancy. Br Med J 1:11–14
Spengler JD, Dockery DW, Turner WA et al. (1981) Long-term measurements of respirable sulfates and particles inside and outside homes. Atmos Environ 15:23–30
Spengler JD, Treitman RD, Tosteson TD et al. (1985) Personal exposures to respirable particulates and implications for air pollution epidemiology. Environ Sci Technol 19:700–707
Tager IB, Weiss ST, Rosner B et al. (1979) Effect of parental cigarette smoking on the pulmonary function of children. Am J Epidemiol 110:15–26
Tager IB, Weiss ST, Munoz A et al. (1983) Longitudinal study of the effects of maternal smoking on pulmonary function in children. N Engl J Med 309:699–703
Taylor B, Wadsworth J (1987) Maternal smoking during pregnancy and lower respiratory tract illness in early life. Arch Dis Child 62:786–791
US Department of Health and Human Services (1984) The health consequences of smoking: chronic obstructive lung disease. US Government Printing Office, DHHS publication no (PHS) 84-50205, Washington, DC

US Department of Health and Human Services (1986) The health consequences of involuntary smoking. US Government Printing Office, DHHS publication no (CDC) 87-8398, Washington, DC
Ware JH, Dockery DW, Spiro A III et al. (1984) Passive smoking, gas cooking, and respiratory health of children living in six cities. Am Rev Respir Dis 129:366–374
Weiss ST, Tager IB, Speizer FE et al. (1980) Persistent wheeze. Its relation to respiratory illness, cigarette smoking, and level of pulmonary function in a population sample of children. Am Rev Respir Dis 122:697–707

25. Environmental Tobacco Smoke and Cancer

G. Pershagen[1]

Introduction

Environmental tobacco smoke (ETS) contains a number of human carcinogens to which passive smokers are exposed. The pronounced effects observed among smokers who are exposed to the same agents at higher doses indicate that cancer risks deserve attention also in non-smokers exposed to ETS. Indeed, since the publication in 1981 of the first reports on increased lung cancer risks in non-smokers married to smokers, there has been a growing interest in the health effects of ETS, both in the scientific community and among the general public. Although cancer induction is probably not the most important effect of ETS from a public health point of view, there is no doubt that the possibility of such effects has had a great impact on the attitudes towards passive smoking and has constituted an incentive for measures to restrict the exposure.

This review focuses on cancer risks in adults following inhalation of ETS. However, cancer risks in children will also be taken up, which to some extent may be related to exposure in utero. In view of the scarcity of experimental data pertaining to ETS and cancer, the main emphasis will be on epidemiologic evidence. Extensive reviews of the health consequences of ETS exposure, including cancer risks, have recently been performed by the National Research Council (1986) and the Surgeon General (1986) of the United States.

Exposure to ETS

Tobacco smoke contains more than 40 carcinogenic agents (International Agency for Research on Cancer 1986). In addition, a large number of other components may be of importance as cocarcinogens or effect modifiers, including many irritants. The complex composition of tobacco smoke and the lack of knowledge of

---

[1]Institute of Environmental Medicine, Karolinska Institute, Box 60208, S-104 01 Stockholm, Sweden.

the role of its specific components in the cancer induction process makes it difficult to identify suitable indicators of the carcinogenic potential of tobacco smoke.

There are major differences in the composition of mainstream and sidestream smoke (National Research Council 1986). As a rule, the amounts emitted from a cigarette are greater in the sidestream smoke. ETS is predominantly an indoor problem, and the exposure levels depend on the amount of tobacco smoked, room size, ventilation, etc. It is evident that smoking is a dominating source of indoor suspended particulates (Spengler et al. 1981; Revsbech et al. 1987). It has been estimated that about 0.1%–5% of the amount deposited from one cigarette in a smoker may be deposited daily in the respiratory tract of passive smokers (National Research Council 1986). For some gaseous components the relative amounts absorbed can be considerably higher, e.g., the equivalent of one to five cigarettes daily for acrolein.

Exposure to ETS can be estimated with biologic markers. Nicotine and its metabolite cotinine are two sensitive and specific markers which may be measured in saliva, serum, and urine. Cotinine has some advantages in comparison with nicotine, including a longer biologic half-life. Measurements of nicotine and cotinine in physiologic fluids of adults show that passive smokers may have from less than one to a few percent of the levels found in smokers (Feyerabend et al. 1982; Jarvis et al. 1983; Jarvis and Russel 1984; Wald et al. 1984; Wald and Ritchie 1984; Lee 1987; Wall et al. 1988). This suggests that the daily exposure to nicotine among passive smokers would correspond to a consumption of about one-tenth to one cigarette. For ETS-exposed children the relative nicotine exposure seems to be even higher (Greenberg et al. 1984; Rylander et al. 1989).

Measurements of nicotine and cotinine in serum and urine of neonates show that the fetus is heavily exposed if the mother smokes (Etzel et al. 1985; Luck et al. 1985). In fact, fetal nicotine levels in serum often exceed corresponding maternal values. Blood carboxy hemoglobin and serum thiocyanate levels are also increased in newborn children of smokers (Meberg et al. 1979; Bureau et al. 1982). Some studies using biologic markers indicate that exposure to ETS in pregnant women may result in fetal exposure (Andresen et al. 1982; Bottoms et al. 1982; Smith et al. 1982).

Experimental Data on ETS and Cancer

Several studies show that inhalation of tobacco smoke can induce tumors of the respiratory tract in laboratory animals (International Agency for Research on Cancer 1986). No consistent effects were seen for tumors of other sites. Cigarette smoke condensate (CSC) induces mouse skin tumors, and its constituents possess tumor-initiating, tumor-promoting, and other cocarcinogenic activities.

No studies are available on ETS inhalation and cancer risks in laboratory animals. In one study, Wynder and Hoffman (1967) applied CSC prepared from sidestream smoke of nonfilter cigarettes to mouse skin for 15 months. A higher tumor response per equivalent dose was indicated than in animals treated with CSC prepared from mainstream smoke of the same cigarettes ($p<0.05$).

Transplacental carcinogenesis has been demonstrated in laboratory animals with a number of compounds in tobacco smoke as well as with CSC (Napalkov 1973; Nikonova 1977; Nicolov and Chernozemsky 1979). Prenatal exposure to transplacental carcinogens may also enhance the effects of postnatal exposure to the same and other carcinogenic substances.

Epidemiologic Data on ETS and Cancer

Cancer in Adults

Lung Cancer

More than 20 epidemiologic studies have now been published on passive smoking and lung cancer in nonsmokers, including a total of nearly 2000 cases. Three studies were of the cohort type (Hirayama 1981; Garfinkel 1981; Gillis et al. 1984) and the rest case-control studies (Chan and Fung 1982; Trichopoulos et al. 1983; Correa et al. 1983; Kabat and Wynder 1984; Buffler et al. 1984; Garfinkel et al. 1985; Koo et al. 1985; Wu et al. 1985; Akiba et al. 1986; Lee et al. 1986; Brownson et al. 1987; Gao et al. 1987; Humble et al. 1987; Lam et al. 1987; Pershagen et al. 1987; Geng et al. 1988; Inoue and Hirayama 1988; Shimizu et al. 1988; Svensson et al. 1989). The studies come from six different countries (China with Hong Kong, Greece, Japan, Sweden, United Kingdom, and United States) and include predominantly women (about 90% of the cases).

The studies published until 1986 have been extensively reviewed (National Research Council 1986; Surgeon General 1986). Table 25.1 summarizes some data from studies on ETS and lung cancer which were published after these reviews. As a basis for pooling the results, only data on spouse smoking in two exposure categories are given, although most studies included more detailed information. All of the eight studies show increased risks associated with living with a smoker, which are statistically significant in four studies. It should be noted that four of the 28 cases in the study by Humble et al. (1987) probably were smokers, but it is not clear from the publication how these distribute with respect to spouse smoking.

Table 25.2 shows overall estimates of relative risks for lung cancer in nonsmokers living with smokers. The eight recent case-control studies give a pooled relative risk of 1.54 with 95% confidence interval of 1.28–1.85. If the studies reviewed by the National Research Council (1986) are also included, the overall relative risk is 1.43 (1.27–1.60). Excluding the study by Humble et al. (1987), which probably contains some smokers, reduces the relative risk to 1.41.

Most of the studies analyzed lung cancer risks in relation to detailed smoking habits of the spouses. With some exceptions, these analyses show positive trends in risk with increased smoking, which are often statistically significant. For other measures of exposure, including smoking habits of other household members and occupational ETS exposure, the results are less consistent. Most analyses of risks associated with specific histologic types of lung cancer, which were

Table 25.1. Summary of recent case-control studies on lung cancer and passive smoking.[a]

Reference	Location	Sex	Spouse smoker		Spouse nonsmoker		Relative risk	Confidence interval (95%)
			Cases	Controls	Cases	Controls		
Brownson et al. 1987	US	MF	57	50	45[c]	81[c]	2.0	1.2–3.4
Gao et al. 1987	China	F	189	276	57[d]	99[d]	1.2	0.8–1.7
Humble et al. 1987[b]	US	MF	20	127	8	165	3.2	1.4–6.7
Lam et al. 1987	Hong Kong	F	115	152	84	183	1.6	1.2–2.3
Geng et al. 1988	China	F	34	41	20	52	2.1	1.1–4.1
Inoue and Hirayama 1988	Japan	F	18	30	4	17	2.3	0.8–6.9
Shimizu et al. 1988[b]	Japan	F	52	91	38	72	1.1	0.6–1.8
Svensson et al. 1989[b]	Sweden	F	21	67	15	52	1.1	0.5–2.3

[a] To facilitate the comparison only data on spouse smoking in two exposure categories are given. Relative risks are estimated from odds ratios and confidence intervals according to the method described by the National Research Council (1986).
[b] Number of cases and controls in different exposure categories estimated from data in the published papers or obtained by personal communication.
[c] Reference category: passive smoke exposure from husband < 3 h/d.
[d] Reference category: lived with smoking husband < 20 years.

performed in about half of the studies, tend to show the strongest association with spouse smoking for squamous and/or small cell carcinomas.

Several types of errors may affect the estimates of relative risks in passive smokers. Confounding by smoking and nondifferential misclassification (of exposure and effect) are probably two of the most important and consistent sources of bias. These biases would be expected to shift the relative risk in opposite directions. All studies based the exposure assessment on information obtained from mailed questionnaires or interviews. With few exceptions, no attempts were made to validate the never smoking status of the study subjects. There is a tendency for smokers to marry smokers. This is evident from those studies referred to in this section, which include both smokers and nonsmokers. In conjunction with some misclassification of smokers as nonsmokers this would tend to produce a spurious association between spouse smoking and lung cancer.

A quantitative assessment of the effect of confounding by smoking on the relative risk estimates in passive smokers rests on several assumptions. Two crucial assumptions are the proportion of those classified as nonsmokers who really have been smokers and the relative risk of lung cancer in this group. The available empirical information is very limited, particularly regarding mis-

Table 25.2. Overall estimates of lung cancer risk in nonsmokers living with smoking spouses.[a]

	"Exposed" cases	Relative risk[b]	Confidence interval (95%)
Review by the National Research Council (1986):			
Ten case-control studies	426	1.24	1.04–1.50
Three cohort studies	251	1.44	1.20–1.72
This review:			
Eight case-control studies	506	1.54	1.28–1.85
Overall for all studies	1183	1.43	1.27–1.60

[a] For two of the 21 studies under review, exposure was only available as smoking by household members (not specified further). Methods for estimation of relative risks and confidence intervals of pooled data were described by the National Research Council (1986).
[b] The "unexposed" reference category normally consists of nonsmokers with a nonsmoking spouse and often also includes unmarried nonsmokers.

classification of smoking for lung cancer cases, and differences in the assumptions may lead to different conclusions regarding the magnitude of the bias (Wald et al. 1986; Lee 1988).

Nondifferential misclassification of exposure to ETS, to which crude exposure measures may contribute, will result in a dilution (underestimation) of the true risks. The principal basis for exposure assessment in most studies was the smoking habits of the spouse. This may be justified by a high reliability of the information (Pron et al. 1988), and it has been shown that nonsmokers married to smokers, on average, have a higher excretion of cotinine than nonsmokers married to nonsmokers (Wald and Ritchie 1984; Lee 1987). It is evident, however, that ETS exposure may result from other sources and that not all smokers smoke at home or in front of their spouses. Empirical data on nicotine exposure (Wald et al. 1986; Gann et al. 1988) suggest that this type of bias can be at least as important as confounding by smoking.

A poor quality in the diagnosis of lung cancer may also lead to an underestimation of the true risks. For example, secondary lung tumors and carcinomas with an unknown primary site appeared in about one-sixth of reported cases of lung cancer on death certificates in the United States (Garfinkel 1981) and in central health registers in Sweden (Pershagen et al. 1987) among female nonsmokers. In many of the studies on ETS and lung cancer there is an inadequate description of the diagnostic criteria.

Analyses of childhood exposure to ETS and lung cancer risks in nonsmokers as well as in smokers have given inconsistent results (Correa et al. 1983; Koo et al. 1985; Sandler et al. 1985a; Garfinkel et al. 1985; Akiba et al. 1986; Gao et al. 1987; Pershagen et al. 1987; Shimizu et al. 1988; Svensson et al. 1989). A major difficulty with the studies in smokers is to fully standardize for differences in smoking by the study subjects.

Other Cancers

Only four studies are available on cancers other than lung cancer in nonsmokers exposed to ETS. Two were cohort studies (Gillis et al. 1984; Hirayama 1984) and the other two case-control studies (Miller 1984; Sandler et al. 1985b). Increased relative risks associated with marriage to a smoker were seen for specific sites and/or for all sites taken together in some of the studies, but no consistent pattern emerged. The results of one study suggested that a combination of ETS exposure during child- and adulthood could be particularly harmful (Sandler et al. 1985c). It is of interest that some of the associations observed with ETS occurred for sites which generally do not show higher cancer risks in smokers.

Cancer in Children

In assessing the cancer risks in children for whom in utero exposure may be relevant, it is of interest to evaluate the data on maternal smoking during pregnancy and childhood cancer risks. This issue has been addressed in a number of epidemiologic investigations, and the data were recently reviewed by Pershagen (1989). There is no consistent increase in overall cancer risk or for specific sites in children exposed to maternal smoking in utero. The only study showing a statistically significant positive "dose-response" relationship was one by Stjernfeldt et al. (1986), who found an increased risk of leukemia in children of mothers who smoked during pregnancy. For ETS exposure the data are very limited. A higher risk of brain tumors was seen in children with mothers who "had lived with a smoker" in one study (Preston-Martin et al. 1982) and rhabdomyosarcoma was related to paternal smoking in another (Grufferman et al. 1982). It should be noted that there was no association between cancer risk and maternal smoking in any of these studies.

Several circumstances have to be considered in the evaluation of the results. All studies were of the case-control type in which the data on smoking during pregnancy were often obtained several years later. As a rule, the mothers provided the information and recall bias cannot be excluded. Confounding factors may also have been of importance (occupational exposures, X-rays taken during pregnancy, perinatal mortality, etc).

Conclusions

Large segments of the population in many countries are exposed to ETS, which means that even small effects of this exposure may be of public health significance. Using biologic markers, it has been estimated that the daily exposure to nicotine in passive smokers may correspond to a dose received from 0.1–1 cigarette in a smoker. The relative exposure for passive smokers in comparison with smokers varies substantially for different components in the smoke. Indirect exposure to constituents in ETS may occur for the fetus in mothers who are exposed to ETS.

There are no data on ETS inhalation and cancer risks in laboratory animals, but one study using the mouse skin cancer model indicated that CSC prepared from sidestream smoke was more potent than condensate prepared from mainstream smoke. Transplacental carcinogenesis has been demonstrated in laboratory animals with a number of compounds in tobacco smoke as well as with CSC.

More than 20 epidemiologic studies have been published on passive smoking and lung cancer. The most common approach in the analysis has been to base the exposure classification on smoking habits of spouses. Taking the studies together, there is a statistically significant increase in the lung cancer risk of about 40% in nonsmokers married to smokers. Higher risks in those married to heavy smokers were often observed, supporting a causal association.

It is probable that a part of the increased risk related to smoking of spouses is explained by confounding by smoking in some study subjects. On the other hand, nondifferential misclassification of the exposure, resulting from crude exposure measures, will give rise to an underestimation of the true risks. The available empirical data are inadequate for a detailed quantitative evaluation of the influence of bias on the observed relative risks. An overall assessment taking into consideration the presence of carcinogens in ETS, dose estimates in passive smokers, dose-response relations in smokers, and the epidemiologic evidence on lung cancer in passive smokers indicates that ETS is of carcinogenic importance.

Only a few studies have been performed on ETS exposure in adults and the risk of malignancies other than lung cancer. Some associations were found, but the data are not consistent and do not permit any firm conclusions. A similar situation holds true for ETS exposure during childhood and cancer risks in nonsmokers and smokers. The available epidemiologic data provide no conclusive evidence of an effect by maternal smoking during pregnancy on the risk of cancer in children, suggesting that cancer risks associated with maternal passive smoking in this situation would be insignificant.

References

Akiba S, Kato H, Blot WJ (1986) Passive smoking and lung cancer among Japanese women. Cancer Res 46:4804–4807

Andresen BD, Ng KI, Iams JD, Bianchine JR (1982) Cotinine in amniotic fluids from passive smokers. Lancet i:791–792

Bottoms SF, Kuhnert BR, Kuhnert PM, Reese AL (1982) Maternal passive smoking and fetal serum thiocyanate levels. Am J Obstet Gynecol 144:789–791

Brownson RC, Reif JS, Keefe TJ, Ferguson SW, Pritz JA (1987) Risk factors for adenocarcinoma of the lung. Am J Epidemiol 125:25–34

Buffler PA, Pickle LW, Mason TJ, Contant C (1984) The causes of lung cancer in Texas. In: Mizell M, Correa P (eds) Lung cancer: causes and prevention. Chemie International, New York, pp 83–99

Bureau MA, Monetta J, Shapcott D, Pare C, Mathieu J, Lippe J, Blovin D, Berthiaume K, Begin J (1982) Carboxyhemoglobin concentration in fetal cord blood of mothers who smoked during labor. Pediatrics 69:371–373

Chan WC, Fung SC (1982) Lung cancer in nonsmokers in Hong Kong. In: Grundmann E (ed) Cancer campaign, vol 6. Cancer epidemiology. Fischer, Stuttgart, pp 199–202

Correa P, Pickle LW, Fonthan E, Lin Y, Haenszel W (1983) Passive smoking and lung cancer. Lancet ii:595–597

Etzel RA, Greenberg RA, Haley NJ, Loda FA (1985) Urine cotinine excretion in neonates exposed to tobacco smoke products in utero. J Pediatr 107:146–148

Feyerabend C, Higgenbottam T, Russel MAH (1982) Nicotine concentrations in urine and saliva of smokers and nonsmokers. Br Med J 284:1002–1004

Gann P, Coghlin-Strom J, Hammond K (1988) Measurement of bias due to misclassification in epidemiologic studies of passive smoking. Am J Epidemiol 128:920

Gao YT, Blot WJ, Zheng W, Ershow AG, Hsu CW, Levin LI, Zhang R, Fraumeni JF (1987) Lung cancer among Chinese women. Int J Cancer 40:604–609

Garfinkel L (1981) Time trends in lung cancer mortality among nonsmokers and a note on passive smoking. J Natl Cancer Inst 66:1061–1066

Garfinkel L, Auerbach O, Joubert L (1985) Involuntary smoking and lung cancer: a case control study. J Natl Cancer Inst 75:463–469

Geng GY, Liang ZH, Zhang AY, Wu GL (1988) On the relationship between smoking and female lung cancer. In: Aoki M, Hisamichi S, Tominaga S (eds) Smoking and health. Elsevier, Amsterdam, pp 483–490

Gillis CR, Hole DJ, Hawthorne VM, Boyle P (1984) The effect of environmental tobacco smoke in two urban communities in the west of Scotland. Eur J Respir Dis 65:121–126

Greenberg RA, Haley NJ, Etzel R, Loda FA (1984) Measuring the exposure of infants to tobacco smoke. N Engl J Med 310:1075–1078

Grufferman SH, Wang H, DeLong ER, Kimm SYS, Delzell ES, Faletta JM (1982) Environmental factors in the etiology of rhabdomyosarcoma in childhood. J Natl Cancer Inst 68:107–117

Hirayama T (1981) Non smoking wives of heavy smokers have a higher risk of lung cancer: a study from Japan. Br Med J 282:183–185

Hirayama T (1984) Lung cancer in Japan: effects of nutrition and passive smoking. In: Mizell M, Correa P (eds) Lung cancer causes and prevention. Chemie International, New York, pp 175–196

Humble CG, Samet JM, Pathak DR (1987) Marriage to a smoker and lung cancer risk. Am J Public Health 77:598–602

Inoue R, Hirayama T (1988) Passive smoking and lung cancer in women. In: Aoki M, Hisamichi S, Tominaga S (eds) Smoking and health. Elsevier, Amsterdam, pp 283–285

International Agency for Research on Cancer (1986) Monographs on the evaluation of the carcinogenic risk of chemicals to humans: tobacco smoking. International Agency for Research on Cancer, Lyon

Jarvis MJ, Russel MAH (1984) Measurement and estimation of smoke dosage to nonsmokers from environmental tobacco smoke. Eur J Respir Dis 65:68–75

Jarvis MJ, Russel MAH, Feyerabend C (1983) Absorption of nicotine and carbon monoxide from passive smoking under natural conditions of exposure. Thorax 38:829–833

Kabat GC, Wynder EL (1984) Lung cancer in nonsmokers. Cancer 53:1214–1221

Koo LC, Ho JHC, Lee N (1985) An analysis of some risk factors for lung cancer in Hong Kong. Int J Cancer 35:149–155

Lam TH, Kung ITM, Wong CM, Lam WK, Kleevens JWL, Saw D, Hsu C, Seneviratne S, Lam SY, Lo KK, Chan WC (1987) Smoking, passive smoking and histological types in lung cancer in Hong Kong Chinese women. Br J Cancer 56:673–678

Lee PN (1987) Passive smoking and lung cancer association: a result of bias? Hum Toxicol 6:517–524

Lee PN (1988) Misclassification of smoking habits and passive smoking. Springer, Berlin Heidelberg New York

Lee PN, Chamberlain J, Alderson MR (1986) Relationship of passive smoking to risk of lung cancer and other smoking-associated diseases. Br J Cancer 54:97–105

Luck M, Nau H (1985) Nicotine and cotinine concentrations in serum and urine of infants exposed via passive smoking or milk from smoking mothers. J Pediatr 107:816–820

Meberg A, Sande B, Foss OP, Stenwig JT (1979) Smoking during pregnancy—effects on the fetus and on thiocyanate level in mother and baby. Acta Paediatr Scand 68:547–552

Miller GH (1984) Cancer passive smoking and non-employed and employed wives. West J Med 140:632–635

Napalkov NP (1973) Some general considerations on the problem of transplacental carcinogenesis. In: Tomatis L, Mohr U (eds) Transplacental carcinogenesis. International Agency for Research on Cancer, Lyon, pp 1–13

National Research Council (1986) Environmental tobacco smoke—measuring exposures and assessing health effects. National Academy Press, Washington

Nicolov IG, Chernozemsky IN (1979) Tumors and hyperplastic lesions in Syrian hamsters following transplacental and neonatal treatment with cigarette smoke condensate. Cancer Res Clin Oncol 94:249–256

Nikonova TV (1977) Transplacental action of benzo(a)pyrene and pyrene. Bull Exp Biol Med 84:1025–1027

Pershagen G (1989) Childhood cancer and malignancies other than lung cancer related to passive smoking. Mutat Res 222:129–135

Pershagen G, Hrubec Z, Svensson C (1987) Passive smoking and lung cancer in Swedish women. Am J Epidemiol 125:17–24

Preston-Martin S, Yu MC, Benton B, Henderson BE (1982) N-nitroso compounds and childhood brain tumors: a case-control study. Cancer Res 42:5240–5245

Pron GE, Burch DJ, Howe GR, Miller AB (1988) The reliability of passive smoking histories reported in a case-control study of lung cancer. Am J Epidemiol 127:267–273

Revsbech P, Korsgaard J, Lundqvist GR (1987) Tobacco smoking and indoor suspended particulate matter. In: Seifert B, Esdorn H, Fischer M, Rüder H, Wegner J (eds) 4th International conference on indoor air quality and climate, 17–21 Aug 1987. Institute for Water, Soil and Air Hygiene, Berlin

Rylander E, Pershagen G, Curvall M, Kazemi-Vala E (1989) Exposure to environmental tobacco smoke and urinary excretion of cotinine and nicotine in children. Acta Paediatr Scand 78:449–450

Sandler DP, Everson RB, Wilcox AJ, Browder JP (1985a) Cancer risk in adulthood from early life exposure to parent's smoking. Am J Public Health 75:487–492

Sandler DP, Everson RB, Wilcox AJ (1985b) Passive smoking in adulthood and cancer risk. Am J Epidemiol 121:37–48

Sandler DP, Wilcox AJ, Everson RB (1985c) Lifetime passive smoking and cancer risk. Lancet i:866–867

Shimizu H, Morishita M, Mizuno K, Masuda T, Ogura Y, Santo M, Nishimura M, Kunishima K, Karasawa K, Nishiwaki K, Yamamoto M, Hisamichi S, Tominaga S

(1988) A case-control study of lung cancer in nonsmoking women. Tohoku J Exp Med 154:389–397
Smith N, Austen J, Rolles CJ (1982) Tertiary smoking by the fetus. Lancet i:1252
Spengler JD, Dockery DW, Turner WA, Wolfson JM, Ferris BJ (1981) Long term measurements of respirable sulphates and particles inside and outside homes. Atmos Environ 15:23–30
Stjernfeldt M, Berglund K, Lindsten J, Ludvigsson J (1986) Maternal smoking during pregnancy and risk of childhood cancer. Lancet i:1350–1352
Surgeon General (1986) The health consequences of involuntary smoking. US Department of Health and Human Services, Washington
Svensson C, Pershagen G, Klominek J (1989) Smoking and passive smoking in relation to lung cancer in women. Acta Oncologol (in press)
Trichopoulos D, Kalandidi A, Sparros L (1983) Lung cancer and passive smoking: conclusion of the Greek study. Lancet ii:677–678
Wald NJ, Ritchie C (1984) Validation of studies on lung cancer in non-smokers married to smokers. Lancet i:1067
Wald NJ, Boreham J, Bailey A, Ritchie C, Haddow JE, Knight G (1984) Urinary cotinine as marker of breathing other people's tobacco smoke. Lancet i:230–231
Wald NJ, Nanchanal K, Thompson SG, Cuckle HS (1986) Does breathing other people's tobacco smoke cause lung cancer? Br Med J 293:1217–1222
Wall MA, Johnson J, Jacob P, Benowitz NL (1988) Cotinine in the serum, saliva and urine of nonsmokers, passive smokers and active smokers. Am J Public Health 78:699–701
Wu AH, Henderson BE, Pike MC, Yu MC (1985) Smoking and other risk factors for lung cancer in women. J Natl Cancer Inst 74:747–751
Wynder EL, Hoffman D (1967) Tobacco and tobacco smoke: studies in experimental carcinogenesis. Academic, New York

26. Risk Assessment for Inhomogeneous Subgroups

B. Schneider[1]

"Risk" has different meanings in our daily language. We speak of risk if a situation could have a negative outcome, e.g., if we could lose money in business or our life in an adventure. In epidemiology such a situation is called a "risk factor" or better "risk condition" (as it is not always clear whether the specific conditions of the situation really "make" the negative outcome). "Risk" is understood quantitatively as the probability or chance for the negative event in presence of the specific conditions. This implies that risk is considered for populations where the specific conditions are present as well as absent. Let us assume, as a specific condition, environmental tobacco smoke (ETS) and, as a negative event, lung cancer (LC); the data layout for risk calculations is given in Table 26.1.

N_1 individuals of the population are exposed to ETS and N_0 are not exposed. From the exposed individuals E_1 show the event (LC) (and $F_1 = N_1 - E_1$ do not). The risk for LC in the exposed group is the ratio $R_1 = E_1/N_1$. Analogous in the nonexposed group, the risk for LC is the ratio $R_0 = E_0/N_0$, if E_0 LCs occurred under the N_0 individuals not exposed to ETS.

As a measure of the risk involved with the condition (ETS), usually the "relative risk" is computed, i.e., the ratio of the risk rate for the exposed individuals to the rate for the nonexposed individuals (reference group). As can be seen in Table 26.1, this ratio can be interpreted either as the ratio of the "observed" number of events (LC) in the exposed group to the "expected" number if the risk ratio of the reference group is applied to the exposed group; or the ratio of the expected number in the reference group to the observed number in this group if the risk ratio of the exposed group is applied to the reference group. The expected numbers are also called "standardized" numbers. For the crude relative risk, either the exposed or the reference group can be used for standardization.

The relative risk is a factor indicating by how much the risk ratio in the exposed group is higher (or lower) than that in the reference group. This figure may be misleading, especially if the risk ratios for both groups are small. A relative risk of 2 or 3 suggests a very high risk, but even for the exposed group the

[1]Hannover Medical School, Institute for Biometry, Konstanty-Gutschow-Str. 18, 3000 Hannover 61, FRG.

Table 26.1. Data layout for crude relative risk.

	Risk (ETS)	Reference (NETS)
Event (LC)	E_1	E_0
No event (NLC)	F_1	F_0
Size (# ind.)	N_1	N_0

Risk rate for ETS: $R_1 = E_1/N_1$
Risk rate for NETS: $R_0 = E_0/N_0$
Crude relative risk for ETS: $CRR_1 = R_1/R_0 = (E_1 \cdot N_0)/(E_0 \cdot N_1) = E_1/[(N_1 \cdot (E_0/N_0)]$
$= [N_0 \cdot (E_1/N_1)]/E_0$
Crude attributable risk for ETS: $CAR_1 = R_1 - R_0$

real risk, i.e., the risk ratio, may be relatively small. A measure which better reflects the magnitude of the risk associated with a specified risk condition is the "attributable risk," i.e., the difference between the risk ratio in the exposed and that in the reference group.

The relationship among the different risk measures can be demonstrated using Hirayama's data (1984a) about LC mortality in nonsmoking women with smoking husbands (Table 26.2). The risk ratio for the exposed is 2.34 per 1000 individuals and for the reference group it is 1.69 per 1000. The (crude) relative risk is 1.38, the (crude) attributable risk is 0.69 per 1000.

It is usually assumed that the attributable risk gives, for each individual of the population, that part of the risk which can be attributed solely to the risk factor. This assumption is only true if the risk ratios for the total population are not confounded or biased by other factors or conditions. In practice, such confounding conditions are always present in populations. The most relevant confounding condition is age. The population has different age groups or strata, and in general the risk ratios for the exposed and reference group vary within the strata. The situation is represented in the general data layout of Table 26.3.

It is assumed that the total population is split up in J subgroups according to one or more confounding conditions such as age. In addition, it is assumed that more than one risk condition (exposure condition) is distinguished (e.g., different times or extents of exposure). Each subgroup is homogeneous with respect to the

Table 26.2. Cancer mortality in nonsmoking women with smoking husbands based on a large-scale cohort study in Japan. (From Hirayama 1984a).

	ETS	NETS
LC	163	37
NLC	69,462	21,858
Size	69,645	21,895

$R_1 = 0.00234$
$R_0 = 0.00169$
$CRR_1 = 1.38$
$CAR_1 = 0.00234 - 0.00169 = 0.00065$

Table 26.3. Data layout for risk conditions C_i ($i = 1, \ldots I$) and reference condition C_0 in a population composed of J strata with levels S_j ($j = 1, \ldots J$).

Stratum	Components of risk rate	Category of risk condition C_1	...	C_i	...	C_I	Reference condition C_0	Relative risk
		.	...	.	...	.	.	
S_1		.	...	.	...	.	.	
		.	...	.	...	.	.	
	Event	.	...	E_{ij}	...	.	E_{0j}	$RR_{ij} =$
S_j	No event	.	...	F_{ij}	...	.	F_{0j}	$(E_{ij}/N_{ij})/$
	Size	.	...	N_{ij}	...	.	N_{0j}	(E_{0j}/N_{0j})
		.	...	.	...	.	.	
S_J		.	...	.	...	.	.	
		.	...	.	...	.	.	
Total population	Event	.	...	E_i	...	.	E_0	$CRR_i =$
	No event	.	...	F_i	...	.	F_0	$(E_i/N_i)/$
	Size	.	...	N_i	...	.	F_0	(E_0/N_0)

confounding condition. Therefore, the risk ratios and relative risks within the subgroups reflect the real risk for the respective individuals. However, the risks among the subgroups are inhomogeneous. Calculating the crude risks from the population totals may produce serious biases. The question is how the risk ratios or relative risks of the different strata can be combined to give a meaningful "overall" measure of the risk for the population.

One approach in solving this question is "standardization." Table 26.4 presents the most common standardized relative risks. Very often the "standardized mortality (morbidity) ratio" is used as a measure of overall risk. As can be seen from Table 26.4, in each stratum (e.g., age group) the observed risk ratios of the exposed and reference groups are standardized to the distribution of the confounding condition (e.g., age distribution) in the respective *exposed* group. This is not the best idea, as Miettinen (1972) has pointed out. A better idea is to standardize to the *reference* group. The result is the "standardized reference risk."

In general, one has to take a relevant standard distribution A_j and standardize all the exposed and the reference risks to this distribution. For cohort studies, an appropriate standard distribution is the distribution of the confounding factor in the population from which the cohort is sampled, e.g., the age distribution of the population.

As can be seen from Table 26.4, the standardized relative risk is a weighted average of the relative risks within the strata. The weights depend on the selected standard distribution A_j and the risk ratios in the reference group.

As "internal" standard distributions, the distribution of the confounding condition in the total cohort or the harmonic mean of the distributions in the exposed and reference group is of special interest. The latter is equivalent to the standardization proposed by Mantel and Haenszel (1959). The relationship among

Table 26.4. Combining the risk ratios of the strata to an overall risk.

Standardized mortality (morbidity) ratio SMR_i (strata distribution in risk population as standard)

$$\begin{aligned} SMR_i &= E_i/\Sigma_j N_{ij}\cdot(E_{0j}/N_{0j}) \\ &= \Sigma_j N_{ij}\cdot(E_{ij}/N_{ij})/\Sigma_j N_{ij}\cdot(E_{0j}/N_{0j}) \end{aligned}$$

Standardized reference risk SRR_i (strata distribution in reference population as standard)

$$\begin{aligned} SRR_i &= \Sigma_j N_{0j}\cdot(E_{ij}/N_{ij})/\Sigma_j N_{0j}\cdot(E_{0j}/N_{0j}) \\ &= \Sigma_j N_{0j}\cdot(E_{ij}/N_{ij})/E_0 \end{aligned}$$

Standardized general risk $SRR(A)_i$ (general strata distribution A_j as standard)

$$\begin{aligned} SRR(A)_i &= \Sigma_j A_j\cdot(E_{ij}/N_{ij})/\Sigma_j A_j\cdot(E_{0j}/N_{0j}) \\ &= \Sigma_j(A_j\cdot(E_{0j}/N_{0j})\cdot RR_{ij}/\Sigma_j A_j\cdot(E_{0j}/N_{0j}) \\ &= \Sigma_j G_j\cdot RR_{ij} \end{aligned}$$

$$\text{with } RR_{ij} = (E_{ij}/N_{ij})/(E_{0j}/N_{0j})$$
$$\text{and the "weights" } \quad G_j = A_j\cdot(E_{0j}/N_{0j})/\Sigma_j A_j\cdot(E_{0j}/N_{0j})$$

Special "internal standards":
Cohort total standard:

$$\begin{aligned} A_j &= N_{ij} + N_{0j}\ (= \Sigma_{i=0..I}N_{ij}) = N_j \\ G_j &= [(N_j/N_{0j})\cdot E_{0j})]/[\Sigma_j(N_j/N_{0j})\cdot E_{0j}] \end{aligned}$$

$$SRR(N)_i = \Sigma_j N_j\cdot(E_{ij}/N_{ij})/\Sigma_j N_j\cdot(E_{0j}/N_{0j})$$

Harmonic mean (Mantel-Haenszel) standard:

$$\begin{aligned} A_j &= (N_{ij}\cdot N_{0j}/N_j\ [= N_{0j}\cdot(\Sigma_{i=1...I}N_{ij})/N_j] \\ G_j &= [(N_{ij}/N_j)\cdot E_{0j})/[\Sigma_j(N_{ij}/N_j)\cdot E_{0j}] \end{aligned}$$

$$SRR(MH)_i = [\Sigma_j(E_{ij}\cdot N_{0j})/N_j]/[\Sigma_j(E_{0j}\cdot N_{ij})/N_j]$$

the various standardized risks is discussed by Ahlborn et al. (1988). The use of these different standardized risks can be demonstrated using the results of Hirayama's (1984b) cohort study of ETS and LC in nonsmoking Japanese women. The age of the women (at study entry) is used as a confounding condition. The age distribution of married women in the Japanese population is used as an exter-

Table 26.5. Hirayama's data stratified to wife's age. (From Hirayama 1984b).

Stratum –wife's age (years)	Category of risk condition (husband's smoking habits) 1–19 cigarettes LC	Size	20+ cigarettes LC	Size	Reference condition nonsmoker LC	Size	Total cohort LC	Size	A_j age distribution in population
40–49	21	17492	21	12615	4	7918	46	38025	42612
50–59	46	15640	31	8814	14	7635	91	32089	29101
60–69	31	10381	10	3793	16	6170	57	20344	15315
70–79	1	671	2	239	3	172	6	1082	4512
	99	44184	64	25461	37	21895	200	91540	91540

Table 26.6. Risk calculations with Hirayama's (1984b) data.

	Risk category 1: 1–19 cigarettes		Risk category 2: 20+ cigarettes	
Crude risks	CRR_1	= 1.32591	CRR_2	= 1.48747
Standardized mortality ratio	SMR_1	= 1.30026	SMR_2	= 1.75154
Standardized reference risk	SRR_1	= 1.36873	SRR_2	= 1.56055
Cohort total standard	$SRR(N)_1$	= 1.35220	$SRR(N)_2$	= 1.59577
Mantel-Haenszel standard (common)	$SRR(MHC)_1$	= 1.37029	$SRR(MHC)_2$	= 1.57941
Mantel-Haenszel standard (single)	$SRR(MHS)_1$	= 1.33123	$SRR(MHS)_2$	= 1.71231
Age distribution of population as standard	$SRR(A)_1$	= 0.97882	$SRR(A)_2$	= 1.30068

nal standard (data from Ahlborn et al. 1988). The age distribution of the population is related to the sample size of 91 540 in Hirayama's cohort. As can be seen from Table 26.5, there are some deviations between the age distribution of the cohort and that of the population. In Table 26.6 the risk calculations are presented for the two risk categories (husband is ex-smoker or smokes fewer than 20 cigarettes per day; husband smokes more than 20 cigarettes per day).

There are marked deviations between the standardized risks for both categories. The lowest relative risks are obtained if the age distribution in the population is used as a standard. In this case, the standardized relative risk in the lower risk category (fewer than 20 cigarettes) is slightly below 1. For the higher risk category (20 and more cigarettes) the standardized mortality ratio is remarkably increased. This shows the influence of the age distribution in the exposed group. The Mantel-Haenszel standard is calculated using the common harmonic mean of all risk categories and the reference as well as the single harmonic mean of the respective risk category and the reference. The first procedure is preferable as the risks for different categories are comparable in this case.

The question of which standardization procedure should be used depends on the aim of the risk calculations. If the risk is to be extrapolated to a population, the distribution of the confounding conditions in this population seems most appropriate. This question plays an important role in the so-called meta-analyses, i.e., for quantitative comparisons of different studies. To make the results of the different studies comparable, they must be standardized to a common external

Table 26.7. Interaction between risk conditions and confounding conditions. RR_{ij} are different for different strata S_j. Example: Hirayama's (1984b) data stratified to wife's age.

Wife's age (years)	Husband's smoking habits: 1–19 cigarettes R_{1j}	RR_{1j}	20+ cigarettes R_{2j}	RR_{2j}	Reference condition nonsmoker R_{0j}	R_{0j}
40–49	0.00120	2.37649	0.00166	3.29524	0.00051	1.00000
50–59	0.00294	1.60399	0.00352	1.91809	0.00183	1.00000
60–69	0.00299	1.15156	0.00264	1.01668	0.00259	1.00000
70–79	0.00149	0.08544	0.00837	0.47978	0.01744	1.00000

Table 26.8. Logistic regression analysis applied to Hirayama's (1984b) data—without interactions: $\log(R_{ij}/(1-R_{ij})) = a_0 + c_1C_1 + \ldots + s_1S_1$.

Main effects	Estimated odds ratio
Husband's smoking habits ($P = 0.049$):	
Nonsmoker	0.765
1–19 cigarettes	1.033
20+ cigarettes	1.265
Age (years) ($P = 0.000$):	
40–49	0.437
50–59	1.047
60–69	1.074
70–79	2.036

standard distribution of the confounding conditions. Unfortunately, in most of the meta-analyses of studies on ETS and health risk this is not done.

As was pointed out in Table 26.4, standardized relative risks are weighted averages of the relative risks in the strata. Such an averaging has a real basis if the relative risks within the strata are similar for the total population. If this is not the case, *interaction* between the confounding condition and the risk conditions is present. I doubt whether in this case any averaged overall risk can be interpreted meaningfully.

Let us investigate the interaction using Hirayama's (1984b) data. In Table 26.7 the relative risks for both risk categories are presented for each stratum. We see remarkable inhomogeneities in the relative risks for different strata. The highest relative risk is in the age group of 40–49 years. With increasing age, the relative risk decreases and, for the highest age group, declines remarkably to below 1. This is the result of the controversial behavior of the risk ratios in the reference and exposed groups. Whereas in the reference group the risk ratio for LC increases with increasing age, in the exposed group there is a decrease, especially for the low-risk category. One can only speculate about the reasons for this. Certainly, a selection effect (which is indicated by discrepancies in the age distributions between the cohort and the population) plays some role. Whether additional effects (e.g., a "competing risk" for other causes of death) are involved cannot be analyzed from the data available.

A method to handle interaction is the *logistic regression analysis*. With this method, it is assumed that the logarithms of the odds ratios for each combination of risk condition and confounding condition is a linear function of these variables. If the variables are qualitative, they are expressed by "indicator variables" (with a value of 1 if the quality of the variable is present, and 0 otherwise). Interactions are included in this model by product terms resulting from variables for risk conditions and confounding conditions.

I applied this technique to Hirayama's (1984b) data (Tables 26.8 and 26.9). If no interactions are included in the model, analysis shows a significant ($P = 0.049$) risk effect of a husband's smoking habits to a woman's LC and a very

Table 26.9. Logistic regression analysis applied to Hirayama's (1884b) data—with interactions: $\log (R_{ij}/(1-R_{ij})) = a_0 + c_1 C_1 + \ldots + s_1 S_1 + \ldots + b_{11} C_1 S_1 + \ldots$

Main effects	Estimated odds ratio
Husband's smoking habits (P = 0.265):	
Nonsmoker	0.990
1–19 cigarettes	0.772
20+ cigarettes	1.309
Age (years) (P = 0.000):	
40–49	0.388
50–59	1.034
60–69	1.060
70–79	2.348

Interactions: [P (before removal) = 0.096]
[P (smoking removed) = 0.028]

	Estimated odds ratios		
Age (years)	Nonsmoker	1–19 cigarettes	20+ cigarettes
40–49	0.499	1.275	1.573
50–59	0.688	1.140	1.275
60–69	0.974	1.174	0.874
70–79	2.992	0.586	0.571

pronounced effect of age (P = 0.000). If interactions are included, the main effect of smoking habits is not significant, but we get a significant interaction effect (P = 0.028) in addition to the significant age effect. The estimated odds ratios for the interactions show very good agreement with the variation of risk ratios in Table 26.7: for younger age groups there is an increase of risk with increasing ETS, for older age groups the risk decreases.

This analysis shows that "risk" of ETS is not homogeneous in the cohort. Its extent depends on age and possibly on other factors. An overall estimate may therefore be misleading for an objective prognosis.

References

Ahlborn W, Tuz H-J, Überla K (1988) Estimating relative risk from heterogeneous strata. Methods Inf Med 27:118–124

Hirayama T (1984a) Cancer mortality in nonsmoking women with smoking husbands based on a large-scale cohort study in Japan. Prev Med 13:680–690

Hirayama T (1984b) Effects of nutrition and passive smoking. In: Mitzell M, Correa P (eds) Lung cancer: causes and prevention. Verlag Chemie, Weinheim, pp 175–195

Mantel N, Haenszel W (1959) Statistical aspects of the analysis of data from retrospective studies of disease. J Natl Cancer Inst 22:719–748

Miettinen OS (1972) Standardization of risk ratios. Am J Epidemiol 96:383–388

Section 7. Automotive Exhaust Emissions and Polycyclic Aromatic Hydrocarbons

27. Epidemiologic Studies of Populations Exposed to Motor Vehicle Exhausts and Polycyclic Aromatic Hydrocarbons

M.B. Schenker[1]

Introduction

Many epidemiologic studies over the last 3 decades have confirmed the association of workplace exposure to polycyclic aromatic hydrocarbons (PAHs) and lung cancer (Lindstedt and Sollenberg 1982). However, most of these studies were of workers with very high exposure from nonmotor vehicle sources (e.g., gas retort and coke oven workers) (Lawther et al. 1965; Lloyd 1971). In addition, exposures in these occupations may also have been qualitatively different from motor vehicle exhausts. More recent epidemiologic studies have focused on workplace exposure to motor vehicle exhaust, with the greatest amount of such work on populations exposed to diesel exhaust. These epidemiologic studies have provided useful data on human risks of lung cancer associated with exposure to diesel exhaust, but have not been useful in evaluating lung cancer risk from gasoline engine exhaust. Population-based studies are limited in their ability to evaluate causal associations of air pollution or PAHs with specific health outcomes, and do not provide useful data for assessing the lung cancer risks associated with motor vehicle exhausts.

An analysis of the epidemiologic studies pertaining to motor vehicle exhaust and lung cancer risk provides useful insights into the value and limitations of epidemiologic studies for assessing the hazard of motor vehicle exhaust inhalation. The contrasts in epidemiologic studies of diesel versus gasoline engine exhausts and occupational versus environmental exposures further illustrate the strengths and weaknesses of epidemiologic studies in evaluating lung cancer risk from chronic low-level exposure.

This paper will review some of the considerations in epidemiologic studies of motor vehicle exhaust and lung cancer risk. Both study design and characterization of exposure for epidemiologic studies will be addressed. Finally, the recent

[1]Division of Occupational and Environmental Medicine, University of California, Davis, CA 95616, USA.

epidemiologic data on motor vehicle exhaust and lung cancer will be summarized and the strengths and limitations of these studies discussed.

Motor Vehicle Exhaust Exposure Assessment

There have been few epidemiologic studies on chronic effects of gasoline engine exhaust, and those that have been conducted have generally not included measurements of PAH exposure levels (Stern et al. 1989). Comments in this section will, therefore, address issues of diesel exhaust exposure assessment. A serious weakness of most diesel exhaust epidemiologic studies has been their limited ability to estimate diesel exposure. This weakness was substantial in a hospital-based case-control study of lung cancer that characterized prior diesel exposure by occupational title (Hall and Wynder 1984). In this study, the United States National Institute of Occupational Safety and Health (NIOSH) classification of diesel-exposed occupations was used. This classification includes such diverse job titles as warehousemen, machine operators, and assemblers in its *high* exposure category (Hall and Wynder 1984). Random misclassification of exposure would be expected to occur with such broad categories and would reduce or eliminate any possibility of observing a small increase in lung cancer risk associated with exposure to diesel exhaust. Failure to account for prior work history would further reduce study power by misclassification of exposure.

Several occupational epidemiologic studies have demonstrated another serious problem of exposure assessment, the absence of documented diesel exhaust exposure or any estimate of possible exposure magnitude if exposure is present. Work *with* diesel-powered equipment cannot a priori be equated with diesel exhaust *exposure*. Truck and bus drivers spend most of their working hours inside the truck cabs or buses, an environment that may not have significant contamination from the engine exhaust. In a study of Swedish automobile repair shops and garages, the levels of benzo(a)pyrene in the workplaces was similar to the levels in the city streets (Lindstedt and Sollenberg 1982). Another problem was demonstrated by the early studies of London bus workers in which the levels of smoke and PAH outside the garages were often higher than those inside the garages, a situation that changed with the improvements in London air quality (Commins et al. 1957). The absence of documented or measured diesel exhaust exposure exists in mortality studies of truck drivers (Luepker and Smith 1978), construction union members (Wong et al. 1985), bus company employees (Edling et al. 1987), and dock workers (Gustafssan et al. 1986). Results from these studies on lung cancer risk associated with diesel exhaust exposure must therefore be interpreted with caution.

Assessment of historical exposure is another weakness of occupational cancer studies, particularly when attempts to *quantitate* lifetime exposures are made. Few industries have accurate historical exposure data, and estimates, when possible, must be made on the basis of very limited data. Many industries using diesel engines have not changed substantially in the past few decades (e.g., railroads, transportation, mining), but it may still be difficult to estimate levels of prior exposures from the limited available historical data (Woskie et al. 1988b). This necessitates several assumptions regarding prior exposures which increase the uncertainty of any quantitative estimates.

Of more importance to current and future studies is the lack of a specific or generally accepted marker of diesel exhaust exposure. Early studies of diesel exposure in the railroads measured a variety of compounds including NO, NO_2, CO, acrolein, and total aldehydes (Woskie et al. 1988b). These are of limited value because of their variable relationship to the particles from diesel exhaust which contain the mutagenic (and carcinogenic) fraction. Some early studies measured total suspended particulates (TSP), which contain the PAH fraction but may also include large concentrations of nonexhaust particles.

The recent studies of diesel exhaust exposure and lung cancer mortality among United States railroad workers focused on respirable particles (RSP), but it was recognized that particles from environmental tobacco smoke (ETS) could significantly contribute to the measured RSP, particularly in the indoor setting (Hammond et al. 1985; Woskie et al. 1988a). This led to the creation of an adjusted respirable particulate (ARP) variable, which corrected for the ETS particulate component in RSP (Woskie et al. 1988a). Environmental chamber studies were used to establish the relationship of ETS to RSP, and the relationship was confirmed in field studies (Hammond et al. 1987).

Several candidate markers, including specific PAHs, have been suggested for diesel exhaust exposure measurement. These include phenanthrene, 1-nitropyrene, elemental carbon, and the ratio of pyrene to nitropyrene. While some of these markers may improve the separation of diesel exhaust from non-diesel sources of particles, they lack specificity for diesel exhaust. Also, concentrations expected in personal samplers would be very small—often less than 1 ng. Another potential marker is mutagenicity of respirable particles from low-volume personal samplers (Kado et al. 1987). Mutagenicity may provide a sensitive measurement of the mutagenic or carcinogenic fraction in diesel exhaust particles but, as with the other markers, it is not specific and other sources such as ETS may contribute to measured mutagenicity (Kado et al. 1987). There is not yet a specific marker of diesel or gasoline engine exhaust of the quality that nicotine is as a marker for cigarette smoke (Hammond et al. 1987).

Finally, the use of biologic markers promises to be a more accurate measure of actual dose than does the air concentration of diesel exhaust components. Urinary mutagenicity has been measured in nonsmokers and associated with ETS exposure (Kado et al. 1987). However, there are numerous potential problems with the use of urinary mutagenicity in populations exposed to motor vehicle exhaust including the kinetics of mutagen excretion, the limits of assay sensitivity, and the effect of dietary factors. DNA adducts have been observed in rats exposed to high levels of diesel exhaust, and further studies of DNA adducts may provide more sensitive and specific markers of PAH exposure and dose from engine exhaust (Wong et al. 1986). The combination of newer and more sensitive laboratory assays with epidemiologic methods (molecular epidemiology) provides hope for more specific evaluation of the risks from occupational and environmental exposures to diesel exhaust.

Epidemiologic Study Design Considerations: Occupational Versus Environmental

Occupationally exposed populations have traditionally been the focus of studies of cancer associated with exogenous agents. A major reason for this focus is the higher exposure concentrations in the workplace setting, but numerous other reasons contribute to this preference. Occupational cohorts can be traced historically and in many industries are very stable (Woskie et al. 1988b). Exposure estimates often exist for current workplace conditions, and for some industries qualitative estimates of historical exposures may be inferred from various sources of data. Finally, appropriate comparison populations frequently exist within the occupational setting, providing useful data for the evaluation of dose-response estimates of cancer risks or rates, including rates among nonexposed populations within the same industry.

The large sample sizes required for adequate study power in occupational or environmental studies of lung cancer and diesel exhaust are a major consideration. Occupational studies of exposure to diesel exhaust suggest a relative risk for lung cancer of <1.5 associated with lifetime workplace exposure (Howe et al. 1983; Garshick et al. 1987, 1988). Assuming a lifetime risk of lung cancer among unexposed subjects of approximately 3% (Schenker and Speizer 1979), to observe an increase of 1.5-fold in this risk would require lifetime follow-up of over 4000 subjects, half of whom were exposed. This translates into approximately 150 expected cases of lung cancer, which is greater than the observed number in most occupational studies. Some of the negative studies of diesel and lung cancer have had substantially smaller sample sizes. For example, there were only 12 lung cancer deaths in a study that compared mortality among underground miners who had worked in mines with and without diesel engines, and there were only six lung cancer deaths in a study of diesel exhaust exposure among bus company employees (Waxweiler et al. 1973; Edling et al. 1987).

Evaluation of lung cancer from environmental exposures has been severely constrained by additional limitations in methodology and exposure assessment. Ambient levels of PAHs are commonly two or more orders of magnitude lower than levels in some workplaces (Lindstedt and Sollenberg 1982; Doll 1978). This necessitates even larger sample sizes for such studies than for the studies of occupational cohorts. The mobility and heterogeneity of the population reduces the sensitivity of studies requiring exposure estimate over decades.

The very small estimated risk from environmental pollution also increases the difficulty of excluding the distinct possibility that observed effects are occurring as a result of confounding by cigarette smoking or even occupational exposures (Hammond and Garfinkel 1980). Differences in smoking habits can exist for different geographic areas, and not be reflected in smoking prevalence. For example, the age at which individuals begin smoking, duration of smoking, and manner of smoking may partially explain the urban-rural difference in lung cancer rates (Doll 1978; Hammond and Garfinkel 1980). Failure to control for such differences may result in failure to observe an effect, or in confounding of results.

Confounding is a serious concern because areas with greater atmospheric pollution may have more residents from lower socioeconomic strata, who are also known to have a higher prevalence of smoking and toxic occupational exposures.

Atmospheric carcinogens arise from multiple sources, and useful measurements of potential exposures to PAHs and other toxic air pollutants are scarce. When good measurements of ambient concentrations exist, estimates of indoor or historical exposure are generally lacking. These serious limitations have prevented scientific consensus on whether air pollution or atmospheric PAHs cause lung cancer (Doll 1978; Hammond and Garfinkel 1980). Another problem is the absence of estimates or measurements for personal exposure to ambient PAHs. The existing studies give, at best, ecologic results useful for future hypothesis testing. The low levels of ambient PAHs and the difficulty in characterizing actual exposures suggest that useful epidemiologic studies of lung cancer risk associated with air pollutants from motor vehicle exhausts or from other sources is unlikely. Estimates of such an effect may be based on extrapolations used in risk assessments, but not directly on the results of epidemiologic studies (Hammond and Garfinkel 1980; McClellan 1986).

Occupational Studies of Lung Cancer and Motor Vehicle Exhaust

Several studies of working populations have evaluated lung cancer risk from diesel exhaust exposure (Raffle 1957; Howe 1983; Wong et al. 1985; Edling et al. 1987; Gustafsson et al. 1986; Garshick et al. 1987, 1988). Many large populations of workers exist with potential long-term exposure to diesel exhaust, especially those using heavy-duty vehicles and equipment such as trucks, ships, railroad engines, and construction equipment. There are few comparable occupations with increased gasoline exhaust exposure, and many of these, such as bridge toll takers and highway workers, have mixed gasoline and diesel exhaust exposure.

Few epidemiologic studies of motor vehicle exhaust-exposed populations adequately address the methodologic and exposure issues discussed above, and there are no adequate studies for evaluating human risk of lung cancer from gasoline engine exhaust. The studies of railroad workers provide the most useful data for assessment of lung cancer risk from diesel exhaust.

A large study of Canadian railway workers used computerized record linkage of employee records to national mortality data to evaluate cause-specific mortality in this population (Howe et al. 1983). Despite the absence of measurement data on duration or level of exposure, these investigators found a small increase in lung cancer risk associated with level of "potential" diesel exposure. The relative risks for lung cancer among workers with "possible" and "probable" exposure to diesel exhaust were 1.20 and 1.35, respectively ($p < 0.05$ for both).

More recently the extensive studies of United States railroad workers by Garshick et al. (1987, 1988) have further explored the association of diesel exhaust exposure and lung cancer among railroad workers. In these studies, personal exposures to diesel exhaust were measured for over 500 workers in 39 jobs from four United States railroads (Woskie et al. 1988a). These jobs were combined into 13 job groups based on similarity of job activities and measured exposures. The 13 categories were further reduced to five career exposure groups that were used in the models of lung cancer risk from diesel exposure. Longitudinal work history data from the railroads confirmed the very stable job histories of workers in this industry.

An historical cohort study (Garshick et al. 1988) evaluated lung cancer risk among 55407 railroad workers aged 40–64 in 1959. There were 1694 lung cancers identified among 19396 deaths to 1980. Lifetime diesel exposure for the risk model was based on accumulated yearly exposures as estimated from annual job titles, with the 4 years before death disregarded. The relative risks (ratio of hazard functions adjusted for age) for lung cancer with 1-4, 5-9, 10-14, and ≥15 years of exposure were 1.20, 1.24, 1.32, and 1.72, respectively ($p < 0.05$).

A similar result was seen in a companion case-control study of United States railroad workers (Garshick et al. 1987). Unlike hospital-based case-control studies of lung cancer, the single industry and complete work histories in this study allowed characterization of prior diesel exposure by occupational titles. Lung cancers ($n = 1256$) occurring over 1 year were matched to controls on age and date of death. Cigarette smoking histories were obtained from next-of-kin, and potential diesel exposure was based on industry work histories and the same industrial hygiene evaluation of railroad jobs and work areas (Woskie et al. 1988a). Using multiple logistic regression to control for smoking and potential asbestos exposure, workers ≤ 64 years of age at death with ≥ 20 years of diesel exposure had a significantly increased odds ratio (OR) of lung cancer (OR = 1.41, 95% CI 1.06–1.88). A dose-response was present with years of diesel exposure.

Current levels of diesel exhaust exposure among these railroad workers, as estimated by the ARP concentrations, were 17–191 μg/m³ (geometric means). These values represented up to a two-thirds reduction in the RSP concentrations because of the contribution from ETS. However, the absence of accurate historical data on exposure has precluded to date an estimate of risk per microgram of diesel particulate.

Summary

Epidemiologic studies that have adequately addressed the issues of study design and exposure assessment provide evidence for the carcinogenicity of diesel exhaust in humans. In general, the negative studies have been seriously deficient in their ability to adequately address the association of diesel exhaust and lung cancer. The consistency of animal studies and epidemiologic data on carcinogenicity of diesel exhaust supports its classification as a probably human carcinogen (Ishinishi et al. 1986). Following long-term workplace exposure to diesel exhaust, a relative risk for lung cancer of approximately 1.4 is observed. The absence of epidemiologic data on human populations exposed to gasoline engine exhaust precludes a similar conclusion, despite the similarity of gasoline and diesel exhaust PAHs, and the evidence for carcinogenicity of gasoline engine exhausts in experimental animals. No conclusions are possible from epidemiologic studies regarding exposure to ambient air pollution or PAHs and lung cancer risk.

Despite the evidence for carcinogenicity of diesel exhaust in human populations, assessment of lifetime exposure to diesel has been limited by the absence of a specific marker for diesel exhaust, and scanty data on historical exposures. These deficiencies have weakened the ability to develop quantitative risk estimates from the epidemiologic data (McClellan 1986). While implementation of changes in public health policy is prudent for workplace exposures to diesel exhaust, the absence of quantitative risk estimates limits the ability to extrapolate workplace diesel exposure studies to populations exposed to gasoline exhaust or to environmental PAHs.

References

Commins BT, Waller RE, Lawther PJ (1957) Air pollution in diesel bus garages. Br J Ind Med 14:232–239

Doll R (1978) Atmospheric pollution and lung cancer. Environ Health Perspect 22:23–31

Edling C, Anjou C, Axelson O, Kling H (1987) Mortality among personnel exposed to diesel exhaust. Int Arch Occup Environ Health 59:559–565

Garshick E, Schenker MB, Muñoz A, Segal M, Smith TJ, Woskie SR, Hammond SK, Speizer FE (1987) A case-control study of lung cancer and diesel exhaust exposure in railroad workers. Am Rev Respir Dis 135:1242–1248

Garshick E, Schenker MB, Muñoz A, Segal M, Smith TJ, Woskie SR, Hammond SK, Speizer FE (1988) A retrospective cohort study of lung cancer and diesel exhaust exposure in railroad workers. Am Rev Respir Dis 137:820–825

Gustafssan L, Wall S, Larsson L, Skog B (1986) Mortality and cancer incidence among Swedish dock workers—a retrospective cohort study. Scand J Work Environ Health 12:22–26

Hall NEL, Wynder EL (1984) Diesel exhaust exposure and lung cancer: a case-control study. Environ Res 34:77–86

Hammond EC, Garfinkel L (1980) General air pollution and cancer in the United States. Prev Med 9:206–211

Hammond SK, Smith TJ, Woskie S, Schenker MB, Speizer FE (1985) Characterization of diesel exhaust exposures for a mortality study. In: Cooke M, Dennis AP (eds) Analysis and biological fate: polynuclear aromatic hydrocarbons. Batelle, Columbus, pp 533–541

Hammond SK, Leaderer BP, Roche AC, Schenker M (1987) Collection and analysis of nicotine as a marker for environmental tobacco smoke. Atmos Environ 21:457–462

Howe GR, Fraser D, Lindsay J, Presnal B, Yu SZ (1983) Cancer mortality (1965–77) in relation to diesel fume and coal exposure in a cohort of retired railway workers. JNCI 70(6):1015–1019

Ishinishi N, Koizumi A, McClellan RO, Stöber W (eds) (1986) Carcinogenic and mutagenic effects of diesel engine exhaust. Elsevier, Amsterdam

Kado NY, Tesluk SJ, Hammond SK, Woskie SR, Samuels SJ, Schenker MB (1987) Use of a salmonella micro pre-incubation procedure for studying personal exposure to mutagens in environmental tobacco smoke: pilot study of urine and airborne mutagenicity from passive smoking. In: Sandhu S, DeMarini D, Mass M, Moore M, Mumford J (eds) Short-term bioassays in the analysis of complex environmental mixtures, vol V. Plenum, New York, pp 375–390

Lawther PJ, Commins BT, Waller RE (1965) A study of the concentration of polycyclic aromatic hydrocarbons in gas works retort houses. Br J Ind Med 22:13–20

Lindstedt G, Sollenberg J (1982) Polycyclic aromatic hydrocarbons in the occupational environment. Scand J Work Environ Health 8:1–19

Lloyd JW (1971) Long-term mortality study of steelworkers. V. Respiratory cancer in coke plant workers. J Occup Med 13:53–68

Luepker RV, Smith ML (1978) Mortality in unionized truck drivers. J Occup Med 20(10):677–682

McClellan RO (1986) Health effects of diesel exhaust: a case study in risk assessment. Am Ind Hyg Assoc J 47(1):1–13

Raffle PAB (1957) The health of the worker. Br J Ind Med 14:73–80

Schenker MB, Speizer FE (1979) A retrospective cohort study of diesel exhaust exposure in railroad workers: study design and methodologic issues. In: Pepelko WE, Danner RB, Clarke NA (eds) Health effects of diesel engine emissions. Proceedings of an International Symposium, 3–5 Dec, Cincinnati, pp 1114–1126 (Environmental Protection Agency-600/9-80-57b)

Stern FB, Lemen RA, Curtis RA (1981) Exposure of motor vehicle examiners to carbon monoxide: a historical prospective mortality study. Arch Environ Health 36(2):59–66

Waxweiler RJ, Wagoner JK, Archer VE (1973) Mortality of potash workers. J Occup Med 15(6):486–489

Wong D, Mitchell CE, Wolff RK, Mauderly JL, Jeffrey AM (1986) Identification of DNA damage as a result of exposure of rats to diesel engine exhaust. Carcinogenesis 7(9):1595–1597

Wong O, Morgan RW, Kheifets L, Larson SR, Whorton MD (1985) Mortality among members of a heavy construction equipment operators union with potential exposure to diesel exhaust emissions. Br J Ind Med 42:435–448

Woskie SR, Smith TJ, Hammond SK, Schenker MB, Garshick E, Speizer FE (1988a) Estimation of the diesel exhaust exposures of railroad workers: I. Current exposures. Am J Ind Med 13:381–394

Woskie SR, Smith TJ, Hammond SK, Schenker MB, Garshick E, Speizer FE (1988b) Estimation of the diesel exhaust exposures of railroad workers: II. National and historical exposures. Am J Ind Med 13:395–404

28. Exhaust-Specific Carcinogenic Effects of Polycyclic Aromatic Hydrocarbons and Their Significance for the Estimation of the Exhaust Exposure-Related Lung Cancer Risk

U. Heinrich[1]

Introduction

When we speak of the carcinogenic effects of polycyclic aromatic hydrocarbons (PAH) and, in particular, of the effects of PAH-containing exhaust emissions and their potential carcinogenic risk to humans, it is almost a failure not to mention Sir Percival Pott, a London physician who, in 1775, was the first to recognize the relationship between scrotal cancer and the occupation of chimney sweep (Pott 1775). Pott assumed that the tumor development was caused by prolonged contact of the skin with coal tar and soot. However, the tumor-inducing effect of coal tar was demonstrated experimentally only 140 years later when malignant tumors developed on the ear of the rabbit when painted with coal tar (Yamagiwa and Ichikawa 1918). The experiment showing that skin cancer also developed after administration of the organic extract of soot particles was conducted only a few years later.

The justification for referring to the observation of Sir Percival Pott also in the context of the discussion on exhaust-specific carcinogenic effects of PAH derives from Pott's opinion that prolonged contact of soot with the skin was necessary for the development of tumors. This remark made more than 200 years ago expressed one of the basic principles in the action of PAH-containing particles and is also significant for the exhaust exposure-related lung tumor induction observed in rats very recently. The residence time or biological half-time, respectively, of the soot particle in the target organ is prolonged because of a compromised clearance mechanism; in respect of the scrotal skin of chimney sweeps, the skin clearance was compromised by the poor hygienic conditions present at that time; as regards the lungs of rats exposed to exhaust gases containing higher concentrations of soot particles, the lung clearance was compromised by the high load of alveolar macrophages with nonmetabolizable carbon black. The target organs and the clearance mechanisms of these two organs are quite different, but,

[1]Fraunhofer-Institut für Toxikologie und Aerosolforschung, Nikolai-Fuchs-Straße 1, D-3000 Hannover 61, FRG.

in both cases, deterioration of the process of soot removal will result in a longer contact time of soot and soot-attached carcinogenic substances with the target cells and, consequently, probably also in a more pronounced genotoxic effect. Without going into too much detail on this subject at this time, it should perhaps be mentioned that this statement holds true only under the following conditions: on the one hand, the potential carcinogenic substances bound to the soot particle must be essentially bioavailable; and, on the other hand, their dissolution rate must not be so high that, because of a very short biological half-time, the greater part of the genotoxic material has already been eliminated before it is able to exert a genotoxic effect.

Returning to the observation of Sir Percival Pott, it is obvious that the development of scrotal tumors in chimney sweeps with this noticeable incidence was only possible through the long contact time of soot with the skin and through an adequate dissolution rate and metabolism of the soot-attached PAH. The scrotal cancer of the chimney sweeps was the first tumor connected with the action of soot particles, but a higher lung cancer risk was also reported for this occupational group more than 150 years after Pott's observation (Kennaway and Kennaway 1936, 1947). The very important question about the alteration of the carcinogenic potency of PAH in the respiratory tract owing to the fact that they are mostly bound to particles will be discussed in more detail later on (p. 304).

Occurrence of PAH and Soot in the Occupational and Ambient Atmosphere

As we all know, PAH and soot are not only produced by the burning of coal, but are also found in all exhaust gases resulting from the incomplete combustion of organic material and from high-temperature heating of coal-derived material. Residential heating systems, diesel and gasoline engines, power plants, and various industries represent the most important anthropogenic sources of PAH-containing exhaust emissions. The known carcinogenic PAH which belong to the PAH with four and more rings are generally adsorbed onto small diameter particles which can be easily inhaled and deposited in the deep lung. Since the burden to humans mostly ensues from inhalation of PAH-containing emissions, we are especially interested in the fate and action of PAH retained in the respiratory tract.

When we talk of the carcinogenic potency of PAH as we find them in ambient air and workplace atmospheres, then we have to realize first and foremost that the carcinogenic or suspected PAH are mostly attached to very fine particulate material of different composition and that, as already mentioned, the effect of the PAH with respect to coal soot can be altered in different ways by the various source-specific materials acting as carriers for the PAH and other organic compounds. Tumors caused exclusively by the action of PAH can only be found under experimental conditions in the laboratory employing pure PAH. In all other circumstances, there is a combination effect of the PAH and their exhaust-specific

Table 28.1. Exhaust-specific modifiers of the PAH effect.

- Binding strength between particle and PAH
- Physical and chemical properties of the PAH carrier particle
- PAH profile (gaseous and particle-bound PAH)
- Particle-attached organic compounds other than PAH
- Gaseous exhaust components

carrier. In most cases, this carrier consists of carbon black particles coated with different amounts of organic material. The surface area and the composition of the soot particle, meaning its mass ratio of carbonaceous core to condensed organic material, appear to be decisive for the carcinogenic potency of the particle-associated PAH.

The adsorption of PAH onto particles is an important determinant for the development of the exhaust-specific carcinogenic effect of PAH, but it is by far not the only one (Table 28.1). Depending on the combustion process and the material burned or heated, the PAH profile emitted may be quite different, leading to different combination effects of the PAH with one another. The organic material, which together with the PAH covers the surface of the carrier aerosol, and also the chemical and physical properties of the carrier particle itself may have an exhaust-specific composition resulting in exhaust-specific toxic and cocarcinogenic effects. The gas phase constituents of exhaust emissions, such as irritant gases, and of various organic compounds represent another possibility to modulate the effect of PAH in the lung in an exhaust-specific way.

In humans, exhaust-specific carcinogenic effects are most likely to occur at the workplace because of exposure to higher concentrations of only one particular exhaust emission. Sources of PAH in the work environment are coke and aluminum production, iron and steel works, iron and steel founding, petroleum processing, coal gasification and liquefaction, shale oil production, asphalt production and use, carbon black production, and combustion engines. Although the PAH in the workplace atmosphere may originate from only two major processes, namely, evaporation during high-temperature heating of PAH-containing material and formation by pyrolysis or incomplete combustion, we cannot simply proceed on the assumption that the PAH-containing exhaust gases of the various workplace atmospheres can be allocated to two categories of carcinogenic potency.

Even if the PAH concentration of the various workplace atmospheres were very similar, it would appear inappropriate to estimate the cancer risk of different exhaust emissions solely on the basis of the known concentration of some PAH. This approach should be applied only if the exhaust emissions in question modulate the action of the PAH in a similar way.

The main sources of PAH-containing exhaust gases in urban areas are domestic heating, motor vehicles, and various industries. Unlike the workplace atmosphere, which is mostly polluted by only one exhaust emission, the ambient air may contain a mixture of various types of exhaust emissions. Because of the mixing of the various exhaust gases, the PAH profile of ambient air sampled at different locations is very similar, except when the samples have been collected near a PAH emission source. The carcinogenic potency of such a mixture of PAH-containing emissions may be governed primarily by the exhaust gases which have the strongest carcinogenic effect and the highest proportionate concentration. But if there is a marked agglomeration of PAH-containing particulates from different exhaust emissions, the resulting carcinogenic potency may change in an unpredictable way; an additive or even synergistic enhancement and an inhibiting effect may likewise develop.

A more reliable evaluation of the PAH-related cancer risk can be achieved with populations living in closer proximity to a PAH source or in an area with only one major PAH emittant, for example, a coke oven plant, an area of domestic coal and wood heating, or an area of high traffic density, provided the carcinogenic potency of the respective PAH-containing exhaust gas has been adequately assessed in various experimental tests.

Animal Experiments Showing Modulating Effects of Exhaust Constituents on the Carcinogenic Potency of PAH

What have the experimental results on the modulating effects of exhaust constituents on the carcinogenic effects of PAH been since Kennaway (1930) and Cook et al. (1933) could demonstrate, for the first time, the tumor-inducing effect of pure PAH in animal experiments? Both used the skin-painting test in mice. Kennaway demonstrated the carcinogenic activity of synthesized dibenz(a,h)anthracene—this was the first recognition of the tumor-inducing activity of a chemical of defined structure—and shortly afterwards Cook et al. succeeded in isolating 7 g of the isomeric compounds benzo(a)pyrene (BaP) and benzo(e)pyrene from two tons of pitch; only the BaP was found to be tumorigenic. Since then, a large number of PAH has been isolated, synthesized, and tested for carcinogenicity. By 1976, more than 30 PAH and several hundred alkyl derivatives of PAH had been reported to have some carcinogenic effects.

However, not all the PAH identified were able to induce tumors. This observation could be explained later by the important recognition that PAH are not in themselves chemically highly reactive but exert their mutagenic and tumorigenic activity through a particular class of alkylating metabolite, namely, the so-called bay region dihydrodiol epoxides (Grover 1986). The formation of this class of metabolite, which represents the ultimate carcinogens of the PAH, is the basic requirement for the development of a PAH-related genotoxic effect. The expression "bay region" refers to the position of the epoxide ring in the "bay" formed by carbons 10, 10a, 10b, and 11, as shown for the BaP bay region dihydrodiol epoxide in Fig. 28.1.

Without discussing the very important subject of PAH metabolism in greater detail, it should be emphasized that the carcinogenic potency of a PAH-containing exhaust emission very much depends on the bioavailability of the PAH

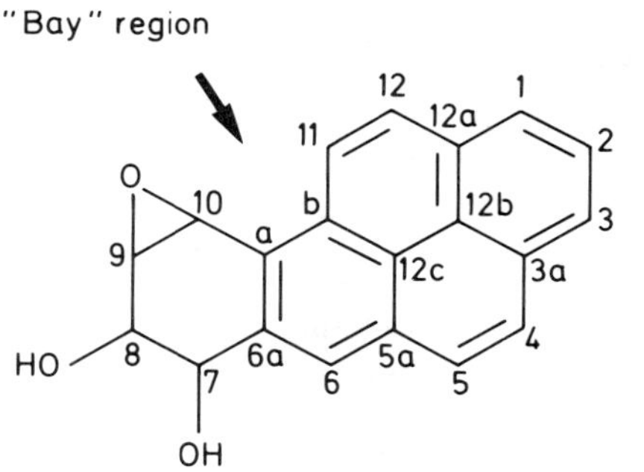

7,8-dihydroxy-9,10-epoxy-7,8,9,10-tetrahydrobenzo(a)pyrene

Figure 28.1. The benzo(a)pyrene bay region dihydrodiol epoxide

and on the activity of the enzyme system which is involved in the metabolic activation of PAH. It is easily conceivable that this enzyme system, which, in conjunction with various types of cytochromes, is also responsible for the detoxification of xenobiotics, can be influenced in its activity by a great number of different events.

This should be kept in mind when considering the various results available from animal experiments indicating interactions of exhaust-specific substances which may contribute to the development of exhaust-specific carcinogenic potencies.

In most of the animal experiments, rodents were used and the test substances were given by epicutaneous, subcutaneous, and intratracheal application or by implantation into the lung (Grimmer 1983). After simultaneous administration of several pure PAH, it could be demonstrated that the tumorigenic effect of the individual PAH was diminished or enhanced depending on the mixture of PAH used.

This means that the carcinogenic potency of a mixture of various PAH cannot be evaluated on the basis of the carcinogenic effect of the individual PAH. Also, PAH which have been shown to be noncarcinogenic can influence the action of carcinogenic PAH.

On testing condensates and extracts from various PAH-containing exhaust emissions and ambient air particulates and various fractions thereof, the PAH-containing fraction has been found to represent most of the carcinogenic activity of the total condensate or extract. But the carcinogenic potency of total condensate and corresponding PAH fraction revealed a marked difference in most instances. A dose-dependent inhibiting effect of the total condensate on the BaP-induced tumor rate was also observed (Pott et al. 1977, 1980; Pott and Stöber 1983; Grimmer 1983). These experiments clearly demonstrated the influence of the non-PAH organic material on the carcinogenic activity of the PAH. But as the organic material, including the carcinogenic PAH, is mostly firmly attached to carbon black particles and has to be dissolved off the particles before it is able to exert an effect, the true exhaust-specific carcinogenic potency of the exhaust could well be different from the potency observed when testing only the particle-free organic material. This effect becomes especially important after inhalation exposure because the extremely small PAH-containing particles of the inhaled exhaust gas are retained in the alveolar region of the lungs for rather a long time.

This particle effect on the carcinogenicity of PAH in the lungs was addressed in a number of studies with hamsters, rats, and mice already published 20–25 years ago. In these studies, particles like hematite, carbon black, and aluminum and titanium oxides were used as carriers for the intratracheal application of BaP into the lungs of the experimental animals. The tumor-inducing effect of BaP instilled intratracheally without particles could be significantly increased when the BaP was firmly attached to the surface of a sufficiently small particle. But also the opposite effect was possible, predominantly with BaP attached to large carbon particles. In this case, the particles were cleared very rapidly from the lungs, and only small amounts of carcinogenic material were released from the carbon particles; therefore, the carcinogenic response in the respiratory tract was correspondingly low (Henry and Kaufman 1973; Farrell and Davis 1974).

The long retention time of BaP adsorbed onto very fine diesel soot particles in the rat lung compared to BaP attached to gallium oxide particles or BaP contained in coal tar aerosols was recently demonstrated by Bond et al. (1986) after short-term inhalation exposure.

Preliminary Results of an Ongoing Inhalation Study Using Various PAH-Containing Aerosols

A very extensive inhalation study has been started to investigate the effect of the carbonaceous core of soot particles on the development of PAH-induced lung tumors. Rats and mice are exposed to carbon black particles devoid of any organic material, to carbon black particles coated with two different amounts of PAH-containing organic material, and to condensed organic material without a carbonaceous core.

Some preliminary results dealing with the development of lung tumors in rats exposed to the differently composed PAH-containing aerosols are reported in the following. The results on mice are described elsewhere (Heinrich et al. 1989).

The exposure in this part of the study was 18 h a day and 5 days a week for 10 months or 43 weeks, respectively, followed by a clean air time which, at the time of reporting, had reached the 57th week. The PAH-containing aerosol was produced as shown schematically in Fig. 28.2 by heating pitch to about 750°C under nitrogen atmosphere and diluting the high-temperature pitch vapor first with equal parts of about 200°C filtered clean air; this prediluted aerosol was further diluted in the ratio of 1:20 with 12°C clean air before entering the inhalation chamber. The mass concentration of the condensation aerosol in the inhalation chamber amounted to about 2.5 mg/m^3, and the corresponding BaP concentration was

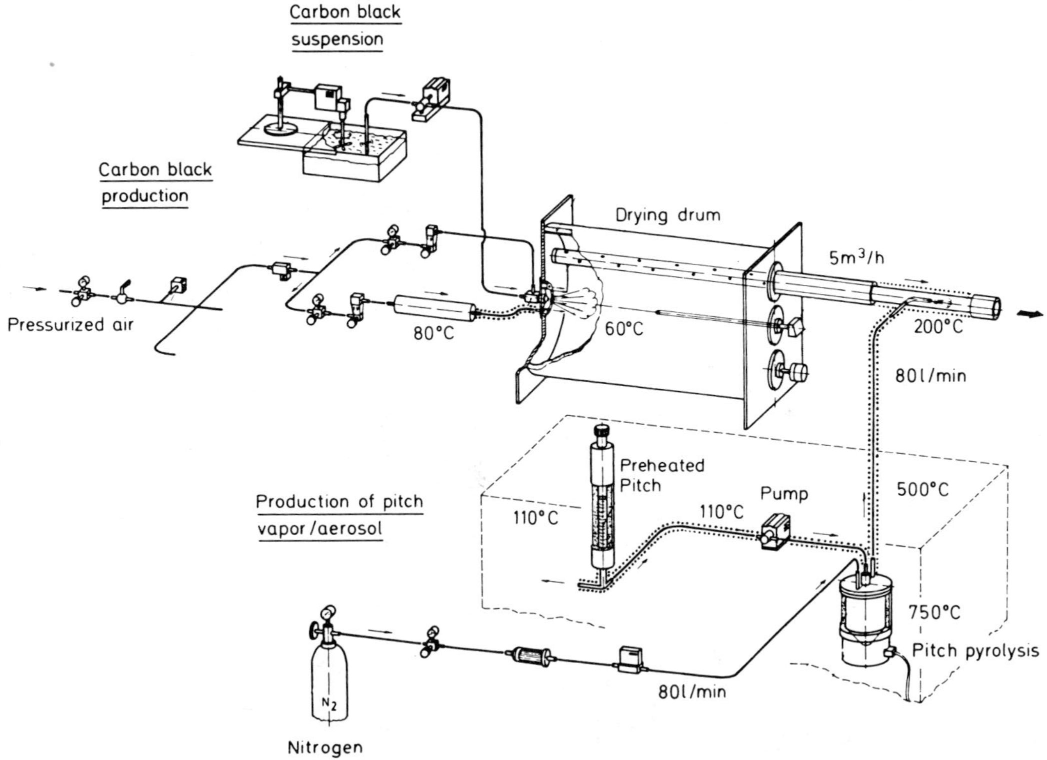

Figure 28.2. Schematic drawing of the experimental set-up used for the production of coated and uncoated carbon black particles as well as carbon black-free pitch vapor condensation aerosol

Table 28.2. Exposure atmospheres used in the ongoing inhalation study.

	Condensation aerosol	2-mg coated carbon black	6-mg coated carbon black	Carbon black particles
Particle concentration (mg/m^3)	2.5	4.5	8.5	6
Carbonaceous core (%)	0	40	70	100
Organic material (%)	100	60	30	<0.02
Benzo(a)pyrene concentration ($\mu g/m^3$)	50	50	50	<0.003

about 50 $\mu g/m^3$. The PAH profile of this exposure atmosphere was very similar to the PAH profile of the coal-derived pitch used for the production of the aerosol.

The same amount of high-temperature heated pitch vapor was mixed at 200°C with 2 or 6 mg/m^3 carbon black particles free of organic material. The mass median aerodynamic diameter of the carbon black particles, produced by nebulizing a suspension of carbon black particles, was below 1 μm; and the surface area of these particles, measured by the nitrogen adsorption method, was about 230 m^2/g particles.

The dilution ratio of pyrolized pitch vapor and aerosol to clean air was the same as mentioned before, and the carbonaceous core of the resulting soot particles represented 40% of the total particle mass in one case and 70% in the other one. The particle mass concentration of the two exposure atmospheres containing two differently coated carbon black particles amounted to 4.5 mg/m^3 and 8.5 mg/m^3, respectively. The concentration of PAH in these two exposure atmospheres containing coated carbon black was the same as in the carbon black-free aerosol atmosphere and, therefore, contained 50 $\mu g/m^3$ BaP as well (Table 28.2).

Another group of animals which is not listed in Table 28.2 was exposed to the carbon black-free condensation aerosol mixed with 5 ppm NO_2, 5 ppm SO_2, and 3 ppm formaldehyde. In addition to the clean air controls, one group was exposed only to the mixture of the irritant gases and one group only to the carbon black particles using the 6 mg/m^3 concentration. Each of the experimental groups comprised 72 female Wistar rats. The mass median aerodynamic diameter of the various particulates used in this experiment was in the range of 0.5–0.9 μm.

The time-to-tumor-rate curves shown in Fig. 28.3 comprise all animals which were sacrificed in a moribund condition and which showed a clear lung tumor at necropsy. The reason for the moribund condition was, in almost all the cases, the development of large lung tumors; therefore, the tumor rates and the mortality rates were very similar. After almost 2 years' experimental time, including 10 months of aerosol exposure, the two exposure atmospheres containing carbon black particles coated with organic material showed a clearly higher carcinogenic potency than the condensation aerosol devoid of any carbonaceous core. "Pitch" in Fig. 28.3 refers to the condensation aerosol produced by pyrolizing pitch containing no carbon black. The carbon black particle which had bound roughly only 30% of its total mass as organic material seems to induce a lower rate of potentially fatal lung tumors than the carbon black particle coated with

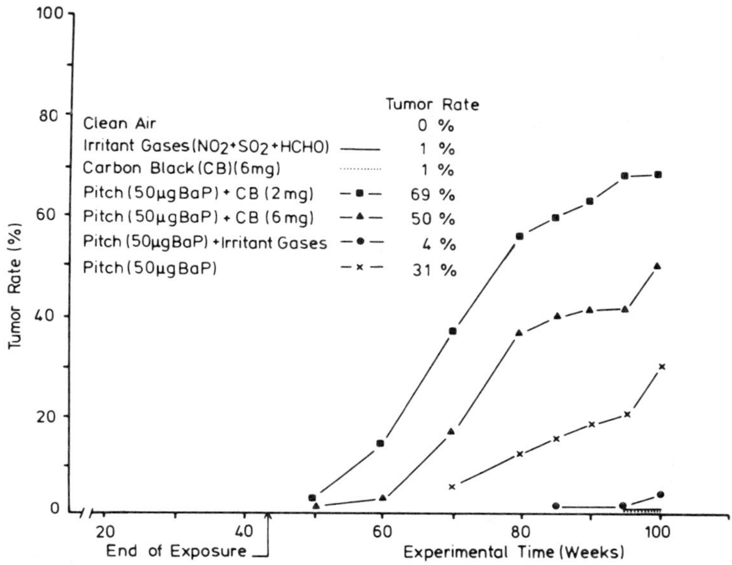

Figure 28.3. Percentage of rats with macroscopically identifiable lung tumors at necropsy sacrificed in a moribund condition at different times after finishing the 43-week inhalation exposure

more than double the amount of organic material; these two experimental groups are termed "pitch (50 µg BaP) + carbon black (6 mg)" and "pitch (50 µg BaP) + carbon black (2 mg)," respectively.

The binding strength between the organic material, including the PAH, and the surface area of the carbon black particle is supposed to be higher when the particle surface is covered with only a thin layer of organic material. Therefore, different dissolution rates of PAH for the differently composed soot particles may play a major role in the development of different tumor rates despite a similar PAH concentration of the exposure atmospheres.

The lower tumor rate of the animals exposed to the aerosol which consisted of condensed organic material without any carbon black particles may be explained by the fast dissolution and elimination of this material by direct absorption through the pulmonary epithelium into the blood. No deposits of particulates could be found in the lungs of the animals exposed to this material.

The amount of carbon black retained in the lungs after 10 months of exposure to 2 and 6 mg/m³ carbon black coated with organic material was 3 and 10 mg, respectively. Although the particle load of the lung after exposure to 6 mg/m³ was more than three times higher compared to the lung load after 2 mg/m³, the macroscopically observed tumor rate of the 6-mg group was clearly lower. The particle load of the rat lung exposed to 6 mg/m³ carbon black free of any organic material amounted to 15 mg per lung after 10 months; up until now, one animal in this

Table 28.3. Exposure to various PAH-containing particulates.

	Pitch aerosol	Coated carbon black		Diesel exhaust[a]	
Particle exposure (mg/m³)	2.5	4.5	8.5	3.5	7.0
Particle dose administration (mg × h/m³) × 10³	9.7	17.4	32.9	14.7	29.4
Concentration of BaP (μg/m³)	50	50	50	<0.05[d]	
Carbonaceous core (% of particle mass)	0	40	70	88	
Organic material (% of particle mass)	100	60	30	12	
Particle mass/lung (mg)	0	3[c]	10[c]	11[e]	21[e]
Lung tumor rate (%)	31[b]	70[b]	50[b]	3.6[f]	12.8[f]

[a] Data from Mauderly et al. (1987).
[b] Macroscopic findings after almost 2 years.
[c] After 10 months of exposure, 18 h a day.
[d] Estimation based on Heinrich et al. (1982, 1986).
[e] After 24 months of exposure, 7 h a day.
[f] After up to 2.5 years.

group has been observed showing a possibly fatal lung tumor at necropsy. However, the mortality rate of this latter group was only 15% at the time of reporting.

Considering all the preliminary results of this inhalation study, it appears to be rather obvious that the development of lung tumors is primarily governed by the presence and bioavailability of particle–attached PAH and not so much by the effects which may be created by a higher load to the lung with particles.

For comparison, the rats which developed a significantly increased lung tumor rate after exposure to diesel engine exhaust containing 7 or 3.5 mg/m³ soot particles had a lung burden of 21 and 11 mg soot, respectively, after a 2-year exposure period. In this study conducted by Mauderly et al. (1987), the animals were exposed 7 h a day, 5 days a week for up to 30 months. The lung tumors were not fatal to the animals, and the tumor rates were 12.8% and 3.6%, respectively. The PAH concentration of the diesel exhaust exposure atmosphere was certainly more than 1000 times lower than that of the coated carbon black atmosphere, and the organic material attached to the diesel soot particle amounted to only 12% of the total particle mass (Table 28.3).

The surprisingly low number of animals with potentially fatal lung tumors exposed simultaneously to the PAH-containing aerosol and the irritant gases in this ongoing inhalation study may be explained by an exposure-related lowering of the breathing frequency, followed by a reduced uptake of PAH-containing particles. Furthermore, the irritating and inflammatory effects of formaldehyde in the nasal cavities may have increased the filtration efficiency of the nose also leading to a lower lung burden. Investigations into particle retention, lung clearance, and mechanical lung function in rats exposed to this mixture of irritant gases are underway, and, therefore, this question will be clarified in the near future.

Conclusions

The preliminary results of this study clearly demonstrate that the carcinogenic potency of the PAH-containing emissions cannot be evaluated simply by their PAH concentrations. The carcinogenic effect of the PAH very much depends on the binding strength of the PAH to the exhaust-specific carrier particle and on the residence time of the PAH-containing particle in the lung. A long residence time provides an extended period for the elution of adsorbed PAH. A short residence time combined with a low dissolution rate of the PAH from the particle will lead to a comparably low carcinogenic effect.

The clearance half-time of diesel soot in the lung is rather long, and, because of the strong adsorption of the PAH onto diesel soot, a continuous dissolution of small amounts of PAH is supposed to occur. The development of PAH-related genotoxic effects may also be supported by the direct uptake of diesel soot by type one epithelial lung cells covering most parts of the alveolar region of the lungs (Barnhart et al. 1980); the penetration of very fine particles into epithelial cells may also occur in the terminal bronchioles. The direct uptake of very small particles by epithelial lung cells, which is likely to occur not just with diesel soot particles alone but also with other very small particulates, enables the whole particle to become involved in the formation of a genotoxic effect. A large surface area and a high reactivity of the particle surface is a good prerequisite for such a particle effect. Furthermore, the release of still reactive PAH metabolites from the diesel soot-laden macrophages and their uptake by other lung cells may be another possibility to expose target cells with diesel soot-derived carcinogens (Palmer et al. 1978). However, the various effects caused by the high load of diesel soot in the rat lung and the resulting consequences for tumor development must not be underestimated.

Finally, it should be mentioned that for PAH-containing emissions which are very similar, especially also in respect of the composition of the PAH-carrying particles, it is by all means reasonable to use the BaP concentration as a measure of the carcinogenic potency. This is demonstrated by the preliminary tumor rates observed macroscopically in the above-mentioned ongoing inhalation study (Fig. 28.4).

Rats were exposed to three different concentrations of the carbon black-free pitch vapor condensation aerosol containing 20, 50, and 90 $\mu g/m^3$ BaP, respectively. Although the evaluation of the tumor rates involved comprised only animals which were sacrificed in a moribund condition, mostly caused by a potentially fatal lung tumor, it is obvious that the lung tumor rates correlate very well with the BaP concentration of the exposure atmosphere. The 20 $\mu g/m^3$ BaP exposure group showed a clearly longer latency period compared to the higher BaP exposure groups, but the lung tumors arose in this group as well.

Acknowledgment. The ongoing inhalation study is supported by the Federal Ministry for Research and Technology (BMFT).

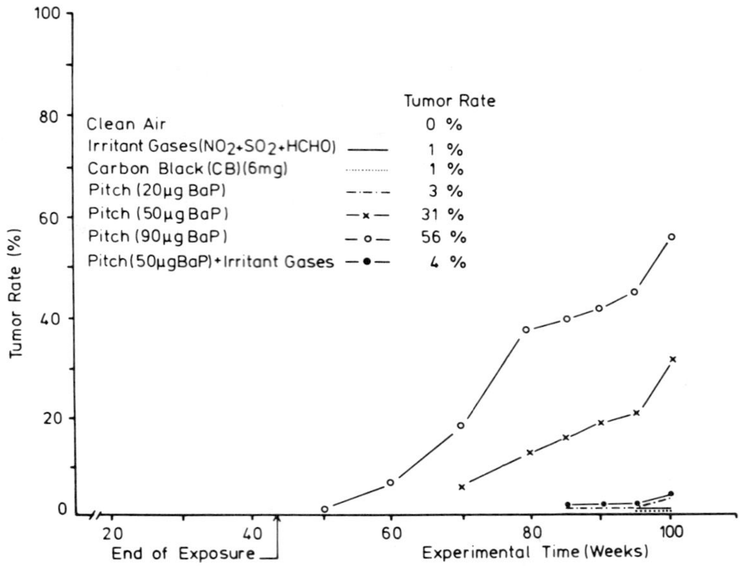

Figure 28.4. Lung tumor rates in rats after 43 weeks of exposure to three different concentrations of carbon black-free pitch vapor condensation aerosol; lung tumor rates increase and lung tumor latency periods decrease with higher BaP concentrations. Lung tumors were recognized macroscopically at necropsy

References

Barnhart MI, Chen S, Puro H (1980) Input of diesel engine exhaust (DEE) particles on the structural physiology of the lung. In: Pepelko WE, Danner RM, Clarke NA (eds) Health effects of diesel engine emissions: proceedings of an international symposium, Dec 3–5, 1979. United States Environmental Protection Agency, EPA-600/9-80-057b, pp 649–671

Bond JA, Sun JD, Mitchell CHE, Dutcher JS, Wolff RK, McClellan RO (1986) Biological fate of inhaled organic compounds associated with particular matter. In: Lee DS, Schneider T, Grant LD, Verkerk PJ (eds) Aerosols. Proceedings of the Second US-Dutch International Symposium on Aerosols. Lewis, Chelsea, MI, pp 579–592

Cook JW, Hewett CL, Hieger J (1933) The isolation of a cancer-producing hydrocarbon from coal tar, parts 1–3. J Chem Soc Trans 1:395–405

Farrell RL, Davis GW (1974) Effect of particulate benzo(a)pyrene carrier on carcinogenesis in the respiratory tract of hamsters. In: Karbe E, Park JF (eds) Experimental lung cancer. Springer, Berlin Heidelberg New York, pp 186–198

Grimmer G (ed) (1983) Environmental carcinogens: polycylic aromatic hydrocarbons. CRC, Boca Raton, FL

Grover PL (1986) Pathways involved in the metabolism and activation of polycyclic hydrocarbons. Xenobiotica 16(10/11):915–931

Heinrich U, Peters L, Funcke W, Pott F, Mohr U, Stöber W (1982) Investigation of toxic and carcinogenic effects of diesel exhaust in long-term inhalation exposure of rodents. In: Lewtas J (ed) Toxicological effects of emissions from diesel engines. Elsevier, New York, pp 225–242

Heinrich U, Muhle H, Takenaka S, Ernst H, Fuhst R, Mohr U, Pott F, Stöber W (1986) Chronic effects on the respiratory tract of hamsters, mice and rats after long-term inhalation of high concentrations of filtered and unfiltered diesel engine emissions. J Appl Toxicol 6(6):383–395

Heinrich U, Fuhst R, Peters L, Muhle H, Dasenbrock C, Pott F (1989) Comparative long-term animal inhalation studies using various particulate matter: objectives, experimental design and preliminary results. Exp Pathol (in press)

Henry MC, Kaufman DG (1973) Clearance of benzo(a)pyrene from hamster lungs after administration of coated particles. J Natl Cancer Inst 51(6):1961–1964

Kennaway EL (1930) Further experiments on cancer-producing substances. Biochem J 24:497–504

Kennaway NM, Kennaway EL (1936) A study of the incidence of cancer of the lung and larynx. J Hyg 36:236

Kennaway NM, Kennaway EL (1947) A further study of the incidence of cancer of the lung and larynx. Br J Cancer 1:260

Mauderly JL, Jones RK, Griffith WC, Henderson RF, McClellan RO (1987) Diesel exhaust is a pulmonary carcinogen in rats exposed chronically by inhalation. Fundam Appl Toxicol 9:208–221

Palmer WG, Allen TJ, Tomaszewski JE (1978) Metabolism of 7,12-dimethylbenz(a)anthracene by macrophages and uptake of macrophage-derived metabolites by respiratory tissues in vitro. Cancer Res 38:1079–1084

Pott F, Stöber W (1983) Carcinogenicity of airborne combustion products observed in subcutaneous tissue and lungs of laboratory rodents. Environ Health Perspect 47:293–303

Pott F, Tomingas R, Misfeld J (1977) Tumors in mice after subcutaneous injection of automobile exhaust condensates. In: Mohr U, Schmähl D, Tomatis L (eds) Air pollution and cancer in man. IARC, Lyon (IARC Sci Publ 16)

Pott F, Tomingas R, Brockhaus A, Huth F (1980) Untersuchungen zur tumorerzeugenden Wirkung von Extrakten and Extraktfraktionen aus atmosphärischen Schwebstoffen im Subkutantest bei der Maus. Zentralbl Bakt I. Abt Orig B 170:17–34

Pott P (1775) Chirurgical observations relative to the cancer of the scrotum. Reprinted in: Natl Cancer Inst Monographs (1963) 10:7–13

Yamagiwa K, Ichikawa K (1918) Experimental study of the pathogenesis of carcinoma. J Cancer Res 3:1–29

29. Molecular Dosimetry of Inhaled Diesel Exhaust

J.A. Bond,[1] J.R. Harkema,[1] R.F. Henderson,[1] J.L. Mauderly,[1] R.O. McClellan,[2] and R.K. Wolff[3]

Introduction

The health effects of diesel exhaust (DE) have been a topic of intensive research during the last decade. DE consists of a mixture of gases, vapors, and soot particles of a respirable size (Cheng et al. 1984). Recent results from several studies (Mauderly et al. 1987; Stöber 1986; Ishinishi et al. 1986; Brightwell et al. 1986) indicate that DE, inhaled chronically at high concentrations, is a pulmonary carcinogen in laboratory rats. The carcinogenicity of inhaled DE was shown to be both dose and time dependent (Mauderly et al. 1987).

It is not clear at this time whether genetic or epigenetic mechanisms or a combination of the two are responsible for the carcinogenic action of inhaled DE. Several studies point to the potential role of the organic chemicals associated with DE soot as causative factors in the carcinogenic response of rodents (reviewed in McClellan 1987). One hypothesis regarding the mechanism for the tumorigenic response is that the organic chemicals must interact with lung DNA to initiate carcinogenesis. In support of this hypothesis is the observation that organic chemicals associated with diesel soot can form DNA adducts in the lung after long-term exposures (Wong et al. 1986). An alternate hypothesis elucidated by Vostal (1986) suggests that the lung tumors in rats exposed to high levels of DE may be due to overloading of the normal lung clearance mechanisms, accumulation of soot particles, and subsequent epigenetic mechanisms (i.e., no direct interaction with genome) for cancer induction. In support of this view are the observations that lung cancer in rodents can be induced by inhalation of particles which are virtually devoid of organic chemicals (e.g., quartz, shale dusts, coal dust, and titanium dioxide), although much higher concentrations of these particles are typically necessary. A third alternative is lung cancer arising from an

[1]Inhalation Toxicology Research Institute, Lovelace Biomedical and Environmental Research Institute, P.O. Box 5890, Albuquerque, New Mexico 87185, USA.
[2]Present address: Chemical Industry Institute of Toxicology, P.O. Box 12137, Research Triangle Park, NC 27709, USA.
[3]Present address: Lilly Research Laboratory, Greenfield, IN 46140, USA.

initiating event involving the organic compounds associated with the DE and the promoting effect of chronic inflammation arising from high concentrations of particles deposited and retained in the lungs.

Our laboratory is exploring the molecular basis for DE-induced pulmonary carcinogenesis in rats by investigating the molecular dosimetry of inhaled DE. The goal of our research is to provide insights into the potential role of DNA adducts in the carcinogenic response to DE. We addressed the following questions:

1. Are lung DNA adduct levels related to exposure concentrations?
2. What is the time course of DNA adduct formation and persistence?
3. Are DNA adduct levels within the respiratory tract increased most in the region in which tumors occur?
4. Would exposure to carbon particles without associated mutagenic organic chemicals also elevate DNA adduct levels?
5. Do species susceptible to exhaust-induced carcinogenesis have higher levels of lung DNA adducts than resistant species?

The research results presented in this paper suggest that the formation of lung DNA adducts by metabolites of soot-associated organic chemicals is probably one of several steps involved in DE-induced pulmonary carcinogenesis in the rat.

Methods

Animals. Female $B6C3F_1$ mice, female Syrian hamsters, male and female F344 rats, and female cynomolgus monkeys were used in these studies. Cynomolgus monkeys were obtained from Charles River Research Primates Corp. (Port Washington, NY) and quarantined for 90 days before they were used. $B6C3F_1$ mice and Syrian hamsters were purchased from Charles River (Wilmington, MA). F344/N rats were born and raised in the institute's barrier-maintained colony. All rodents were 11–15 weeks of age at the start of exposures and were continuously housed in individual wire cages within exposure chambers (H1000 and H2000, Hazleton Systems, Aberdeen, MD). Monkeys were housed in individual cages in the ITRI colony and were moved to individual cages within the exposure chambers (two cages per H2000 chamber) during exposure hours. All animals were allowed to acclimate to the exposure chambers for 2–3 weeks before initiation of particle exposures. Water (automated system) was supplied *ad libitum* during the acclimation and exposure periods. Food was provided outside exposure hours. All animals used in these studies had no clinical signs of respiratory disease prior to exposure.

Exposure Systems. Details of the DE exposure system have been reported (Mokler et al. 1984). Briefly, exhaust was generated by 1980 Model 5.7-liter Oldsmobile V-8 engines mounted on test stands and connected via automatic transmissions to dynamometers and flywheels to simulate a medium sized passenger car operated on continuously repeating United States Federal Test Procedure urban cycles. Fuel meeting USEPA certification standards was used (D-2 diesel control fuel, Phillips Chemical Co., Borger, TX) (Cheng et al. 1984). The exhaust was routed through a standard automotive muffler and tailpipe, diluted 10:1 with filtered air, and then diluted to the final concentration. Exposure chamber atmospheres were measured by daily filter samples for particles and by weekly bag samples for carbon monoxide, carbon dioxide, hydrocarbon vapors, nitrogen oxides, and ammonia. The particle size distribution was measured using a Lovelace multijet cascade impactor and a parallel flow diffusion battery operated in series (Cheng et al. 1984).

Carbon black (Elftex 12, Cabot Corp., Boston, MA) was obtained as a dry powder (elementary particle size, 37 nm; specific surface area, 43 m^2/g). The carbon black was diluted with filtered air and was dispersed using a Wright dust feeder (BGI, Inc., Waltham, MA) operated at 16 psi using a flow rate of 40 L/min. The powder was compressed into a Wright dust feeder cup for 2 h at 1500 psi, using a hydraulic press. The particle exposure concentrations were measured using daily filter samples. The particle size distributions were measured as described above.

Animals were exposed for 6–7 h a day, 5 days a week for up to 12 weeks to filtered air (controls), diluted DE (0.35–10 mg/m^3), or carbon black (3.5 or 10 mg/m^3).

Animal Sacrifice and DNA Isolation. Rodents were killed by either carbon dioxide asphyxiation or a lethal injection of sodium pentobarbital. Monkeys were killed by an intravenous injection of sodium pentobarbital and exsanguinated by cardiac puncture. Respiratory tract tissues were removed as previously described (Bond et al. 1988). DNA was isolated by a modification (Gupta 1984) of the Marmur procedure (Marmur 1961). The DNA concentration was estimated spectrophotometrically (1 A_{260} = 50 μg DNA/ml).

^{32}P-Postlabeling Assay. [$\gamma^{32}P$]Adenosine triphosphate ([$\gamma^{32}P$]ATP) was synthesized by the method of Johnson and Walseth (1979), as described by Reddy and Randerath (1986), except that the reaction cocktail was freshly made for each assay. The specific activity of [^{32}P]ATP was determined using a known amount of 2′-deoxyadenosine-3′-phosphate (Reddy and Randerath 1986).

The nuclease Pl procedure was used to separate DNA adducts (Reddy and Randerath 1986; Bond et al. 1988). The chromatographic conditions to separate DNA adducts and the methods for isolation and quantitation of DNA adducts were described previously (Bond et al. 1988).

Results and Discussion

The exposures did not cause overt signs of toxicity and there were no significant exposure-related differences in body weight (data not shown). Gross observations

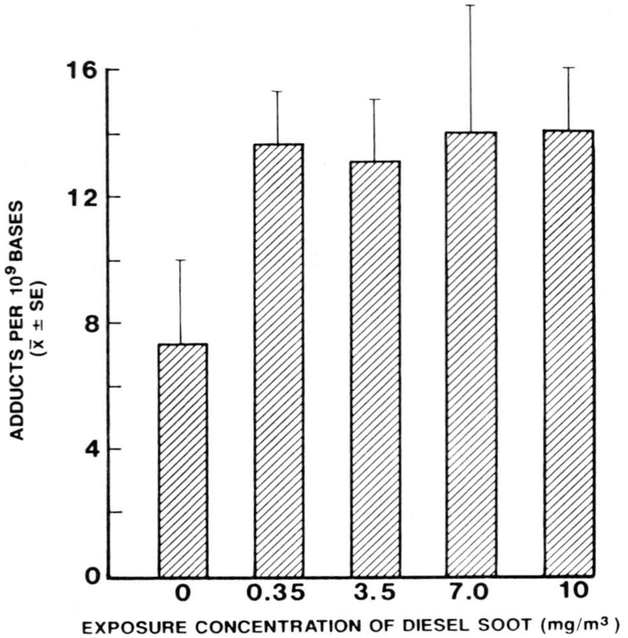

Figure 29.1. Effect of DE concentration on DNA adduct levels in rat lung

of the lungs of animals exposed to DE and carbon black showed that the lungs were darkly pigmented, while the lungs from sham-exposed animals were not.

Are Lung DNA Adduct Levels Related to Exposure Concentrations?

Groups of rats were exposed to 0, 0.35, 3.5, 7.0, or 10 mg soot/m^3 for 12 weeks, and DNA adducts were measured in lungs. The results are summarized in Fig. 29.1. Levels of DNA adducts in lungs of rats exposed to the different exhaust concentrations were similar and were about 14 adducts per 10^9 bases. Levels of lung DNA adducts in exhaust-exposed rats were nearly twice as high as in sham-exposed rats.

These data demonstrated that DNA adduct formation in lungs was independent of exhaust exposure concentration at the exhaust exposure levels tested. Bond and Mauderly (1984) reported on the capacity of isolated perfused rat lungs to metabolize chemicals associated with DE and showed that lung metabolism of these chemicals could be saturated at high concentrations. One explanation for the results reported here is that, at the exhaust concentrations tested, the lung enzymes responsible for metabolism of soot-associated chemicals to metabolites that bind to DNA were saturated. Furthermore, the lack of an exposure-response relationship for DNA adduct formation and the finding that DNA adducts were increased at an exposure level which did not increase lung tumor incidence (0.35

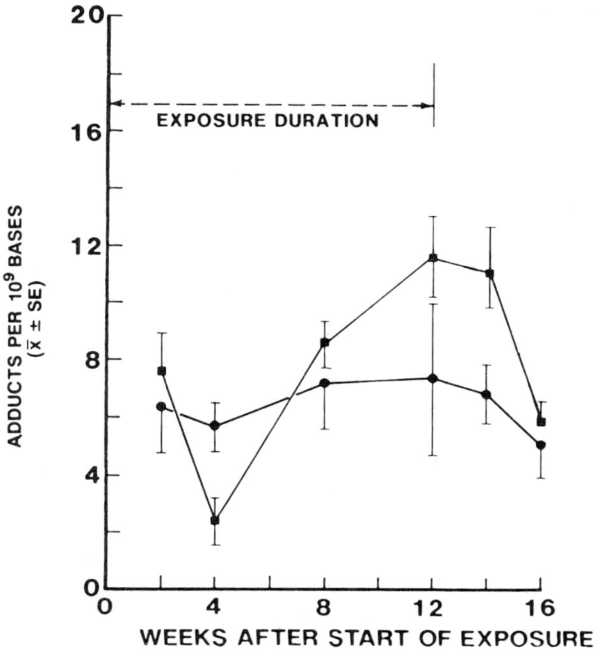

Figure 29.2. Time course for formation and persistence of DNA adducts in lungs of rats exposed for up to 12 weeks to air (*circles*) or to 7 mg soot/m³ (*squares*)

mg/m³; Mauderly et al. 1987) suggest that other factors are also important in exhaust-induced carcinogenicity.

What is the Time Course for DNA Adduct Formation and Persistence?

Experiments were designed to determine whether the amount of adducts increased linearly with time during a 12-week exposure and whether steady-state levels of adducts could be predicted after a chronic exposure to DE. Rats were exposed to 7 mg soot/m³ for up to 12 weeks and were sacrificed at 2, 4, 8, 12, 14, and 16 weeks after the start of exposure. The results are summarized in Fig. 29.2. Lung DNA adducts accumulated slowly during the 12-week exposure and were highest at the end of exposure. In contrast, lung DNA adducts in sham-exposed rats remained relatively constant throughout the 12-week exposure. DNA adduct levels declined rapidly after the termination of exposure. By 4 weeks after the end of exposure, adduct levels were not significantly different from those of controls. The observation that repair of adducts was faster than adduct formation suggests that steady-state levels of DNA adducts would be reached during long-term exposure to DE. It will be important to understand the kinetics of formation and repair of *individual* adducts to elucidate the relationship of DNA adduct formation to carcinogenesis.

Table 29.1. Regional distribution of DNA adducts in the respiratory tract of rats exposed to diesel exhaust[a]

Tissue	Total adduct formation[b]	Net adduct formation[b,c]
Maxilloturbinates	3.8 ± 1.8	2.6
Ethmoturbinates	5.3 ± 1.2	5
Trachea	2.3 ± 2.2	0
Mainstem bronchi	1.2	0
Axial airway	1.1	0
Peripheral tissue	18 ± 10	8.4

[a] Values represent the mean ± SE, $n = 2$. Values with single numbers are from only one group of rats.
[b] Adducts expressed per 10^9 bases.
[c] Net adduct formation represents the difference between adducts per 10^9 bases in exposed animals and adducts per 10^9 bases in comparable control groups.

Are DNA Adduct Levels Within the Respiratory Tract Increased Most in the Region in Which Tumors Occur?

The studies of Wong et al. (1986) did not provide information on the regional dosimetry of DNA adducts within the lung. Thus, experiments were designed to determine the location of DNA adducts in the respiratory tract following exposure of rats to DE and to determine if the highest levels of DNA adducts were located in the region of the respiratory tract where exhaust-induced tumors were found. Rats were exposed for 12 weeks to DE at 10 mg soot/m³. The rats were then sacrificed and different regions of the respiratory tract (maxilloturbinates, ethmoturbinates, trachea, left mainstem extrapulmonary bronchus, intrapulmonary axial airways, and peripheral lung) were removed and analyzed for DNA adducts (Bond et al. 1988). DE-induced adducts were found only in the peripheral lung tissue and nasal tissue, and total levels of DNA adducts were highest in the peripheral tissue (Table 29.1). These data clearly indicate that the total level of DNA adducts and the level of exhaust-induced adducts are highest in the region of the rat respiratory tract where tumors are formed following exposure to a carcinogenic concentration of DE (Mauderly et al. 1987). This observation, while not proving causality between DNA adducts and tumors, lends support to the hypothesis that adduct formation and subsequent tumor formation are related.

Would Exposure to Carbon Particles Without Associated Mutagenic Organic Chemicals Also Elevate DNA Adduct Levels?

In order to investigate the effect of particle exposure on DNA adduct levels, groups of rats were exposed either to DE or carbon black at 0, 3.5, or 10 mg/m³ for 12 weeks. At the end of exposure, lungs were analyzed for DNA adducts. The DE soot particles had about a 30% solvent-extractable organic content, whereas the carbon black particles had only about 0.04% solvent-extractable organic content. High exposure levels (10 mg/m³) of both carbon black and DE increased

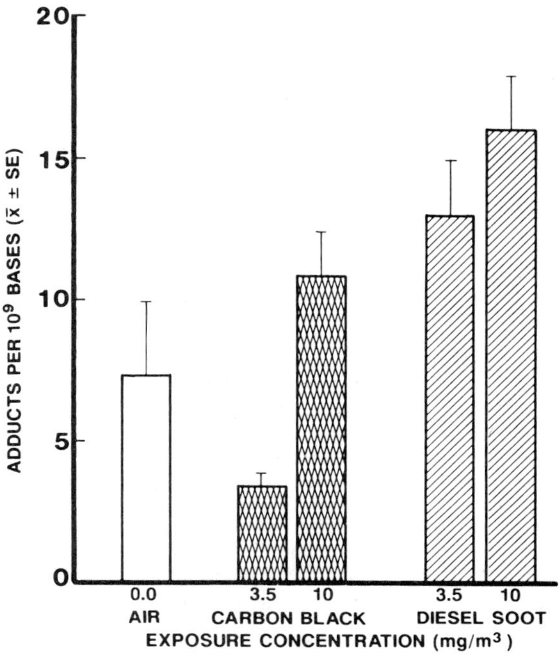

Figure 29.3. Effect of a 12-week particle exposure on DNA adduct levels in rat lung

DNA adduct levels, although adduct levels in DE-exposed rat lungs were about 30% higher than carbon black-induced adducts (Fig. 29.3). However, at an exposure level of 3.5 mg/m^3, only DE was found to increase DNA adducts (~ 50% above control values). These studies indicate that high levels of carbon particle exposure can increase DNA adduct formation, although the levels are lower than those of exhaust-related adducts. It is possible that the small content of organic chemicals associated with carbon black may have desorbed from the particles and was metabolized to metabolites that bind to DNA.

It is unlikely that particles themselves play a role in the genetic mechanisms of DE-induced carcinogenicity. Rather, the role of particles in the carcinogenic response is probably related to the chronic inflammation they induce. Henderson et al. (1988) recently reported a chronic inflammatory response to DE as characterized by dose-dependent increases in inflammatory cells, cytoplasmic and lysomal enzymes, and protein in brochoalveolar lavage fluid. The degree of inflammatory response was related to the lung burdens of particles. As part of the inflammatory response, neutrophils and macrophages are stimulated to release mediators of inflammation as well as injurious oxygen radicals. These oxygen radicals could be important in cytotoxicity or even cause DNA damage as part of the overall carcinogenic response to DE.

Table 29.2. DNA adduct formation in lungs of different species exposed to diesel exhaust.

Species	Exhaust-induced tumors	Weekly CxT[a] (mg h/m³)	Diesel particle lung burden[b] (mg/g lung)	Lung DNA adducts (% control)
Mouse	No	250	0.95 ± 0.05	94
Hamster	No	250	0.63 ± 0.07	110
Rat	Yes	300	1.8 ± 0.1	160
Monkey	Unknown	250	4.2 ± 0.4	180

[a]Weekly concentration × time product; exposures were 12 weeks.
[b]Values are means ± SE.

Do Species Susceptible to Exhaust-Induced Carcinogenesis Have Higher Levels of Lung DNA Adducts Than Resistant Species?

The purpose of this experiment was to investigate species differences in the formation of DNA adducts in the peripheral lung during exposure to DE. This study used four animal species: mice and hamsters (species resistant to exhaust-induced carcinogenesis); rats (species susceptible to exhaust-induced carcinogenesis); and monkeys (no data available on tumorigenicity). All species were exposed to DE at 8.1 mg soot/m³ for 12 weeks. The results (Table 29.2) revealed that species resistant to exhaust-induced cancer also had no increase in peripheral lung DNA adducts above that measured in sham-exposed animals. In contrast, there was a 60%–80% increase in peripheral lung DNA adducts in rats and monkeys. We hypothesize, based on the similarities in the degree of elevation of adducts above controls in monkeys and rats, that monkeys may also develop tumors following long-term exposure to high concentrations of DE.

In summary, the results of these studies suggest that the formation of lung DNA adducts by metabolites of soot-associated organic compounds is one step in the initiation of DE-induced pulmonary carcinogenesis in the rat. The data that support this statement include the observation that DNA adduct levels are highest in the region of the respiratory tract where tumors occur. Furthermore, the observation that mice and hamsters (species resistant to DE-induced cancer) do not have elevated levels of DNA adducts and that rats (species susceptible to DE-induced cancer) do have elevated adduct levels also lends support to this statement.

However, other factors must also be involved in DE-induced carcinogenicity. The lack of an exposure-response relationship and the finding that adducts were increased in rats at an exposure level which did not increase lung tumor incidence (0.35 mg/m³; Mauderly et al. 1987) support this notion. It is possible that the role of particles in the carcinogenic response to DE may be related to their ability to induce an inflammatory response in lungs. The inflammatory response, coupled with an increase in lung cell turnover induced by DE (Mauderly et al. 1983), may be important factors in the progression to tumors following exposure to DE.

Acknowledgments. The authors gratefully acknowledge the technical assistance provided by numerous members of the institute's technical staff throughout the conduct of these studies. The authors also acknowledge the critical review and discussions with a number of our colleagues. This research was supported by the Office of Health and Environmental Research, US Department of Energy, under contract No. DE-AC04-76EV10103, in facilities fully accredited by the American Association for Accreditation of Laboratory Animal Care.

References

Bond JA, Mauderly JL (1984) Metabolism and macromolecular covalent binding of ^{14}C-1-nitropyrene in isolated perfused/ventilated rat lungs. Cancer Res 44:3924–3929

Bond JA, Wolff RK, Harkema JR, Mauderly JL, Henderson RF, Griffith WC, McClellan RO (1988) Distribution of DNA adducts in the respiratory tract of rats exposed to diesel exhaust. Toxicol Appl Pharmacol 96:336–346

Brightwell J, Fouillet X, Cassano-Zopi AL, Gatz R, Duchosal F (1986) Neoplastic and functional changes in rodents after chronic inhalation of engine exhaust emissions. In: Ishinishi N, Koizumi A, McClellan R, Stöber W (eds) Carcinogenicity and mutagenicity of diesel engine exhaust. Elsevier, Amsterdam, pp 471–485

Cheng YS, Yeh HC, Mauderly JL, Mokler BV (1984) Characterization of diesel exhaust in a chronic inhalation study. Am Ind Hyg Assoc J 45:547–555

Gupta RC (1984) Nonrandom binding of the carcinogen N-hydroxy-2-acetylaminofluorene to repetitive sequences of rat liver DNA in vivo. Proc Natl Acad Sci USA 81:6943–6947

Henderson RF, Pickrell JA, Jones RK, Sun JD, Benson JM, Mauderly JL, McClellan RO (1988) Response of rodents to inhaled diluted diesel exhaust: biochemical and cytological changes in bronchoalveolar lavage fluid and in lung tissue. Fundam Appl Toxicol 11:546–567

Ishinishi N, Koizumi A, McClellan RO, Stöber W (1986) Carcinogenicity and mutagenicity of diesel engine exhaust. Elsevier, Amsterdam

Johnson RA, Walseth TF (1979) The enzymatic preparation of [α-^{32}]ATP, [α-^{32}-P]GTP, [α-^{32}P]cAMP, [α-^{32}P]cGMP, and their use in assay of adenylate and guanylate cyclases, and cyclic neucleotide phosphodiesterases. Adv Cyclic Nuecleotide Res 10:135–167

Marmur JA (1961) Procedure for the isolation of deoxyribonucleic acid from microorganisms. J Mol Biol 3:208–218

Mauderly JL, Benson JM, Bice DE, Carpenter RL, Evans MJ, Henderson RF, Jones RK, McClellan RO, Pickrell JA, Redman HC, Shami SG, Wolff RK (1983) Life-span study of rodents inhaling diesel exhaust: results through 30 months. In: Marshall TC, Guilmette RA, Byers RL (eds) Annual report of the Inhalation Toxicology Research Institute, LMF-107, pp 305–316

Mauderly JL, Jones RK, Griffith WC, Henderson RF, McClellan RO (1987) Diesel exhaust is a pulmonary carcinogen in rats exposed chronically by inhalation. Fundam Appl Toxicol 9:208–221

McClellan RO (1987) Health effects of exposure to diesel exhaust particles. Annu Rev Pharmacol Toxicol 27:279–300

Mokler BV, Archibeque FA, Beethe RL, Kelly CPJ, Lopez JA, Mauderly JL, Stafford DL (1984) Diesel exhaust exposure system for animal studies. Fundam Appl Toxicol 4:270–277

Reddy MV, Randerath K (1986) Nuclease Pl-mediated enhancement of sensitivity of ^{32}P-postlabeling test for structurally diverse DNA adducts. Carcinogenesis 7:1543–1551

Stöber W (1986) Experimental induction of tumors in hamsters, mice and rats after long-term inhalation of filtered and unfiltered diesel engine exhaust. In: Ishinishi N, Koizumi A, McClellan R, Stöber W (eds) Carcinogenicity and mutagenicity of diesel engine exhaust. Elsevier, Amsterdam, pp 421–439

Vostal JJ (1986) Factors limiting the evidence for chemical carcinogenicity of diesel emissions in long-term inhalation experiments. In: Ishinishi N, Koizumi A, McClellan R, Stöber W (eds) Carcinogenicity and mutagenicity of diesel engine exhaust. Elsevier, Amsterdam, pp 381–396

Wong D, Mitchell CE, Wolff RK, Mauderly JL, Jeffrey AM (1986) Short communication: identification of DNA damage as a result of exposure of rats to diesel exhaust. Carcinogenesis 7:1595–1597

PART IV

Interfaces Between Scientific Judgement and Prudent Environmental Health Policy

30. Risk Assessment as an Instrument of Environmental Policy

M. Fischer[1]

The existence of finite concentrations of no-threshold carcinogenic substances in the environment forces policy makers to follow a strategy of keeping exposure "as low as reasonably achievable" (ALARA principle). The answer to "What is reasonable?" can, in principle, be found from a balance of risks and benefits.

In medicine the health benefits from a therapy or from preventive measures (X-ray, vaccination), to which there is no alternative, justify considerable health risks from these measures, and balancing these risks and benefits is part of the medical service.

Outside the medical field, national programs for the peaceful use of nuclear energy were the first instance of establishing a broadly accepted balance between certain social benefits (increased energy independence, less environmental pollution, economic growth) and specific risks from enhanced radiation or very rare but disastrous failures.

To balance health risks against the benefits from economic growth is difficult. Myers et al. (1981) attributed the increase of longevity in modern high-energy societies (1–10 kW per capita) partially (10% of a total increase of 35 years) to this energy consumption and compared it to the calculated loss of life expectancy (less than 0.1 years) due to cancers and other deleterious health effects from the generation of that energy.

The example is cited in order to illustrate a possible approach in reasoning about carcinogenic environmental pollutants. If the processes generating them are necessary for the survival of a society and their emission cannot be reduced to zero, one has to accept finite concentrations in the environment, i.e., finite risks to the population. Although these risks are usually too small to be empirically detectable, they may still be of political concern, and society may question the benefits of accepting them.

Practical decisions in this field involve several carcinogens of different potency from various sources, changing technologies, and varying economical and geographical settings. A common numerical expression of risks could considerably support environmental policy making by reducing the complexity of toxicological

[1]Bundesgesundheitsamt, Postbox 33 00 13, D-1000 Berlin 33, FRG.

or epidemiological findings, while at the same time translating them into the language of decision making (Fischer 1984; WHO Europe 1987 p. 3). The scientists in public administration have a twofold responsibility in this process of translation:

- Not to misuse the authority of science when giving advice to the government
- To put the questions from the political process into a form which can be answered by scientific means

The widely adopted methods of risk quantification provide a proper framework to cope with this responsibility.

Formulas for Low-Dose Risk Extrapolation

We are primarily interested in an expression for the probability of dying from cancer due to a lifetime exposure to an environmental carcinogen X. This probability is made up of two terms, the spontaneous or background term p_0 and a dose-dependent term $p_1\ (D_x)$:

$$p(D_x) = p_0 + (1 - p_0)\, p_1\ (D_x) \tag{1}$$

For the moment we only assume that D_x is positively related to the environmental concentration [X]. In the next step—restricting ourselves to epidemiological data for the moment—the probabilities p and p_0 are related to the number of cases in the population under observation by

$$\begin{aligned} \text{Obs}\ (X) &= p(D_x)\ \text{N}_\text{A} \\ \text{Exp} &= p_0\ \text{N}_\text{A} \end{aligned} \tag{2}$$

where N_A is the number of persons in the population; $p_1\ (D_x)$ can then be calculated from

$$p_1\ (D_x) = (\text{Obs}(X) - \text{Exp}) \,/\, (1 - p_0) N_A \tag{3}$$

which is the form most appropriate for cohort studies. If—as in case/control studies—the relative risk is used (Obs (X) = $\text{RR}_x \cdot \text{Exp}$) one gets

$$p_1\ (D_x) = p_0\ (RR_x - 1) \,/\, (1 - p_0) \tag{4}$$

Because of $p_0 \ll 1$, the term $1 - p_0$ is often omitted in the denominators in 3 and 4.

The terms on the right-hand sides of 3 and 4 are known from observations usually at workplaces [Obs (X), RR_x] and from cancer statistics (Exp, p_0). In order to make an extrapolation one needs the dose D_x as a function of concentration $[X]$ in air. In many cases the best available approximation is

$$D_x = \text{average workplace concentration} \cdot \text{breathing rate} \cdot \text{duration of exposure at the workplace}$$

with only a few dose points available in each study. Owing to the latency times of occupational cancers, the workplace concentrations refer to situations several decades back in time and may only be "guestimates." Keeping all this in mind, one should not be surprised that a plot of the data—which is only a section of the complete dose-response curve—can often be fitted by a straight line, at least in the lower part of the observational range. There is an infinite set of smoothly increasing positive functions which all satisfy the following conditions:

- They come numerically close to the sectional fit in the observed dose interval.
- They become zero for vanishing dose.
- They approach 1 for very large doses.

In risk extrapolation one often picks one of them by the following *ansatz*:

$$p_1 (D_x) = 1 - \mathrm{Exp}\,(-\mathrm{K}D_x) \tag{5}$$

where K is a positive constant. This formula has the nice property of being linear for small doses:

$$1 - \mathrm{Exp}\,(-\mathrm{K}D_x) \cong \mathrm{K}D_x \tag{5'}$$

whenever $\mathrm{K}D_x \ll 1$, which is the case in environmental situations where $\mathrm{K}D_x$ is below 10^{-3}, as will be seen later. This *ansatz* characterizes the whole dose-response curve by one single parameter K which is closely related to the "unit risk," as will now be shown in the next step of this section. Using 4 and 5' we get for one specific workplace concentration: $\mathrm{K} = p_0\,(\mathrm{RR}_{\mathrm{work}} - 1) / D_{\mathrm{work}}$ and with $D_x = [X] \cdot$ lifetime exposure

$$p_1 (D_x) = p_0\,[X]\,(\mathrm{RR}_{\mathrm{work}} - 1) / \mathrm{D}_{\mathrm{work}}\,(\text{lifetime exposure})^{-1} \tag{6}$$

Defining the unit risk by $p_1 = \mathrm{UR}[X]$, we get (WHO Europe 1987, p. 14):

$$\mathrm{UR} = p_0\,(\mathrm{RR}_{\mathrm{work}} - 1) / [X]_{\mathrm{work}}\ \mu\mathrm{g/m}^3 \cdot 8/24 \cdot 240/365 \cdot \text{years of employment}/70 \tag{7}$$

where the same breath rate has been assumed for workers as in the general population.

Our *ansatz* 5 is equivalent to the "one-hit (and one-stage) model;" more refined models are possible and have been suggested in order to account for the fact that carcinogenesis is a multistage process. In a brilliant review Zeise et al. (1987) recently discussed these more sophisticated models and their biological content. Their conclusion is that, at best, very carefully conducted animal experiments, including accumulation of many more data on pharmacokinetics and cell dynamics, would allow one to discriminate, on the basis of biological reasons, between different models, but that these data are still lacking. It should be intuitively clear that the risk curve calculated from a multistage model always falls below the linear extrapolation for low doses because, if several hits or stages are needed to induce cancer, the probability per unit of the carcinogenic agent must be smaller than in the one-hit case. Therefore the one-hit model is conservative, i.e., on the safe side [for a critical discussion of so-called superlinear models see Zeise et al. (1987)] and should for this reason be preferred for regulatory purposes.

Table 30.1. Unit risk for carcinogenic air pollutants calculated from epidemiological studies (see text for details).

Pollutant	Unit risk $[\mu g/m^3]^{-1}$		Reference
	Value	Range	(WHO Europe 1987)
Acrylonitrile	2×10^{-5}	1.5 – 1.7	p. 42
Benzene	8×10^{-6}	3 – 8.1	p. 53
BaP[a]	9×10^{-2}	0.11×1 – 8.7	p. 113
Vinyl chloride	4×10^{-6}	0.5 – 30[f]	p. 166
Arsenic	8×10^{-3}	3.1 – 7,5[g]	p. 178
Asbestos[b]			
Mesothelioma	2×10^{-5}	0.5 – 2	p. 193
Lung cancer[c]	2×10^{-5}	0.04 – 1.6	p. 195
Cadmium	2×10^{-3}	1.8 – 92[h]	p. 207
Chromium-VI[d]	1×10^{-1}	0.11 – 1.3	p. 228
Nickel	6×10^{-4}	1.5 – 5.9	p. 293
Radon (EER)[e]	2×10^{-4}	0.7 – 2.18	p. 335

[a] BaP as indicator for urban PAHs.
[b] The unit is $(100 \text{ critical fibers/m}^3)^{-1}$ measured with SEM.
[c] Calculated for smokers, the unit risk for nonsmokers would be lower by a factor of ten.
[d] It is very difficult to measure chromium-VI separately in the environmental aerosol.
[e] The unit is $(\text{Bq/m}^3 \text{ EER})^{-1}$.
[f] The highest value from human data is 4×10^{-6}.
[g] In a recent paper Becher and Wahrendorf (unpublished data) calculated the unit risk from 19 different studies and obtained a range of $(0.4 - 34)\times10^{-3}$.
[h] The highest value from human data was 2.3×10^{-3}.

We finish this section with a list (Table 30.1) of unit risks for several air-borne carcinogens. The listing was part of a WHO-coordinated effort by many experts over 3 years towards *Air Quality Guidelines for Europe* (WHO Europe 1987).

We have given the unit risk value to one significant digit only. The ranges are not statistical confidence intervals but result from different interpretations of the data, especially the exposures. In some cases, an upper bound in the statistical sense was calculated and used for the extrapolation. Before these results are used for regulatory purposes, an international convention on these procedures should be reached in order to free the discussion on the EEC or OECD level from purely technical details.

According to our present knowledge, the list seems to be fairly complete with two exemptions: diesel soot and environmental tobacco smoke (ETS). It is hoped that the material presented in this volume will further clarify these matters. Man-made mineral fibers could be a further candidate for inclusion in the list.

The organic substances listed are clearly genotoxic, and therefore there is good reason to apply the no-threshold hypothesis in these cases. The same holds true for ionizing radiation from radon and its daughters. Several epidemiological studies on the carcinogenic effects from indoor radon exposure are now being carried out, opening a way to empirically test the risk extrapolation.

With respect to the other inorganic materials in our list, the evidence for genotoxicity is less clear, so there is no general reason for assuming no-threshold. Furthermore, "asbestos" designates a whole family of fibrous minerals, and the metallic materials exist in many different compounds in the atmospheric aerosol, the potency of which turns out to be different according to carefully conducted toxicological studies (see WHO Europe 1987 for details). So the unit risk attributed here to the "substance family" as a whole may actually represent the most dangerous of its members.

A further piece of pessimism baked into the unit risks derived from epidemiological studies is due to the fact that the populations studied not only smoked a lot, but were also simultaneously exposed to a lot of air pollutants at their workplaces and even during the rest of their lives, because urban air pollution was much more pronounced during the 1940s and 1950s when these people were active. Therefore possible synergisms are—to a certain degree at least—included in these unit risks.

In summing up, one can be sure that the real cancer risk from an exposure to the pollutants listed will not be larger than the value calculated by using the unit risk figure. How much less it will actually be is, however, impossible to tell—otherwise one could improve the risk extrapolation. In this sense, risk extrapolation makes the best use of existing knowledge.

Limitations of Quantitative Risk Assessment

If risk extrapolation is to be more than curve fitting, it has to be applied with sound biological reasoning, especially if only animal data are available. Thoughtless and schematic applications have discredited the method and led to a renewed discussion of its usefulness (see Fischer, 1984, for a short summary of earlier discussion). In revising its *Guidelines for Drinking-Water Quality* WHO has recently stated the basic assumptions which should apply if a risk extrapolation is to be made (WHO 1988; see also Preuss and Ehrlich 1987). We cite from WHO (1988):

Carcinogenicity

The initiating event in the process of chemical carcinogenicity is often considered to be the induction of a mutation in the DNA of a somatic cell. In some cases the chemical acts directly on the DNA, in other cases it may require activation by a metabolic system. This process theoretically has no threshold and is the basis for mathematical models for low dose extrapolation.

There are, however, some substances which are capable of producing cancer in animal studies but which cannot be shown to have genotoxic activity. In these cases the mode of action is probably either cytotoxicity in the target organ, with the resulting tissue repair increasing the possibility of spontaneous mutation, the promotion of previously initiated cells, or the deregulation of metabolism. In such cases there is likely to be a threshold dose and the use of low-dose extrapolation models would not be appropriate.

The linearized multistage model, which is formally equivalent to our *ansatz* 5, is generally recommended for extrapolation by WHO: "If suitable pharmacokinetic data are available and their use can be justified, they should be incorporated into the low-dose extrapolation model."

Regional Variations in Exposure and Risk

The concentration of a pollutant varies across a country according to the localization of sources and the meteorological factors of atmospheric transport, as shown by a map (Fig. 30.1a) of the vinyl chloride concentrations in the Netherlands. If this information is combined with population density data, information is obtained on the cumulative fraction of the population exposed (Fig. 30.1b). For regulatory purposes, however, we would like to know the cumulative fraction as a function of risk and we would like to compare the risks from several pollutants. This can easily be done by using the product

$$x = \mathrm{UR}_x[X]$$

as the common variable as shown in Fig. 30.2. In order to construct this figure for the Federal Republic of Germany, various data had to be used:

- The distribution of radon has been measured in a sample of about 6000 households in an official program (Schmier 1984). His data have been redrawn by using the relationship [EER] = 0.5 [Rn] for the equilibrium equivalent radon concentration (McLaughlin 1988). The International Commission on Radiological Protection (ICRP) reference distribution is shown for comparison (Jacobi et al. 1987).
- In 1984 the Länderausschuss Immissionsschutz (LAI; State Commission on Air Pollution Control), a focal point of clean air policy in our country, appointed a special working group to develop guidelines on carcinogens in ambient air. This group also collected available data on average concentrations in rural areas, cities, large conurbations, and in the vicinity of large emitters. No correlation with population density has yet been made by the working group, and the data provided in Fig. 30.2 are the result of my own guesswork in order to illustrate their possible use in policy making (I thank the chairman of this group, Mr. Schmitt, for permission to use the data here). Also included are data from the Netherlands (MVROM 1984) and from a survey on indoor air pollutants by Krause et al. (1987). A similar approach has been used by Stolwijk (1987) with data from the United States concentrating primarily on the indoor situation. For the sake of clarity, his data have not been included in Fig. 30.2. Because the geographical distribution of radon concentrations is completely uncorrelated to that of the other pollutants we have to keep radon separate in the following discussion.

In the Federal Republic of Germany about 15% of the population live in rural areas. The pollution levels (excluding radon) recorded there mark the background

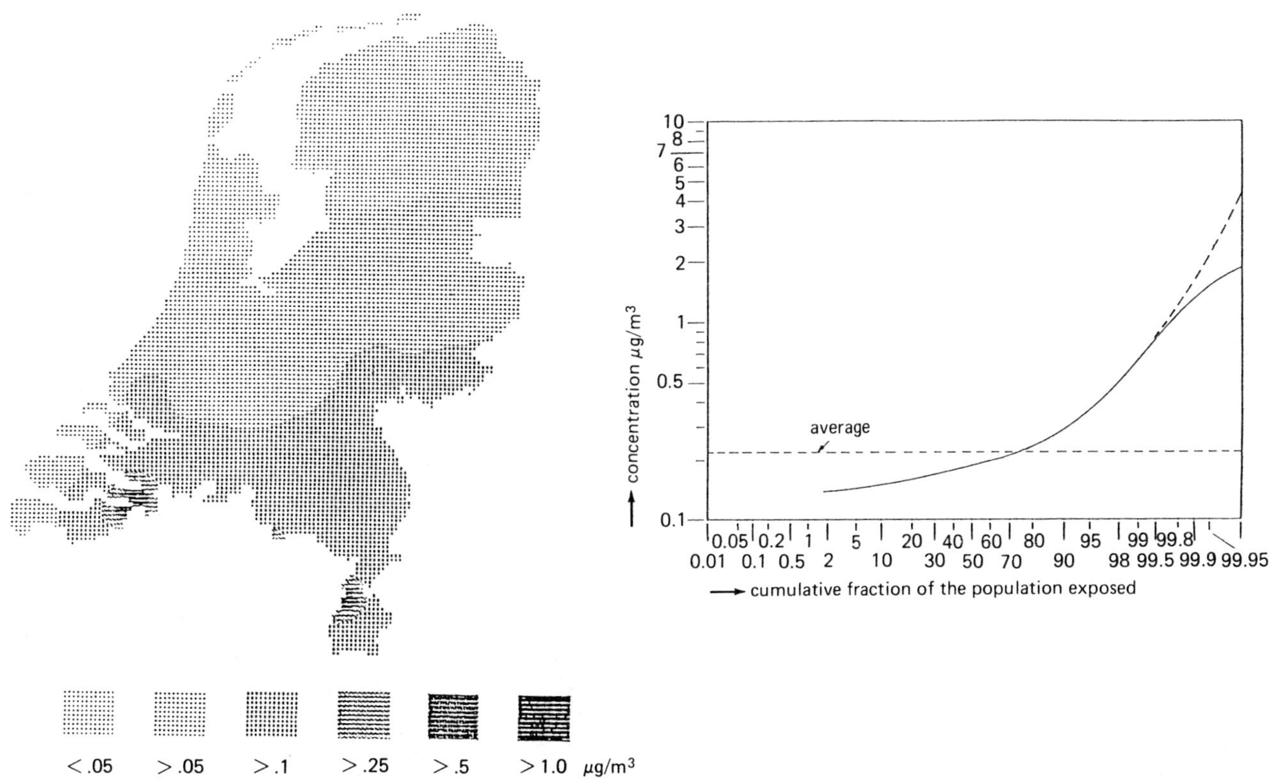

Figure 30.1. a Distribution of vinyl chloride concentrations in The Netherlands. **b** The cumulative frequency distribution shows what fraction of the population is exposed up to a certain concentration. (From MVROM 1984)

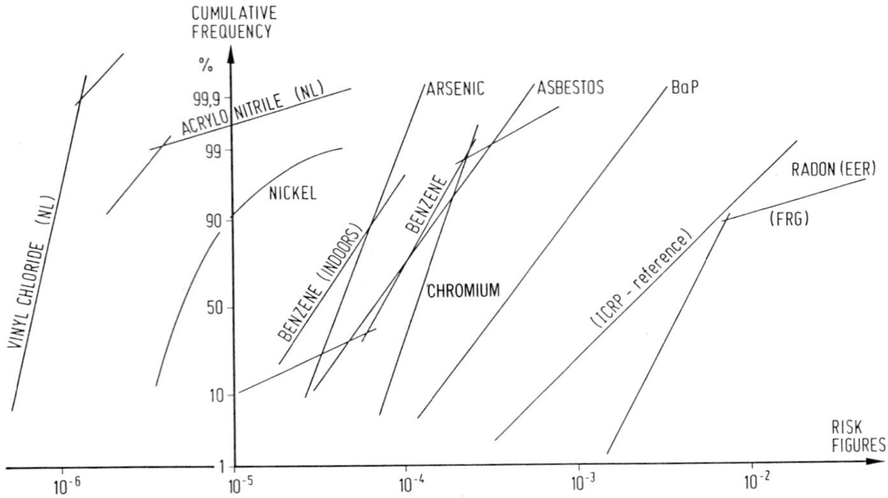

Figure 30.2. Cumulative frequency distribution showing the exposure to various airborne carcinogens in the Federal Republic of Germany. Source: see text

levels in the sense that these concentrations are geogenic, the result of long-range transport or of the emissions from many small, widely dispersed sources, like heating. In any case, these concentrations are more or less below the level of social concern and control.

While the mixture of pollutants will be roughly the same in all rural areas, this is certainly not the case in those places where the highest concentration of one specific carcinogen is found. This means that the fraction of the population with the highest exposure to, say, arsenic will most probably not be the same group of persons who are most severely exposed to, say, vinyl chloride. But it is the bulk of the population, living in cities and conurbations, which faces both massive and multiple pollution because the crowding of people correlates positively with that of industry, traffic, and nonindustrial fossil fuel consumption. We therefore concentrate on this part of the population which in our graph is in the band between 15% and approximately 99%.

The risk figures, also called individual risks, should primarily be considered as an index combining the concentration of a pollutant and its potency. They can be added because we can safely assume that simultaneous carcinogens act independently at environmental concentrations.

Looking at Fig. 30.2, three groups of pollutants can be formed according to the level of risk they pose. Radon is the single most risky component (Kerr 1988). At the 50% level its risk figure is several times that of all other carcinogenic air pollutants together (ETS not included). The individual risk from radon ranges from 2×10^{-3} (at 15%) to possibly 2×10^{-1} (at 99%). In the next group the risk figures are all below 2×10^{-3} and mostly in the range 10^{-4}–10^{-3}. In this group we find the "classical" pollutants: polycyclic aromatic hydrocarbons (PAH) (indicator benzo-a-pyrene, BaP, 1.8×10^{-4} – 1.8×10^{-3}) asbestos (3×10^{-4} – 4.3×10^{-5}), chromium (2.4×10^{-4} – 8.2×10^{-5}), benzene 10^{-5} – 2×10^{-4}), and arsenic (10^{-4} – 2.8×10^{-5}). The aforementioned working group has not yet considered diesel, but preliminary calculations show that diesel soot would range between PAH and asbestos (M.J. Csicsaky, personal communication). With respect to benzene, the indoor concentrations in both the United States (Stolwijk 1987) and northern Italy (De Bartoli et al. 1985) were clearly higher than those in ambient air. Nickel and, if the Dutch data are representative of the situation in the Federal Republic of Germany, acrylonitrile and vinyl chloride belong to a third group with risk values below 10^{-5}.

Proposing Guidelines for the Total Risk Level

Many of our concepts in risk policy have been developed in radiation protection parallel to and in close interaction with the progress of nuclear technology. The natural dose rate and its variations were so high if compared to the normal release from nuclear power stations that the acceptability of the latter was never seriously questioned by the administration.

When regulating carcinogenic chemicals in the environment, politicians and administrators face a much greater variety in sources: (a) residues from industrial activities terminated long ago (waste dumps, or *Altlasten* in German), (b) old plants in declining industries which are often localized in areas with high unemployment; (c) small, but numerous sources such as cars, private heating, and asbestos products. Even though environmental legislation requires the application of the "best technical means" for the minimization of carcinogenic emissions, there is often no way to enforce their implementation with today's administrative instruments.

This situation could be improved if a guide-value for the total risk level in any region were introduced. This value would cover all carcinogenic air pollutants other than radon. There is no natural level for orientation, but, if we consider the values for rural areas, the guide-value would be somewhere in the range $(1–5) \cdot 10^{-4}$. Because only the sum of risk values from several pollutants would be specified, their individual concentrations could vary from region to region. Radon is to be excluded because its high background value would make the index insensitive to all changes in the other substances.

The index could be used to localize regions of high risk value and to set priorities for appropriate countermeasures. Substances and source groups which are of relevance in many regions could be identified and appropriate federal programs initiated. Locally dominating risk sources would be identified as such, and specific strategies could be discussed.

More formalized decision-making tools can be used in the analysis of concrete projects. Cost-effectiveness of emission control in petrochemical plants can be used to evaluate the impact of different technical measures on the risk level in the vicinity of the source (MVROM 1984). The weathering of asbestos cement tiles and sheets contributes to the asbestos aerosol. Fischer (1989) has analyzed proposals for the total or partial elimination of this source in terms of their health benefits (avoidance of cancers due to asbestos) and their health costs (loss of life in the work force necessary for the realization of these proposals). He argues that one should consider the net benefit (in terms of health) of labor-intensive programs for the elimination of environmental carcinogens.

References

DeBortoli M et al. (1985) Measurements of indoor air quality and comparison with ambient air: a study on 15 homes in northern Italy. Office for Official Publications of the European Communities, Luxemburg

Fischer M (1984) Die Risikoabschätzung als Entscheidungshilfe. In: Fischer M, Meyer E (eds) Zur Beurteilung der Krebsgefahr durch Asbest. MMV Medizin, Munich, p8ff

Fischer M (1989) On risks related to risk reduction. In: Proceedings, symposium on management of risk from genotoxic substances in the environment, Stockholm (1989)

Jacobi W et al. (1987) Lung cancer risk from indoor exposures to radon daughters. In: ICRP (ed) Radiation protection. ICRP publication no. 50. Pergamon, New York

Kerr AR (1988) Indoor radon: the deadliest pollutant. Science 240:606–608

Krause C et al. (1987) Occurrence of volatile organic compounds in the air of 500 homes in the FRG: proceedings of a symposium series, vol 1. Indoor Air '87, Aug 1987, Berlin (West). Institute for Water, Soil, and Air Hygiene, Berlin

McLaughlin J (1988) Radon in indoor air. In: Commission of the European Communities (ed) European concerted action: indoor air quality and its impact on men. COST project no 613. Report no 1. Commission of the European Communities, Luxemburg

MVROM – Ministerie van Volkshuisvesting, Ruimtelijke Ordening en Milieubeheer (1984) Criteria document on vinyl chloride. MVROM, Lucht 34, AC's Gravenhage, The Netherlands

Myers DK et al. (1981) Carcinogenic potential of various energy sources: proceedings of a symposium series. IAEA-SM-254/02. International symposium on health impacts of different sources of energy, 22–26 June, Nashville

Preuss PW, Ehrlich AM (1987) Assessment of health risks associated with noncriteria air pollutants. APCA J 37(7):784–791

Schmier H (1984) Die Strahlenexposition in Wohnungen durch die Folgeprodukte des Radon und Thoron. Paper presented at the information meeting of the Federal Ministry of the Interior on information and results on radiation protection research in the Area of Responsibility of the Interior Ministry, 30 May 1984, Wissenschaftszentrum Bonn. Report of the Ministry of the Interior, pp 37–54

Stolwijk JAJ (1987) The determination of health effects of indoor air pollution: proceedings of a symposium series, vol 4. Indoor Air '87, Aug 1987, Berlin (West). Institute for Water, Soil, and Air Hygiene, Berlin

WHO Europe (1987) Air quality guidelines for Europe. WHO regional publications. European series no 23. World Health Organization, Copenhagen

WHO (1988) Guidelines for drinking-water quality. World Health Organization regional bureau of Europe, Copenhagen

Ziese L, Wilson R, Crouch EAC (1987) Dose-response relationships for carcinogens: a review. Environ Health Perspect 73:259–308

31. The Perils of Prudence*

P.W. Preuss,[1] V.J. Cogliano,[1] and P.D. White[1]

Introduction

Risk assessment is an integral part of decision-making at the United States Environmental Protection Agency (US EPA). In some US EPA programs, no regulatory decisions are made without considering risk, often in the form of a quantitative risk assessment. Partly because of its widespread use, quantitative risk assessment has come under increasing scrutiny, particularly the way assumptions are made and uncertainty is expressed: often, risk assessments ignore or gloss over uncertainty, and confine the expression of uncertainty to a qualitative discussion at the end of the document. In response to these concerns, the US EPA is reexamining the assumptions in its previous risk assessments, especially the assumptions in the cancer assessments, and is searching for new ways to incorporate and describe uncertainty.

The current discussion about assumptions and uncertainty in risk assessment is partly the result of policy decisions made over a decade ago. These decisions were used to formulate an approach to calculate reasonable worst-case estimates and plausible upper bound estimates to the risk (Albert et al. 1977; Interagency Regulatory Liaison Group 1979; US EPA 1976). These policy decisions left unclear what these estimates mean and how uncertainty should be described. They also left unclear the relationship of the plausible upper bound (or reasonable worst case) estimate to the true risk[2] and to more central estimates of risk.

---

*Disclaimer. The views expressed in this paper are those of the authors and not those of the US Environmental Protection Agency.

[1]Office of Research and Development, US Environmental Protection Agency, 401 M St. S.W., Washington, DC 20460, USA.

[2]The EPA guidelines (US EPA 1986) state: "It should be emphasized that the linearized multistage procedure leads to a plausible upper limit to the risk that is consistent with some proposed mechanisms of carcinogenesis. Such an estimate, however, does not necessarily give a realistic prediction of the risk. The true value of the risk is unknown, and may be as low as zero."

As currently practiced, the reasonable worst case/plausible upper bound approach is intended to provide a risk estimate that is conservative (i.e., the risk in the situation assessed is not likely to exceed the calculated value), but not unduly prudent. This means that the assessment is based on a series of conservative, but not necessarily worst-case, assumptions. When definitive scientific data are lacking, "fallback" assumptions, based on scientific judgement and science policy, are often used. As a result, scientists in the United States and other countries often disagree about which assumptions are appropriate to use. Unfortunately, these differences are generally not reflected in the assessment and are often presented only as a part of the qualitative discussion of the uncertainty. The "masking" of these uncertainties has tended to fuel the flames of the controversy and has helped to foster the notion that the assumptions used result in a wholly unrealistic, overly conservative assessment.

Agencies like the US EPA are mandated to protect public health and the environment. Consequently, as a matter of policy, they are likely to err on the side of prudence and, therefore, to select the assumptions that give rise to a more prudent risk estimate. To avoid misinterpretation of assessments by decision-makers and the public, it has become critical to develop a rationale and approach that will promote understanding about how the key assumptions and uncertainties affect an assessment (National Research Council 1983; Office of Science and Technology Policy 1985; Finkel 1987; Environ 1988). Several groups in the EPA and elsewhere have tried to represent the uncertainty in assessments by using various extrapolation models (including all the different assumptions and uncertainties of the models themselves) to provide a range of estimates to the decision-maker. These approaches have not succeeded because they are too confusing: they present a complex and bewildering range of risk estimates without providing either a way to understand the basis for the range or a rationale to select a narrower subset of the range.

To address the assumptions/uncertainty problem, we are currently trying to develop a new way to express uncertainty and apply it in the decision-making process with the following goals:

- The discussion of uncertainty should be quantitative as well as qualitative.
- The quantitative uncertainty should be expressed using relatively few, clearly derived, numbers.
- The approach should be applicable to many different kinds of assessments.

Development of the Dual Risk Estimate Approach

To meet these goals, we have been exploring ways to develop a "more central" estimate of risk that parallels and supplements the usual plausible upper bound estimate. This central estimate would provide a risk value that the assessor believes is likely to be closer to the "true risk" than the plausible upper bound estimate, but that would still protect public health.

A rationale for a more prudent central estimate could be developed as follows. A central aim of a regulatory agency is to reduce aggregate cancer risks to exposed populations through a lengthy series of decisions regarding exposures to carcinogens. In principle, the long-term performance of the agency in cancer risk assessment could be evaluated by comparing the aggregate actual risk to the population (following the agency's decisions) to the aggregated estimated risk. Of course, actual risk is not in general ascertainable, so that an appropriate standard would be whether the agency can be confident that the aggregate actual risk is unlikely to exceed the estimated risk. Central risk estimates could then be constructed to be only as prudent as necessary to meet this criterion. Therefore, more central estimates can be less conservative than plausible upper bound estimates because the assessor is now willing to tolerate the potential for underestimation of a single risk as long as long-term performance continues to be prudent. It should be noted that, if the potential risk from one or two carcinogen exposures is especially high, then the assessor may need to be particularly prudent in these instances if long-term performance is to remain prudent overall.

As with plausible upper bound estimates, the protectiveness of the assumptions made for the central risk estimate would most likely be based on scientific judgment, since precise statistical tools could probably not be applied.

The above discussion has focused on risk to a population. In considering individual risk, the concern is not the long-term prudent performance of the government, but, rather, assurance to an individual that a specific exposure is adequately controlled. In other words, the individual is concerned not only about how high a risk may be, but also about the likelihood that the true risk is in fact as high as the estimate. Therefore, a prudent individual would be interested in both a reasonable worst-case estimate as well as a more central estimate of risk.

Based on this rationale developed for both population risk and individual risk, we propose adding another dimension to cancer risk assessments to demonstrate explicitly some of the uncertainties. Specifically, we suggest calculating a more central estimate in addition to the plausible upper bound that is currently estimated. The central estimate would be derived by changing a few of the potency/dose-response and exposure assumptions, such as:

- Use of the most sensitive species and sex.
- Combining benign and malignant tumors.
- Using surface area as a scaling factor when extrapolating from animals to humans.
- The often used assumption that exposure lasts 24 h a day for 70 years.
- The often used assumption that exposure occurs at the point of highest concentration of the pollutant.

This dual calculation of the risk would make it relatively easy to understand how a few basic assumptions affect the risk estimate. At the same time, this dual estimate should help to confine disagreement about assumptions and uncertainties to those few cases where it really makes a difference, which would allow the system to proceed with the next steps (e.g., formulating regulations) for all others.

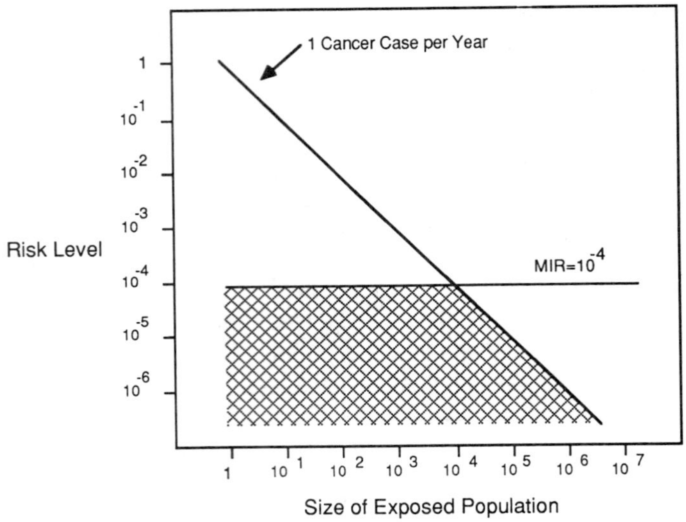

Figure 31.1. Possible criteria for decisions about acceptability of risk. The two criteria used in this example bound an area of acceptable risk and of unacceptable risk. Changes in the magnitude of the two criteria chosen would change each of the areas under the lines *MIR*, maximal individual risk

Figures 31.1 and 31.2 show how the two-risk estimates could be applied to determining whether the risk is acceptable for specific chemicals. Two hypothetical criteria for acceptable risk are shown in Fig. 31.1: a maximal individual risk of 1×10^{-4}, and a population risk of one excess cancer case per year. (These lines can each be moved in both directions to meet any desired policy.) The intersection of these two lines (criteria) defines ranges or zones of acceptable and unacceptable risks that might be used in a health-based regulatory decision. Figure 31.2 superimposes onto the risk criteria the plausible upper bound and a more central estimate for three hypothetical chemicals:

Case a. Since both risk estimates fall into the zone defined as "acceptable," the decision would be that the risk is acceptable.

Case b. Since both risk estimates fall into the zone defined as "unacceptable," the decision would be that the risk is unacceptable.

Case c. Since one estimate falls into the "acceptable" zone and the other into the "unacceptable," further analysis would be required to better define and understand the uncertainty in that assessment.

Clearly, the major hurdle to the successful application of a dual-estimate scheme is defining the criteria by which a more central estimate should be selected.

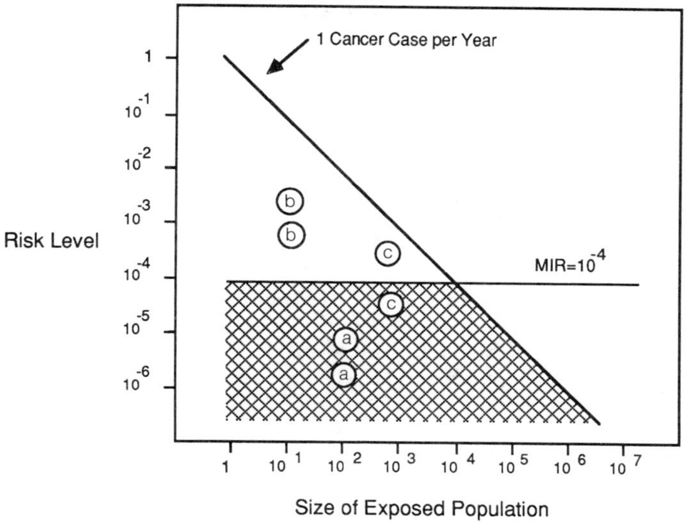

Figure 31.2. An example of superimposing dual risk calculations of plausible upper bounds to risk and more central estimates of risk on Fig. 33.1

Examples of Issues Affecting Risk Estimates

Five examples are provided below to illustrate how more central risk estimates might be derived. (These examples are meant to illustrate specific points; the data have not been fully analyzed, so that the results presented here cannot be used for other purposes.) The first four examples concern questions related to multiple estimates of potency; the fifth addresses the variation obtained in exposure estimates.

Potency

Benzene

In 1985, the US EPA considered several alternatives in estimating the cancer potency of benzene. These alternatives were based on three epidemiologic studies, two estimates of exposure, and two mathematical models. Recently, an industry-sponsored analysis (Clement Associates 1988) has proposed a new potency estimate that challenges several of the US EPA assumptions. Table 31.1 lists the estimates derived by changing individual assumptions. The final estimate proposed, 1.43×10^{-4}, represents the aggregate of these changes. It is clear from the table that the major change is a 25-fold decrease between the last two estimates, which results from application of the hypothesis that two molecules are required to induce the single molecular event that causes the cell transition. Their other proposed changes seem less critical.

Table 31.1. Influence of the modification of US EPA's assumptions on the unit risk estimate.

Modification	Unit risk estimate (per ppm in ambient air)
Combined EPA estimate	2.60×10^{-2}
Model restricted to absolute risk-weighted cumulative dose form and Rinsky cohort	1.76×10^{-2}
Two additional years of observation (1979-1981) added	1.71×10^{-2}
Job code errors corrected	1.84×10^{-2}
New statistical methodology employed alternative definition of biological latency distribution and maximal likelihood estimate of the transition rate parameter using time to tumor survival analysis	3.19×10^{-3}
New definition of diseases induced by benzene and the estimation of the background rate of these diseases using US vital statistics data and a two-stage model	3.48×10^{-3}
Two molecules of a benzene metabolite are required to induce the single molecular event that causes the cell transition	1.43×10^{-4}

Note: Unit risks can be transformed to a mg/kg/day basis by using the formula 1 mg/kg/day = 1.093 ppm for a 70-kg person breathing 20 m^3 of air per day.

Chloroform

Because there were no studies of chloroform inhalation, the US EPA estimated the cancer potency for chloroform inhalation using a 1976 gavage study that showed significant increases in the incidence of liver tumors in male and female mice (National Cancer Institute 1976). In 1985, Jorgenson et al. showed that chloroform in drinking water significantly increases the incidence of kidney tumors in male rats.

It is not clear which of these two studies is most appropriate for estimating potency. While mice appear to be more sensitive to chloroform than rats, the validity of gavage studies using corn oil has been questioned. Similarly, while administration in drinking water seems more appropriate than gavage, the low-fat rat diets in the Jorgensen et al. study were not typical and may have affected the results. Two alternative potency estimates have been suggested:

1. A potency of 0.08 per mg/kg/day based on the geometric mean of potencies calculated from male and female mice in the gavage study. (This value indicates that at low doses the risk would equal this value multiplied by the lifetime daily average exposure in mg kg^{-1} day^{-1}.)
2. A potency of 0.006 per mg/kg/day calculated from the kidney tumors in male rats in the drinking water study.

This issue has been very difficult to resolve, and the order of magnitude difference can significantly influence the acceptability or unacceptability of risk in specific situations.

Dieldrin

Thirteen studies have been judged suitable for estimating the cancer potency of dieldrin (US EPA 1987). All thirteen show significant increases in the incidence of liver tumors in male or female mice in a variety of strains. However, because several studies in rats, albeit mostly inadequate studies, have not shown any increased tumor incidence, questions have been raised about the relevance of the mouse tumors to other species. The estimates obtained from the thirteen studies are very close, ranging from 7 to 55 per mg/kg/day, with a geometric mean of 16.

Several alternative potency estimates may be considered:

1. Use of the most prudent estimate, i.e., 55 per mg/kg/day.
2. Use of the geometric mean from all positive mouse studies, i.e., 16 per mg/kg/day.
3. No estimate, based on discounting the mouse liver tumors entirely.

This analysis shows that it makes little difference which mouse liver data set is chosen, and that there would be little value from any additional studies in mice. The most significant question is the relevance of the mouse data to other species. If relevance is not established or assumed, no alternative estimate is possible since the only other relevant data are negative.

Pentachlorophenol

In a 1988 draft report, the National Toxicology Program showed that technical-grade pentachlorophenol significantly increases the incidence of liver carcinomas and (rare) adrenal pheochromocytomas in male mice and liver carcinomas and (rare) hemangiosarcomas in female mice.

Again, the relevance of the mouse liver tumors has been questioned. Also, the relevance of the pheochromocytomas in mice has been questioned because the relevance of pheochromocytomas in rats has been debated and little is known about this tumor in mice. Finally, the validity of using the male mouse data has been questioned because many of the control mice died early. For these reasons, several alternative upper bound potency estimates have been suggested:

Estimate 1: Accept the Validity of All tumors. Base the potency on the combined liver and adrenal tumor incidence in male mice, the more sensitive sex.

Estimate 2: Address Questions about Control-Group Mortality. Use the same tumors as in Estimate 1 but apply a time-to-tumor analysis, which effectively discounts early deaths.

Estimate 3: Discount Adrenal Tumors. Base the potency on only the liver tumors in male mice.

Estimate 4: Discount Adrenal Tumors and Liver Tumors. No relevant response in male mice is left, so base the potency on hemangiosarcomas in female mice.

Table 31.2. Estimated individual risk.

Chemical		"Upper" estimate	"More central" estimate	Ratio
Benzene[a]		2.6×10^{-5}	1.4×10^{-7}	~200
Chloroform[b]		8×10^{-5}	6×10^{-6}	~ 10
Dieldrin[b]	a	5.5×10^{-2}	1.6×10^{-2}	~ 3
	b	5.5×10^{-2}	?	?
Pentachlorophenol[b]	a	1.3×10^{-3}	9×10^{-4}	~ 1
	b	1.3×10^{-3}	1×10^{-4}	~ 10
2,3,7,8-TCDD[c]	a	4×10^{-3}	2×10^{-5}	~200
	b	2×10^{-4}	1×10^{-6}	~200

[a] Assume exposure to 1 ppm in air.
[b] Assume exposure to 1 μg kg^{-1} day^{-1}.
[c] Exposures as calculated in the text.

Preliminary analysis yields 1.3 per mg/kg/day for Estimate 1, a slightly higher value for Estimate 2, 0.9 for Estimate 3, and 0.1 for Estimate 4. This indicates that the issues about control-group mortality or the relevance of adrenal tumors do not substantially affect the potency estimate, but that the issue about the relevance of liver tumors does. The relevancy of the hemangiosarcomas in female mice is not in question, however, so Estimate 4 is relatively "firm," indicating that it would be inappropriate to discount the risk entirely.

Exposure

2,3,7,8-TCDD (Dioxin)

The recent US EPA draft report (US EPA 1988) on estimating exposures to 2,3,7,8-TCDD considered the multiple pathways that could lead to human exposure to dioxin. For each source and pathway a series of assumptions was made to estimate exposure. For example, the report estimated individual risks for rural families raising and consuming beef from a field adjoining a site contaminated with dioxin. This case required a variety of estimates concerning:

- The quantity of soil eroded from the contaminated site.
- The fraction of this eroded soil placed on the field where cattle graze.
- Degradation rates for 2,3,7,8-TCDD in surface and deeper soils.
- A "distribution factor" relating 2,3,7,8-TCDD levels in beef raised on the site to the soil concentration.
- The quantity of beef consumed by individuals.
- The fraction of an individual's beef consumption coming from the contaminated source.
- The years lived at the site.

The authors made reasonable worst-case and more central assumptions for each of these input factors. Based on these assumptions, they arrived at individ-

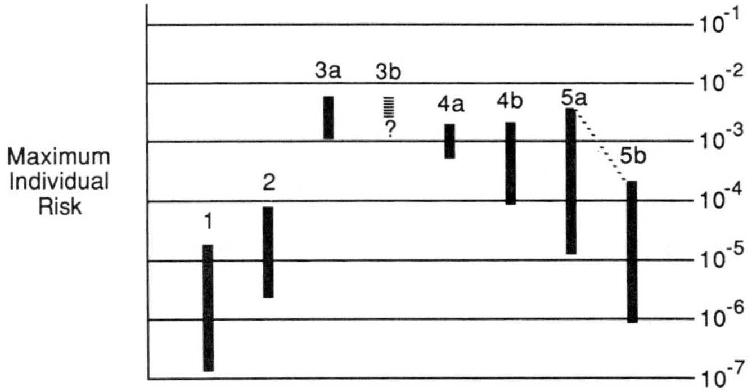

Figure 31.3. Examples of dual risk calculations for five chemicals on a graph of possible individual risk levels. *1*, Benzene; *2*, chloroform; *3*, dieldrin; *4*, pentachlorophenol; *5*, 2,3,7,8-TCDD

ual exposure estimates of 2.2×10^{-2} and 1.3×10^{-4} ng/kg/day for the reasonable worst-case and more central estimate, respectively, for a site with 1 ppb of 2,3,7,8-TCDD in the soil. (It is important to note that the reasonable worst-case estimate was based on high-end assumptions for several, but not all, these factors.) These exposures correspond to individual lifetime cancer risks of 4×10^{-3} and 2×10^{-5} using the US EPA upper bound cancer potency factor for 2,3,7,8-TCDD of 0.16 per ng/kg/day. Had the recently considered cancer potency factor of 0.01 per ng/kg/day been used (US EPA 1988), the estimated risks would have been 2×10^{-4} and 1×10^{-6}, respectively.

Estimating Individual Risk

The various estimates discussed previously may be used to estimate individual risk. Table 31.2 shows the upper and more central estimate of individual risk assuming an exposure of 1 ppm in air for benzene, 1 μg/kg/day for chloroform, dieldrin, and pentachlorophenol, and the exposures as calculated above for 2,3,7,8-TCDD. The results are plotted in Fig. 31.3 to illustrate the way in which these calculations can be used to help judge the acceptability of risk. Assuming, for the purpose of discussion, that a policy decision had been made to use an individual risk of 10^{-4} as a benchmark for acceptable risk, the following can be said about each assessment:

Benzene

Both risk estimates are less than 10^{-4}, so the risk in this instance would be considered acceptable. The potency calculations suggest that the most important issue to the risk assessment is whether one or two molecules are required to induce the single molecular event that causes the cell transition.

Chloroform

Both risk estimates are less than 10^{-4}, so the risk in this instance would also be considered acceptable. Under other exposure scenarios, the difference in the potency estimates might be a critical factor, so resolution of some of the issues raised by the two bioassays (gavage vs. drinking water, corn oil vehicle, diet) might be very useful.

Dieldrin

Both risk estimates in Case a exceed 10^{-4}, so the risk in this instance would be considered unacceptable. The issue raised in Case b, concerning the relevance of mouse data and the negative results in rats should be resolved since a conclusion that the substance poses no risk to humans would obviously have a significant impact on the assessment.

Pentachlorophenol

Both risk estimates exceed 10^{-4}; the risks in this instance would therefore be considered unacceptable. Analysis of the potency estimate indicates that the relevance of the liver tumors in male mice is an important issue, but that the issues related to early deaths in the control groups and the adrenal tumors are of less importance quantitatively.

2,3,7,8-TCDD

Both estimates for 2,3,7,8-TCDD bridge the 10^{-4} risk line, and therefore the risks cannot be considered acceptable or unacceptable without further analysis. In both Case a and Case b, there is an approximately 200-fold difference between the upper and more central risk estimates due to the different assumptions about exposure. Additional work to better define the exposure assessment would be very useful. The difference between Cases a and b is derived from the 16-fold difference in the potency estimate. This difference can be important in certain circumstances as well, and deserves further clarification.

Conclusion

In this paper we have proposed an approach to risk assessment that entails calculating reasonable worst case/plausible upper bounds and more central estimates of potency, exposure, and risk. The central estimate is derived by changing a limited number of key assumptions. These dual calculations can serve a number of ends, including:

- Providing the decision-maker and the public with a more quantitative appreciation of the scientific uncertainty in the assessment, and an appreciation for how various assumptions influence risk estimates.
- Providing a mechanism for incorporating uncertainty into decisions about acceptability of risk.
- Providing a means of defining the research that would most usefully reduce some of the key uncertainties in a risk assessment.

References

Albert RE, Train RE, Anderson E (1977) Rationale developed by the Environmental Protection Agency for the assessment of carcinogenic risks. J Natl Cancer Inst 58:1537–1541

Clement Associates Inc (1988) Quantitative re-evaluation of the human leukemia risk associated with inhalation exposure to benzene. Final Report submitted to the US EPA

Environ Corp (1988) Towards better risk assessment and risk management. Prepared for American Petroleum Institute

Finkel AM (1987) Uncertainty, variability, and the value of information in cancer risk assessment. Thesis, Harvard University, Cambridge

Interagency Regulatory Liaison Group (IRLG) (1979) Scientific basis for identification of potential carcinogens and estimation of risks. J Natl Cancer Inst 63:245–267

Jorgenson TA, Meierhenry EF, Rushbrook CJ, Bull RJ, Robinson M (1985) Carcinogenicity of chloroform in drinking water to male Osborne-Mendel rats and female $B6C3F_1$ mice. Fund Appl Toxicol 5:760–769

National Cancer Institute (1976) Report on the carcinogenesis bioassay of chloroform. PB-264018. National Technical Information Service, Springfield, VA

National Research Council (1983) Risk assessment in the federal government. National Academy Press, Washington

Office of Science and Technology Policy (1985) Chemical carcinogens; a review of the science and its associated principles. US Federal Register, March 14, 1985, pp 10 371-10 442

US Environmental Protection Agency (US EPA) (1976) Interim procedures and guidelines for health risk and economic impact assessments of suspected carcinogens. Federal Register 41:21402–21405

US Environmental Protection Agency (US EPA) (1986) Guidelines for carcinogen risk assessment. Federal Register 51:33992–34003

US Environmental Protection Agency (US EPA) (1987) Carcinogenicity assessment of aldrin and dieldrin. EPA/600/6-87/006 Office of Health and Environmental Assessment, Washington DC

US Environmental Protection Agency (US EPA) (1988) A cancer risk-specific dose estimate for 2,3,7,8-TCDD. EPA/600/6-88/007Aa Office of Health and Environmental Assessment, Washington DC

US Environmental Protection Agency (US EPA) (1988) Estimating exposures to 2,3,7,8-TCDD. EPA/600/6-88/005A Office of Health and Environmental Assessment, Washington DC

32. Science, Ethics and Public Policy

G. Rose[1]

Measuring the Risk

Observation Takes Precedence over Theory

In an area of northwestern England, close to the Sellafield nuclear reprocessing plant, there occurred an excess of childhood leukaemia which is unlikely to have been due to chance. Epidemiologists and radiation experts have been sharply divided on its cause. The epidemiologists, on the one hand, have a lot of experience of phenomena whose mechanisms they cannot explain, going back to the days of the smoking/lung cancer association, and beyond; and they have been impressed by the facts that radiation causes leukaemia and that Sellafield emits more radiation than any plant in western Europe. The epidemiologists also point out that the real dose received by individuals can only be guessed, not known, and that the tissue susceptibilities of young children are also uncertain. This leaves them unimpressed by the argument of the radiation and environmental experts: they tend to exonerate radiation, saying, in effect, "Radiation-induced leukaemia would be inexplicable here, so it did not happen".

If there has to be a choice between trusting observation or reason, it is surely safer to trust observation; for reason is so often based on incomplete information on transfer routes, doses and responses. The argument that what cannot be explained cannot occur is weak, for one can never exclude the explanation that has not been considered. In estimating risks, *direct observations of evident health effects should take priority over theoretical expectations*.

Policy Depends on Level of Absolute Risk

Table 32.1 taken from the American Cancer Society study (Hammond et al. 1979), shows the well-known multiplicative interaction between asbestos exposure and cigarette smoking. The relative risk of lung cancer associated with

[1]Department of Epidemiology and Population Sciences, London School of Hygiene and Tropical Medicine, Keppel Street, London WC1E 7HT, UK.

Table 32.1. The relative risks of death from lung cancer according to exposure to asbestos and cigarette smoking. (From Hammond et al. 1979).

		Cigarettes	
		No	Yes
Asbestos	No	1.0	10.9
	Yes	5.2	53.2

asbestos exposure is similar in non-smokers and smokers (about fivefold). Relative risk measures aetiological force, and it is often (as here) generalisable. This makes it the researcher's preferred measure of risk; but it is in this instance highly misleading as a guide to policy, quite failing to identify the smoker's far higher absolute increment of risk when also exposed to asbestos.

A further shortcoming of measures of relative risk is that they take no account of the size of the background risk. Thus, in the case of asbestos exposure, a fivefold increase in the risk of mesothelioma would still be a very small risk, whereas a fivefold increase in lung cancer is far more serious. A 20% increase in the risk of a common condition is a public health disaster (especially if exposure is widespread), but a 1000% increase in haemangiosarcoma of the liver will involve only a handful of cases. *Policy decisions should always be based on absolute, not proportionate, measures of excess risk.*

Individual or Population?

Medicine is traditionally concerned with the health problems of individuals (Rose 1985). Applying this to environmental medicine leads to the recognition that no individual should face an unacceptable risk. The law reinforces this view, since it is aggrieved individuals—not communities—who take employers or polluters to court. So it is that legal and industrial limits on environmental pollution are designed to avoid such levels as would cause identifiable risk to the most exposed individuals.

The policy of controlling conspicuous dangers to individuals is bound to be gratifying. The regulating authority has taken visible action. If this does its job, then there are no more complaining individuals or lawsuits. And the workforce (or the public) can relax, because they feel exculpated and safe, having been assured that exposure was someone else's problem (and anyway, it has been dealt with).

There is of course nothing wrong with setting limits to toxic exposures, such that identifiable harm to individuals is avoided—so long as it is not assumed that that is necessarily sufficient. Figure 32.1 shows the distribution of radiation doses accumulated over their working lives by employees of the United Kingdom Atomic Energy Authority (Beral et al. 1985). It is widely agreed that cumulative

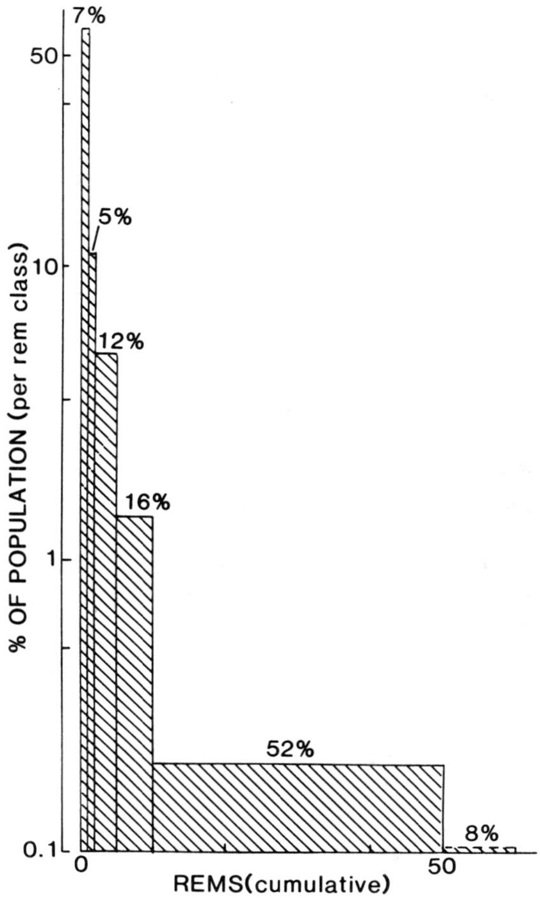

Figure 32.1. Percentage distribution of cumulative radiation dose among employees of the United Kingdom Atomic Energy Authority

doses exceeding 50 rems (500 mSv) are unacceptably risky; and happily these levels were uncommon, accounting for only 8% of total exposure. Most of the total radiation fell on the large numbers of workers who individually received only very small doses. In fact, 40% of the total exposure arose among men whose life-time dose was no more than 10 rems (100 mSv); this is not a level to scare anyone, nor could it be avoided without great difficulty. But if the relation of radiation exposure to cancer risk is linear, with no threshold effect, then this distribution of doses also represents the distribution of induced cancers. Control efforts confined to preventing high individual exposures could touch at most 8% of the problem, since the latter arises mainly from exposure of a very large number of people to an individually inconspicuous risk.

A similar situation applies to all environmental or industrial toxic hazards where the dose-response curve has no threshold. Whenever this is the case, effective control can be achieved only by reducing total emissions, not by simply preventing the occurrence of excessive levels.

The implication of this argument is that control policy should reflect the estimated total *population-attributable(excess)risk*, not simply the attributable risk among heavily exposed individuals. This is easily said, but hard to establish; for the population effects of environmental exposures depend critically on the shape of the dose-response curve, and on whether or not the toxic effects occur only above a certain threshold. A large number of people exposed to a low risk may potentially generate much more disease than a small number with high exposure; but generally we cannot measure this directly, so that our only estimates of what happens with exposure to low doses come from extrapolating downwards from the effects of high-level exposure.

It is hard to think of any instance where a relative risk below 1.5 has been shown convincingly to be causal, for at these low levels it is always possible that the association has some other explanation. This problem has bedevilled, for example, interpretation of the studies of passive smoking, where the possibility cannot be wholly excluded that the associations arose because of unrecognised confounding or through biased reporting of exposure.

We reach then an impasse: conspicuous harm to individuals can be established but it may not be the main problem, whereas an inconspicuous risk for the population or workforce as a whole might be much more serious, but it cannot be established. If we then choose to make the "worst case" assumption of a linear, threshold-free dose-response curve, then we ought to control total emissions, down to that level at which the costs of a further reduction would not be justified in relation to supposed benefit (recognising that zero risk implies zero emissions). Any policy that confines itself to control of high exposures only ought frankly to recognise the public health uncertainties involved: it may be touching only the margin of the problem.

From Scientific Evidence to Policy Decisions

Certainty is Unattainable

It is usual for those who oppose some action to point out that the matter is still uncertain and that more research is needed; and they are quite right. Public health, like the rest of medicine, proceeds by a series of fallible judgments, any of which may need to be revised in the light of further experience; but the public continues to be misled by the experts and the media into believing that policies should be founded on proof. In fact, there is no such thing as certainty: the word ought to be banned! And yet experts continue to say, "A nuclear accident could not happen here," or "The smell from this chemical plant cannot possibly endanger health", and so on. Conversely, we who are enthusiastic for preventive medicine often seem to be equally free of doubts about our advice.

Everyone – public and experts alike – feels threatened by uncertainty; so we tend to deny its existence, by exaggerating the evidence and oversimplifying the policy decision. Lack of proof is not in itself a sufficient objection to action, and it can be perfectly proper to implement a policy whilst still pursuing research which might lead to its modification.

Legislative control of air pollution by smoke is a case in point. In Britain (and elsewhere) laws were passed to prohibit the emission of black smoke from city chimneys. This measure was expected to have a dramatic effect on particulate pollution but little or no effect on sulphur dioxide concentrations: to control the latter would have been far more expensive and socially disruptive. It seemed, nevertheless, a reasonable decision to take, even though the benefits to respiratory health were not at all sure. In the event, subsequent research showed evidence of large health benefits; but it could have turned out otherwise, and then the policy would have had to be reconsidered.

What Level of Evidence Is Required?

Once we have moved from a rigid position of expecting proof before action, it becomes necessary to decide what level of evidence—short of proof—is to be required. Put broadly, the evidence should be appropriate to the decision, and, in particular, to its costs and its benefits, and to the consequences of making the wrong decision. This matching of evidence to policy decisions cannot be achieved when scientists and policy-makers do not communicate, which is so often the case. Academic work is presented in a form suitable for other academics to read: editors and programme committees expect it, and our whole training equips us for that mode of thinking and presentation.

The problem goes further back, to the stage of study design and analysis. We have been taught that a good study tests a hypothesis, so that the conclusion can be expressed by statistical significance. *P* values are of little use to decision-takers; nor (for reasons given earlier) are estimates of relative or proportionate effects. Decision-taking calls for magnitude estimates, effects being measured in absolute units.

In applied scientific research the investigator who plans a study should always ask, "To what specific policy decision will my findings apply?" Only if the application of the findings is first defined will the design be appropriate as a subsequent guide to policy. This implies that epidemiologists and other medical scientists need to become more aware than at present of public health policy choices and the information needed to guide them.

Constructing the Balance-sheet

Every policy decision must consider various items of benefit and costs, each of which needs to be quantified. This factual basis for decisions is most conveniently displayed in the form of a balance sheet, in which costs are expressed in relation to one unit of the main intended benefit. In regard to control of passive smoking, for example, the unit of benefit might be one lung cancer prevented: the balance sheet would show the various costs (and also other benefits) involved in achieving it. Similarly for control of an industrial exposure, it would be necessary to formulate the costs to workers and the industry of preventing one case of the related disease. In every case, policy rests on quantitative information, since no physical benefit is worth unlimited costs.

Who Decides?

The measurement and display of all relevant benefits and costs is helpful; but scientific evidence does not make the decision. To say that "Smoking causes cancer" is a factual, scientific statement. To say, "People should not smoke" is an opinion, based on assumptions about the values assigned to health and to smoking. Medical science can produce the balance sheet; but the decision depends on the values assigned to each of the items, which are all expressed in different kinds of unit.

Doctors and other technical experts have their own opinions on these values, but these may not be shared by the people who have to live with the decisions. Health, for example, tends to be given a higher valuation by doctors than by the public, especially if the latter are healthy at the time. Medical scientists take long-term risks very seriously; but the public, and politicians, are much more concerned with short-term results. Epidemiologists express forcibly their anxiety about population-attributable risk, which is an invisible concept; but the public (and the media) worry about personal threats, or perceived threats. In particular, the public feel threatened by invisible hazards (like radiation, or other pollutants whose presence they cannot recognise) or by hazards which are imposed on them and are hence unavoidable (like industrial emissions), more than by recognisable and avoidable exposures (like tobacco smoke); and they have a special fear of cancer.

Faced with these divergent sets of values, it cannot be said that one set is right and another wrong. It is a matter of choice, not of science or logic. Neither technical experts on the one hand, nor regulatory authorities and governments on the other, have the right to impose their own choices. They are the servants of the public.

The Role of Experts

Environmental issues, such as those discussed in this volume, are all highly technical, and the public cannot be expected to understand the technicalities. This puts the experts in a powerful position, and we are in constant danger of overstepping our authority: we are only technical experts, not experts on values or political decisions. This gives the public an ambivalent attitude towards us. We are trusted, because we know a lot and therefore speak with authority; and we are mistrusted, because people sense (often correctly) that we confuse our technical expertise with the right to decide what is best for them.

There is an analogy between our role in public health policy and the role of doctors in advising patients. Society, like the patient, invites health experts to give advice, both as technical experts and because our advice should be more disinterested than that of the powerful commercial and political pressures to which they are also exposed. Our responsibility is to try to give a correct and balanced account of the scientific and technical issues, with the aim of enabling the public to make choices that are as well-informed as possible. This is our first task.

Sometimes we shall also be asked (as patients ask their doctors) to advise on what we think the public ought to do about the scientific facts; and here we are in danger of overstepping our limited authority as experts, by incorporating into our recommendations our own particular set of values. We need to be much more aware of this danger than is commonly the case, and to distinguish clearly between when we speak as experts (with authority) and when we are presenting our personal views on policy.

Experts are also called on to advise governments, regulatory authorities and managements. This can again create a difficult confusion of roles. Our role as members of official advisory committees is to speak as experts on behalf of the public health; but the official and political members of such groups are likely to be responding to wholly different pressures, relating to the economy, politics and public relations. These are important and legitimate issues; but where they conflict with the interests of the public health, then the medical experts must be careful to keep their distance.

The Role of Government

Government's first duty in regulating environmental health is to protect the individual's freedom of choice. Individuals who do not wish to inhale noxious fumes should not have to do so, whilst those who, for example, wish to smoke should have the opportunity to do so. Neither governments nor managements have the right to impose constraints on people simply because they are believed to be good for them. If we really believe in freedom of choice, then we must respect it consistently.

This principle is not generally accepted in safety legislation, which is often paternalistically protective. For myself, I was strongly tempted to stop wearing my car seat-belt when this became legally compulsory! I wished to protest at an unwarranted infringement of freedom, even though I believe that wearing a seat-belt is eminently sensible.

Ethical decisions are rarely simple, and what we need is not more ethical rules (which inhibit judgement), but rather clear thinking and exposure of the key questions and conflicting interests. If everyone could choose, as an individual, whether or not to be exposed to a particular health hazard, then the choice could (and should) be left to the individual. Unfortunately many situations are more complex. If industrial plants emit pollutants, then local residents cannot help inhaling them. If workers have no reasonable alternative to employment in a toxic environment (including sharing an office with smokers), then individual freedom of choice cannot operate. Where exposure is collective and unavoidable, only a collective control policy can be effective.

In this situation society, in effect, authorises government (or management), as its agent, to take and enforce a collective decision. Thus if most people want, for example, smoke pollution or nuclear power to be banned, then government—whether it likes it or not—should implement such decisions. Manifestly this does not always happen: at present in many countries the people assign a much higher value than governments to environmental control.

Governments may seek to justify their undemocratic behaviour by pointing out that the public is unaware of, or does not understand, the complex issues involved. This may often be true (although it is also true that the divergence of opinions may arise from disagreements on values and objectives); but even where serious ignorance is a factor, governments have only themselves to blame for their secretiveness and for a failure of public education.

Fundamental ethical principles do not change, but their practical outworkings do. Societies grow less and less willing to leave major policy decisions to expert authorities and power groups. Increasingly they expect to be informed and to be consulted; and, since they are the ones who must live with the consequences of those decisions, that position is surely to be welcomed.

References

Beral V, Inskip H, Fraser P, Booth M, Coleman D, Rose G (1985) Mortality of employees of the United Kingdom Atomic Energy Authority, 1946–1979. Br Med J 291:440–447

Hammond EC, Selikoff IJ, Seidman H (1979) Asbestos exposure, cigarette smoking and death rates. Ann NY Acad Sci 330:473–490

Rose G (1985) Sick individuals and sick populations. Int J Epidemiol 14:32–38

33. Recent Progress and Problems in Setting and Applying Occupational Exposure Standards

D. Henschler[1]

Introduction

The history of occupational exposure limits dates back more than one century (Lehmann 1866). They were introduced in the interest of employers as well as employees: employers were trying to preserve work forces, and employees were striving for social security. From the very beginning of establishing exposure limits, both observations on humans and animal experiments were used to explore the borderline between the state of health and the first signs of intoxication; the standards were then formulated as maximal tolerable concentrations in the air at the workplace. The philosophy behind standard setting was to exploit fully the range of concentrations between zero and a determinable threshold of toxic effects so that the limit value was identical with, or closely below the lowest effective concentration. Some experimental evidence was provided for the existence of "thresholds", as for instance with hydrocyanic acid, a completely reversible toxicant (Flury and Heubner 1919). For almost 6 decades, these occupational exposure limits prevailed as the only type of standard to protect human health. They were entirely health-based and intended to serve as recommendations for industrial hygiene inspectors, employers and employees.

The original concept was, however, questioned and, to some extent, diluted by the introduction of other standards from the middle of this century onwards, and by conceptual differences underlying most of these additional limit values. Also, a variety of challenges by criticism and scientific innovations has been influential on the establishment and use of occupational exposure limits, and has enforced a series of re-evaluations. Some of these will be dealt with in order to characterise the present status, research needs and future developments of this essential instrument in the protection of worker's health.

[1]Institute of Toxicology, University of Würzburg, Versbacher Str. 9, D-8700 Würzburg, FRG.

Introduction of New and Different Standards

Radiation standards, maximal levels of food additives and residues, average daily intake values, ambient air standards, drinking water standards for contaminants were all introduced within a short period of time in the late 1940s and 1950s. Many others of minor importance followed, such as maximal concentrations of impurities in technical products, residual contents of monomers in plastic materials, oxidative derivatives in foodstuffs, nitrosation products in drugs and chemicals, etc. The process of creating new standards is continuing at a fast pace. One is tempted to envisage a real flood of new standards. These new entries apply to completely different sectors of man's environment, and to protection of widely varying objects, including non-human organisms and inanimate materials. Of necessity, the philosophies to justify and strategies to establish these standards are different. For instance, when discussing the first radiation standards, radiobiologists were well aware of the non-existence of no-effect levels, and introduced as a handicap a political element, a tolerance level above background radiation exposure from other sources, for instance in the form of the "doubling dose concept" (Ehling 1987) which implicates a doubling of the normal mutation rate as being acceptable. By this, they anticipated a decision which normally should have been taken by government or parliament. This procedure was completely alien to the occupational standard setters of those days. Food standards, on the other hand, could not take advantage of observations in humans and were hampered by many uncertainties in chronic animal studies. Instead, the levels were based on large safety margins, and thus moved into what was soon called the area of non-toxicology. Similarly, ambient air standards have to take into account the youngest and oldest, and increased sensitivity from disease states which are more or less undefined or undefinable.

More important seems to be a fundamental difference in the possibilities of health surveillance. Whilst in the occupational field we have a closed system in that the populations exposed may be kept under permanent medical control allowing for the early detection of untoward reactions due to inappropriate standards, or non-compliance to a well-set tolerance level, no such surveillance is practicable in other fields which constitute open systems, e.g., ambient air standards, food standards and, of course, today's radiation standards. In other words: the appropriateness of standards in these fields can hardly be proven by the occurrence or non-occurrence of health effects.

And yet politicians and administrators feel called upon to unify and simplify the criteria for setting and applying the ever increasing bulk of environmental standards, simply because of the ease and convenience in the administrative handling of standards, and the facilitation of jurisdiction. In other words: science is replaced by administration. This oversimplification of a rather complex area has had negative influence on the traditionally well-established occupational exposure standards.

Recommended Versus Legally Fixed Standards

Occupational exposure limits have been taken, from the very beginning, as recommendations or guidelines for occupational physicians and hygienists, engineers and analytical chemists. They should not be used like physical constants. The argumentation for this has been: there is large interindividual variability in sensitivity to toxic reactions, imprecision of methodology of determining thresholds, and—very important even today—an insufficient data base for setting a reliable limit.

However, legislators have always been keen on converting these recommended values into legally fixed standards. This is a change from the flexible to an inflexible system. The motivation behind these efforts is facilitation of intervention in operations of risk management on the one hand, and the discomfort of juridicial people with an oscillating biological reagibility on the other. The misconception of taking compliance with a standard as safe, and non-compliance as proof of impairment of health is still haunting court trials worldwide. This misuse of occupational standards, particularly in handling claims for compensation, falsifies the very aim of standards: prevention. It is a persistent challenge to the system. Nonetheless, some countries have already introduced legal standards; others work with both systems such as, for instance, the United States. The dilemma of such a situation can easily be identified: inflexibility does not only characterize the handling of legal standards already introduced but, even more so, the process of their introduction: it takes decades to establish just a few (Federal Register, 1988). The predominant disadvantage of legally fixed standards is, however, that new information and scientific progress cannot be incorporated immediately. The long delay before the adaptation of a legal standard to a new situation is to the detriment of the exposed workers. The European Community has not yet taken decision to introduce a flexible (annually revised lists of standards) or an inflexible system (legal standards); it is a struggle of differentiation versus simplification or, put in a simpler way: science versus administration.

Health-Based Versus Socioeconomically Feasible Standards

Originally, occupational exposure limits were entirely health based. Some countries or social systems introduced additional criteria: technical and analytical feasibility, and social and economic acceptability. This is one of the major reasons for the differences encountered in the standards of different countries. If criteria other than health are taken into consideration, the process of standard setting needs the involvement of representatives of non-scientific organisations such as trade unions, delegates of industrial and consumer organisations, regulatory authorities, and maybe others. The United Kingdom has recently introduced a tripartite system for occupational standard setting. In this case, the main issue for achieving a compromise is how much health, or how much deduction from complete health a society is willing to accept in balancing health against economy. Two difficulties will arise in these negotiations: who is able to evaluate costs resulting from compliance with certain standards, now and in the future? Who is in a position to quantify reliably impairment of health in the range of standards under discussion? Some of these issues have certainly not yet been resolved convincingly in the existing systems for standard setting.

There are three disadvantages in a switch from health-based to socioeconomic standards. One is an unavoidable demolition of the consistency of our system of evaluation of health effects. No lay-man will understand why with one chemical a little more, with another a little less irritation of mucous membranes is regarded as acceptable. Once a certain deduction from health has been agreed upon, the next case will be easier to handle, and so forth. And which member of the scientific community will set a scale for acceptable deductions from health? The ethical implications connected with this problem are self-explanatory. The other drawback is to be seen in major differences in the economic structure and interests in different countries. In addition to this, technical performance of protective devices, historical experience and training of personnel, as well as the manpower available to control adherence to standards differ widely from country to country. This complicates efforts towards an international unification of standards which is a legitimate desire for the facilitation of commerce.

A third problem may be expressed by the question: who is to take responsibility for what? The toxicologist is ignorant in the fields of technology and economy, and vice versa, and the politician who should take the final responsibility is ignorant throughout. To overcome these complex problems, some countries have introduced a two-step procedure for standard setting: scientific evaluation as the first step, and negotiation of a compromise as the second. Holland (Arbeidsinspectie 1989), the Federal Republic of Germany (Deutsche Forschungsgemeinschaft 1989) and Sweden (Swedish National Board of Occupational Safety and Health 1987) are exercising this system in principle, with some minor variations. Although the tendency prevails to keep the two steps separate and independent, every organisation has been confronted with pressures and with efforts to interfere in step 1 with arguments from step 2. The only way out of this obvious dilemma is to make every discussion, argumentation and recommendation completely transparent (Henschler 1987a). This transparency may save the credibility of all the parties involved in the whole process.

Definition of Health: Doctor's Diagnosis Versus Analytical Parameters

There now seems to be an overwhelming agreement that the World Health Organization's definition of health, which includes social well-being, is not applicable for occupational exposure standard setting because most factors which influence an individual's well-being can hardly be identified among the sequelae of chemical exposure. Also, the earlier Russian view of changes in conditioned reflexes being relevant indicators of health effects from chemical exposure now seems to be moving towards the traditional concept which has governed standard setting in Western countries. However, more recent developments have induced changes in the definition(s) of health and provided new instruments with ever increasing precision to detect deviations from the norm—a process which is continuing with increasing velocity reflecting scientific progress in all medical disciplines. The more we learn about the biochemical and biophysical mechanisms underlying pathophysiological reactions, the more analytical parameters will be found and introduced to measure imbalances at ever earlier stages and even prestages. Many of these early indicators are not necessarily representative of pathological states; rather, they are reversible deviations from the normal average which just tells you the variance within a steady-state compensatory system. They may be indicative of an organism's capability to defend itself against, rather than being damaged by, a chemical exposure.

One of the consequences of this fascinating development is the proposal by scientists and the public as well to reduce standard-setting procedures to the basis of these sensitive analytical parameters, rather than taking an integrated view and diagnosis of what is normal and what is illness. The result of these proposals is, of course, that exposure limits have to be lowered. This may or may not be a matter of concern. More relevant seems to be the fact that by introducing this type of standard based just on analytical parameters, we are in danger of replacing an integrated health surveillance system by automatism, and thereby may overlook more relevant signs and symptoms which may develop independently of the supposedly most sensitive parameter.

Occupational Carcinogens: Are Health-Based Standards Definable?

The nature of carcinogenic effects, the irreversibility of primary lesions in particular, and the failure to demonstrate unequivocally the existence of thresholds even in large-scale animal experimentation stands against the establishment of health-based exposure limits for this type of compound. Progress in elucidating the mechanisms of chemical carcinogenesis have nourished the expectation that thresholds will be definable for certain types of chemical carcinogens: "epigenetics" or "promotors". In the meantime, it has become clear, however, that a precise differentiation between genotoxic and epigenetic, or between initiators and promotors is questionable. At least for some sorts of promotive effect, a mutation is also involved. So the hope that health-based limits may be introduced for some carcinogens is turning out to be fallacious (Henschler 1987b). Therefore, the risk evaluation of carcinogens continues to be forced to quantify the risks (Richardson 1988) as a basis for reasonable risk management measures.

One approach towards quantification has recently been introduced in the EEC Scientific Advisory Group. Their classification system has been adopted from the list of the Federal Republic of Germany, and comprises human carcinogens (category 1), animal carcinogens (category 2) and suspected carcinogens (category 3). The last group has long been increasing rapidly due to incoming information from short-term mutagenicity tests. If long-term animal bioassays are lacking, the dissolution of suspicion and ensuing classification in a higher or lower group is impossible. There are, however, some compounds listed in category 3 with which animal experiments have been performed but only with borderline results. A repetition of these experiments will not change the data base and are therefore not recommended. The suggestion has been made to form a subgroup for these compounds which obviously have a very low carcinogenic potency, the significance of which cannot be evaluated for human beings. One prerequisite for inclusion in this subcategory is the absence of evidence for genotoxicity in any relevant testing system. At present, this suggestion is exposed to discussion in the scientific community. If a new subgroup is established, this may facilitate the regulation of a number of compounds in the occupational area.

Air Monitoring Versus Biological Monitoring

Monitoring of concentrations of occupational toxicants in the air of working premises by analytical methods has served as the classical instrument in controlling exposure and protecting workers' health. Previous methodological shortcomings such as lack of sensitivity, specificity and velocity in response to monitoring machinery have been overcome by the marvellous development of new physical methods such as gas chromatography, mass spectrometry, high-performance liquid chromatography, nuclear magnetic resonance and high-performance detectors in dynamic flow systems. With only a few exceptions, analytical chemistry now provides methods to meet the requirements for the control of present exposure limits. Air monitoring covers air-borne exposure and is also the essential instrument in the control of the efficiency of technological protective measures.

Long field experience has shown that uptake of occupational toxicants by routes other than inhalation may contribute considerably to the body burden, mostly through skin absorption and ingestion. The proportion of uptake through alternative pathways depends on the physical state of the compound and the type of technical operation used. The need to monitor these parapulmonary entries was, however, not recognised before the early 1960s. Biological monitoring procedures are a suitable means of filling this gap. They were introduced in the early 1980s, the Federal Republic of Germany being the first country to recommend and officially use biological exposure limits in 1981 (Deutsche Forschungsgemeinschaft 1981).

These methods bring an additional advantage: they identify individuals at higher or at lower risk than average due to genetically or otherwise determined capabilities to incorporate, metabolically activate or inactivate chemicals to a greater or lesser extent, or to repair toxic lesions more or less efficiently. The most fascinating progress has been made in detecting primary lesions of the genetic material by mutagens and carcinogens even at levels far below those known to cause cancer (Farmer et al. 1987). The elaboration and introduction of sophisticated methods to detect DNA damage constitutes by far the most important progress in identifying health effects from occupational chemicals. Used in context with prospective epidemiological studies, it may be the most promising method to quantify carcinogenic risks.

Some circumstances inhibited the early application of field studies on a broader scale. The validation of these methods seems to be more difficult than usual, in part because of high background levels of poorly defined origin, and in part because of concern brought forward by employees and their professional organisations: the detection of an unfavourable genetic disposition may open the way to an individual's discrimination. There have been recommendations to boycott the introduction of these methods for screening purposes. This would be a bad solution because studies of this kind will be done anyway, and certainly under less stringent control than if they were based on a valid consensus of the scientific community. Biomonitoring of genotoxic interactions of chemicals with DNA in human populations is not only one of the most fascinating innovations in chemical carcinogenesis, but also a great step in the progress towards exposure control and protection of workers' health.

The Future: Hope or Scepticism?

Occupational exposure limits have enjoyed a long probationary period. Much criticism has been raised, and more will be raised in years to come. The concept of zero tolerance will continue to haunt the minds of a disappointed minority of our prosperous society. Will the system of occupational exposure limits persist? I predict it will, because it has to, since no realistic and practicable alternative is in perspective. There may be changes in the procedures for establishing standards, and in the way they are applied. It is daring to make predictions regarding these issues, except for one: the more scientific backing they obtain, the better their credibility will be.

References

Arbeidsinspectie (1989) Nationale MAC-lijst. Zevende Druk, Voorburg

Deutsche Forschungsgemeinschaft (1989) MAK-Werte und BAT-Werte. VCH, Weinheim

Deutsche Forschungsgemeinschaft (1981) MAK-Werte und BAT-Werte 1981. Boldt, Boppard

Ehling U (1987) Quantifizierung des strahlengenetischen Risikos. Strahlenther Onkol 163:283–291

Farmer P, Neumann HG, Henschler D (1987) Estimation of exposure of man to substances reacting covalently with macromolecules. Arch Toxicol 60:251–260

Federal Register (1988) Dept. of Labour, OSHA, 29 CFR Part 1910, Air Contaminants; Proposed Rule. June 7, 1988, pp 20960–21393

Flury W, Heubner W (1919) Über Wirkung und Entgiftung eingeatmeter Blausäure. Biochem Z 95:249–256

Henschler D (1987a) Exposure limits: history, philosophy, future development. Ann Occup Hyg 28:79–82

Henschler D (1987b) Risk assessment and evaluation of chemical carcinogens – present and future strategies. J Cancer Res Clin Oncol 113:1–7

Lehmann KB (1866) Experimentelle Studien über den Einfluß technisch und hygienisch wichtiger Gase und Dämpfe auf den Organismus. Arch Hyg 5:1–125

Richardson ML (ed) (1988) Risk assessment of chemicals in the environment. Royal Society of Chemistry, London

Swedish National Board of Occupational Safety and Health (1987) Occupational exposure limit values. Swedish National Board of Occupational Safety and Health, Stockholm

Some Reflections on the Symposium*

D.V. Bates[1] and R.O. McClellan[2]

We thought it would be more entertaining, and more interesting, if we adopted a format whereby each of us says something, then the other comments on it, and goes on to make an additional point. We really need one of those clocks they use at international chess games to record the time, but I think we can dispense with that.

Dr. Bates: I would like to kick off by reminding you that we said the purposes of this symposium were threefold:

1. To review data on specific substances. We have done that.
2. To advance our understanding of how the integration of diverse data should be addressed, and how disparate data from animal exposures and human epidemiology should be approached. We have touched on those issues.
3. To consider what general lessons may be derived from a review of particular materials. This is what we will touch on.

This was not designed as a "consensus" conference, but as we have gone along, I expect many of you will have tried to keep some kind of score. I find it very interesting that if, for the eight substances we have considered, we look at the strengths of animal data, the strengths of our understanding of mechanisms, and the strengths of epidemiological data, what we see is a spectrum, in each area going from very strong to almost zero. For instance, in the case of animal data, it is fair to say that for benzene and diesel exhaust this is very strong, but then it tails off to environmental tobacco smoke where it is absent. In relation to mechanisms, we have heard very interesting new approaches (particularly in regard to diesel exhaust). This has great promise for the future. But again, we go all the way from that to substances for which we have very little understanding of mechanism. As far as epidemiologic studies are concerned, we have some standard mortality ratios (SMR) as high as 5 and 7 (in the case of arsenic), but for others these are in the range of

*The following paper reproduces the essence of the discussion that concluded the symposium on which this volume is based. It has been reproduced in this form to indicate how these two scientists viewed the results of this meeting.

[1]Department of Health Care and Epidemiology, University of British Columbia, School of Medicine, Vancouver, British Columbia, Canada.
[2]Chemical Industry Institute of Toxicology, Research Triangle Park, North Carolina 27709 USA.

perhaps three for polycyclic hydrocarbons, and for a group of substances, such as man-made fibers, the SMRs are about 1.4. One of the points worth making is that you find the high SMRs in cases where our knowledge of exposure is relatively good; and the low SMRs are for situations (like environmental tobacco smoke) where our actual knowledge of exposures is very approximate. This raises the interesting question of whether in some of these instances, the "weak association" is partly to be explained by very imprecise exposure data, so that although the factor may be fairly strong, because the exposure data is very weak, it comes out with an SMR of 1.4. I don't think that in a conference of this kind you can resolve that question; but I think our symposium, by illustrating the relative strengths and weaknesses of our three kinds of information, certainly raised very interesting questions.

Dr. McClellan: Let me offer some comments first on your comments. I found it interesting that you made your assessment of the relative quality of the data by focusing on the methodological approach used: animal investigations, mechanistic research, and epidemiological studies. I would have preferred that you had taken an agent-oriented approach. At the beginning of the Symposium I noted that one of our challenges was to break down the barriers between the various methodologies or discipline-oriented approaches and focus on the problem, i.e., to what extent does a given agent pose a health risk to people. We may have an interest in improving methodology but our ultimate interest is in solving problems.

I was encouraged by our progress this week in bringing together people from different disciplines and with different orientations. I sensed improved dialogue between people from the different areas and, on some problems, a real integration of information.

Let me now turn to a second issue you raised, that of the relative cancer risk posed by exposure to various agents. No doubt there is some truth in your contention that the low relative risk values developed for some agents is related to poor exposure data. That in fact, if we were able to better measure exposure, it might be feasible to segregate out a smaller and more highly exposed subpopulation that would end up with a greater relative risk. It is important to keep in mind that the relative assignment of more cancers to the highly exposed subpopulation will leave fewer cancer cases in the lesser exposed subpopulations.

To state the obvious, changing exposure group assignments does not change the total number of cancer cases. Better exposure group assignments will help us detect low potency carcinogens. I believe it is important to recognize that it is just such agents for which we need data from laboratory animal bioassays and mechanistic studies to complement the meager epidemiological data base.

Dr. Bates: I have no difficulty with that. I wrote down a one liner, which is that epidemiology is the "Till Eulenspiegel" of the world of toxicology. We all know what happened to him. But I think that John Bailar raised an interesting point early in our discussion, which had to do with how you move from an association to a cause-and-effect conclusion. I recently had occasion to review this, and went back to Sir Austin Bradford Hill's presidential address to the Royal Society of Medicine in 1965, in which this was the burden of his talk. It is worth rereading that paper, because in it he sets out the factors that he finds helpful in going from an association to a cause-and-effect conclusion. He does this very modestly; these are not presented as laws, although they are sometimes written as if they were rigid criteria. He also makes the point that Geoffrey Rose made, namely that public policy should not always wait for the mass of evidence to constitute definitive proof. He foresaw, I think, the situation we are now in, the sense that we have a variety of substances with an SMR of about 1.4; and we have the problems of public policy decisions. He would not be surprised that we are trying to decide if the strength of epidemiologic data really warrants a shift of public policy. I hope that he would agree that what actually happens in an open society is that as the data accumulates (exaggerated often by the media but sometimes minimized by interested parties), we reach a certain point when the burden of proof shifts. Perhaps by 1965, it was proper to say that the burden of proof about lung cancer and cigarettes had shifted. We now were required to be told by the industry how the persistent association between cigarette smoking and lung cancer could be explained except on a causal basis. What confounder could have influenced so many different studies in the same direction? So I think that in our approach to the issue, the discussion we have been having is basic to our individual judgement of where the burden of proof now rests. Geoffrey Rose also reminded us that the scientist is not in any sense the sole authority on that; indeed in some senses it is always the community, as represented by its legitimate representatives, that must eventually decide where the burden of proof lies.

Dr. McClellan: In addition to establishing causation, I think it is increasingly important to establish quantitative estimates of the potency of carcinogenic agents. It is clear that an absolutely risk-free world is not achievable; we are going to have to control and manage risks. This includes developing reasonable estimates of the potential health costs of exposure to various agents as a means of assigning priorities for using societal resources to control the risks. This requires quantitative estimates of cancer risks for various exposure scenarios.

I see a particular need for developing improved approaches to understanding and assessing the cancer risks of new materials. In these situations it is not possible to obtain epidemiological information because human exposures have been very limited and/or recent. This requires that the carcinogenic risk assessments be accomplished using data from studies on laboratory animals or on sub-animal systems, i.e., macromolecules, cells, and tissues including materials obtained from both humans and laboratory animals.

One point that was re-emphasized to us this week is that we have relatively few known human carcinogens. I feel strongly that it would be appropriate to revisit some of the known human carcinogens with a view to seeing how well estimates of risk, developed from contemporary animal bioassays and mechanistic studies, agree with those developed from epidemiologic studies. This approach should help develop improved approaches to estimating human health risks for materials for which we hopefully will never gain human carcinogenic experience. Although it is always exciting to be working with new materials, I am confident that a structured approach to studying some of the well-known human carcinogens will yield substantial dividends.

Dr. Bates: Yes, always of course with the proviso that you don't insist that action not be taken until we understand mechanisms. I'm sure that will not be your interpretation.

Dr. McClellan: No, I would not argue that point with regard to new materials. However, I would suggest that we need to have a range of possible regulatory actions for new materials. It should not be absolutely zero exposure.

Let me turn to another issue, the identification of susceptible individuals as discussed by Dr. Curtis Harris earlier in this symposium. This is an area where advances can be facilitated greatly by having human material available. Indeed, it would be most advantageous to have available material, both cancerous and "normal" tissue, from individuals in which the cancers were presumably caused by different agents. At the present time, much of our information has been developed on lung cancers induced by our most notorious carcinogen, cigarette smoke. Hopefully, our past and current efforts to minimize occupational and environmental exposure to carcinogens (ignoring for now our nation's poor record in dealing with cigarette smoking) will pay off with fewer and fewer cancers attributable to such exposures. That suggests that we should give high priority to banking human material whenever we think we can establish a causation linkage. This material should prove especially valuable as we make further improvements in molecular and cellular techniques for understanding mechanisms of carcinogenesis.

Dr. Bates: I also hope that you ended the week with a rather warmer feeling toward epidemiologists. I would like to address that issue because we did have a lot of "wringing of hands." In fact, at one point I noted that Peter Lee's comment almost suggested that epidemiologists should move into the field of literary criticism. That's a wonderful world, as those of you who read the *New York Review of Books* will appreciate. It is a world of no numbers; of bias elevated to an art form; in which nonrandom selection is imperative; and polemic is acceptable provided it is witty and devastating. I don't think we are quite at that point. Our problem is that epidemiologic studies are the only tools we have to address the problems. That is a very real issue, I think. If you really want to know if there is any evidence of an effect on the human population, you are back with the same dilemmas and difficulties we have been wrestling with.

The emphasis doesn't have to be on amplifying the SMR of 1.4 so that it is 14 (though that would be desirable if it were possible); what we have to do is to follow Bradford Hill's advice: namely, examine all the data; replicate studies on different populations; and finally be prepared to come to a judgement. One difficulty arises when we have to convert statistical data into words. John Bailar and I had to give this problem some thought when we were working together on a committee of the National Academy of Sciences. We could identify examples in which what had been at fault was the choice of words to describe data, which were inappropriate to the actual statistics. The first error in communication that occurs is when the scientist chooses a form of words inappropriate to the actual data he or she has collected. It can be inappropriate either way; it may minimize or it may exaggerate.

Dr. McClellan: I have been working all week to reestablish my relationships with epidemiologists. However, some of us still have some differences of opinion. For example, you noted "epidemiologic studies are the only tools we have to address the problems." I disagree; epidemiologic studies are options that are not available to us for many materials because human exposure has been limited and/or too recent. Thus, there is no basis for conducting epidemiologic studies. We must use alternative approaches. For other materials, human exposure occurred but it is simply not possible to adequately establish a gradient of exposure. A gradient of exposure is an essential ingredient of any epidemiologic investigation.

I concur with the view expressed by one of our speakers earlier this morning that, in terms of identifying human carcinogens, epidemiologic techniques are faced with diminishing returns. This arises from our control of human exposures.

The aspect of epidemiology that I am most excited about relates to identifying susceptible subpopulations, and individuals within larger populations. I suggest this may require the development and introduction of some new techniques for detailed analysis of material from individuals. Ultimately, I think we will try to draw linkages between a health outcome in an individual and some prior molecular events. Obviously, the only way we can hope to have a "fingerprint" of those prior molecular events is if we have human material, as I have previously discussed.

A related issue I believe we need to start addressing relates to understanding mechanisms of action in laboratory animals and people. At what level of action must we demonstrate common mechanisms of action to feel comfortable in making extrapolations across species? Is it appropriate to make extrapolations based on exposure-dose-response relationships based on the amount of a material or metabolite in a given tissue? Or would we require evidence of modification of DNA in a similar manner in the laboratory animal species and people? I think we have some real interesting issues to face as we push consideration of "dose" down to the molecular level.

Dr. Bates: Yes, I would agree with that. I was fascinated by the diesel exhaust data. I had a feeling that we were being taken gently by the hand into a completely new approach to animal experimentation. The arsenic data was very interesting too, because although the human data have been known since 1948, only in the past 10 years have we begun to understand the metabolic changes which are going to determine animal sensitivity to arsenic as a carcinogen. There is great promise in that work.

I wonder if we are going to reach the dilemma in our deepening understanding of mechanisms, that we reach when we can measure things with greater and greater sensitivity. Are there ways in which this understanding is going to make our life more difficult, rather than simplify it?

Dr. McClellan: Without question, life is going to be more difficult in terms of the kinds of issues we must face. We had a glimpse of that in the exchange between Professor Henschler and Dr. Vostal over the appropriateness of effecting control based on biological phenomena rather than exposure. It is my personal opinion that we must always use exposure limitations as a means of attempting to limit health outcomes. However, we need to recognize that the advances in sensitivity and specificity in measurement technology that have been so prominent in exposure characterization in the last several decades are now upon us in measuring biological structure and function. For example, we are on the threshold of measuring mutation spectra attributable to "background" and a range of chemicals. I am certain our ability to measure will far outstrip our ability to understand what the measurements mean. Indeed, I suspect we may have at least as much difficulty in understanding, interpreting, and advising on the measuring of individual to individual variations, absent any changes relatable to chemical exposures. Without question, a lot of research will be needed before we can advise a worker on what the measurements mean relative to her or his future well-being. Last night, I suggested to the medical director of a major company that the time was at hand when he would need a genetic counsellor to address these issues for both workers and management.

Dr. Bates: I would like to ask you how you react to the very large epidemiologic studies, such as that on diesel exhaust involving more than half a million people, or the very large cohorts in the man-made fiber studies. We are now able to handle these enormous numbers in ways we could not conceive of 20 years ago. Are you encouraged by those examples of the ability of epidemiology to study such large populations?

Dr. McClellan: Yes, I am encouraged to see these types of studies being conducted. I am pleased that some agencies have found it appropriate to fund these large efforts and investigators have been willing to invest many years of effort. It is worthwhile and appropriate because important problems are being addressed. Hopefully, we will see more of these type efforts.

It is critical that we continue to try to convince both funding agencies and investigators that these are worthwhile efforts. For example, it is my opinion that we should be mounting several very large studies now focusing on late-occurring effects of exposure to ozone and copollutants. The cost would be large in absolute terms, but strikingly small relative to the expenditure that will be made on ozone control strategies.

On a related issue, I believe we need to continue to improve our ability to integrate and interpret data from multiple studies, the conduct of metaanalyses, if you will. This is especially critical for materials of low carcinogenic potency or when low exposure concentrations are involved. A good example involved the integrated analysis of studies of environmental tobacco smoke.

A related matter that I believe deserves additional attention is the availability of data from studies for reanalysis and possible reinterpretation by individuals other than the original team that collected the data. I feel strongly that at some appropriate interval, after a study has been completed and published, the original data should be made available to others. This would seem to be warranted based on the expense and importance of many of the studies being conducted today. So that my epidemiology colleagues do not feel that I am unfairly pointing the finger at them, I would offer the same suggestion relative to long-term animal studies.

Dr. Bates: Yes, I think that is an interesting issue; I think it is the subject of an editorial recently in the *American Journal of Epidemiology*. When large studies are done, the question is appropriate. But I am a little hesitant to endorse the concept of some sort of "Good Housekeeping Label." One would need to be assured that the data base could not be changed or manipulated by someone with an interest in doing so, and then be "reinterpreted." When I was a member of an NIH study section on epidemiology, we debated a number of times whether the grant should be conditional on the investigators making their data base available to others. I don't think we ever made it a condition. To pursue the earlier thoughts I had about literary criticism, it seems that in the field it is customary to assume that the individual is badly motivated. I think that epidemiologists are right to assume that their colleagues are not motivated in that way.

We have to remember also that major epidemiologic studies undergo very rigorous scrutiny before they are funded; when the data are presented for publication, there are expert statisticians like John Bailar to act as vultures. So the system has some built-in strengths.

Dr. McClellan: I do not mean to suggest that the investigators involved with epidemiologic or animal studies are not capable and well-intentioned. They certainly are. What I am suggesting is that I believe something is to be gained when large data sets are analyzed and interpreted by more than one investigator or group of investigators. I agree with you that much is gained when the protocols for studies of all kinds are rigorously reviewed before they are initiated. It is my opinion that we need to seek additional vehicles for improving experimental protocols during the planning stage.

Dr. Bates: I think that one of the things that should be said is that animal studies and epidemiologic studies continually reinforce each other. That is why the kind of discussions we have been having are very necessary when a sufficient body of data exists in relation to a specific substance.

Dr. McClellan: I strongly agree that the studies utilizing the various approaches reinforce and complement each other. I am optimistic that we can do even more to facilitate the integration of information. One approach we need to exploit better is to give consideration during the research planning stage to how epidemiologic, animal bioassays, and mechanistic studies can complement each other. This process will be greatly facilitated when individuals in the various disciplines, i.e., epidemiology, toxicology, experimental pathology, biochemistry, molecular biology, etc., are educated to better appreciate the potential role of each of the disciplines in solving problems.

Dr. Bates: In relation to epidemiology, I think there is a lot of discouragement. The NIH study section I was on was only able to fund about 18% of the applications submitted to it. This is depressing. There is a problem in getting the political side to understand the importance of epidemiologic data. There are difficulties in getting enough money for training; but the low level of funding I have referred to is a disincentive to bright students to select epidemiology as a career; it is intensely competitive.

Dr. McClellan: I agree with you that we need to continually be concerned about obtaining adequate funding for research of the kind we have been discussing this week. I am especially concerned that "tight-funding" situations may encourage individuals to promote their own discipline at the expense of multidisciplinary problem-solving approaches. The formal sessions of this symposium have been very successful in reviewing and integrating the contributions of the various disciplines. Moreover, many informal discussions between the participants have identified potential collaborative projects which will further advance our ability to understand and assess carcinogenic risks of occupational and environmental exposures to chemicals and radiation. I am optimistic that if we revisit the topics of this symposium in a decade, we will all be pleased with the progress made using the emerging techniques of molecular biology in studies of cells, tissues, laboratory animals, and people.

On behalf of all the participants, I extend a note of thanks to Dr. Mohr and his Hannover colleagues for their fine hospitality. It was an enjoyable and scientifically stimulating week.

Reference

Hill AB (1965) Presidential address. The environment and disease: association or causation? Proc R Soc Med 58:295–300

Index